Energy Storage

Robert A. Huggins

Energy Storage

 Springer

Prof. Dr. Robert A. Huggins
Stanford University
Department of Materials Science & Engineering
94305-4034 Stanford California
USA

ISBN 978-1-4899-9031-0   ISBN 978-1-4419-1024-0 (eBook)
DOI 10.1007/978-1-4419-1024-0
Springer New York Heidelberg Dordrecht London

Printed on acid-free paper

Springer is part of Springer Science+Business Media (www.springer.com)

Preface

Introduction

Energy is necessary for a number of reasons, the most basic and obvious involve the preparation of food and the provision of heat to make life comfortable, or at least, bearable. Subsequently, a wide range of technological uses of energy have emerged and been developed, so that the availability of energy has become a central issue in society.

The easiest way to acquire useful energy is to simply find it as wood or a hydrocarbon fossil fuel in nature. But it has often been found to be advantageous to convert what is simply available in nature into more useful forms, and the processing and conversion of raw materials, especially petrochemicals have become a very large industry.

Wood

Wood has been used to provide heat for a great many years. In some cases, it can be acquired as needed by foraging, or cutting, followed by simple collection. When it is abundant there is relatively little need for it to be stored. However, many societies have found it desirable to collect more wood than is immediately needed during warm periods during the year, and to store it up for use in the winter, when the needs are greater, or its collection is not so convenient. One can still see this in some locations, such as the more remote communities in the Alps, for example. One might think of this as the oldest and simplest example of energy storage.

It was discovered long ago that it is possible to heat wood under oxygen-poor conditions such that some of its volatile constituents are driven off, leaving a highly porous carbon-rich product called charcoal. Charcoal has a higher heating value per unit weight than the wood from which it was produced, approximately

$30,400$ kJ kg^{-1}, instead of $14,700$ kJ kg^{-1}. Thus, it is more efficient to store and to use to produce heat. This is an example of the conversion of a simple fuel into one with a higher energy value before storage.

Fossil Fuels

Coals

Natural deposits of carbon were also discovered long ago, and it was found that they can likewise be readily burned to produce heat. These solid carbon-rich materials are often described as various types of coal, with different energy contents. The lowest energy content form is called peat, followed by lignite (brown coal), subbituminous coal, bituminous coal, and then hard coal, or anthracite. Their approximate specific energy contents are shown in Fig. 1.

The harder forms have sufficient energy contents that it is economical to not only store them, but also transport them to other locations. Coals constitute the largest fossil fuel resource in the world and are now the most important energy source in a number of places. Where it is available, coal is the least expensive fuel, less than oil or natural gas.

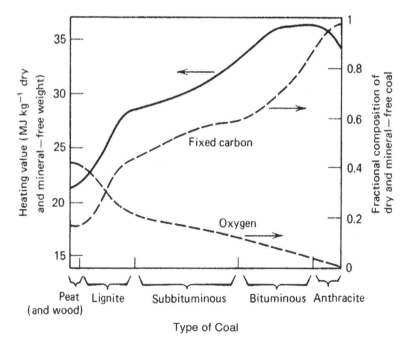

Fig. 1 Energy content and carbon, and oxygen contents of the different types of coal. Based on US DOE data

As in the case of wood, coals can be heated in an air-poor environment to produce a more valuable product, called coke. Coke can then be burned to produce more heat per unit weight and volume than the coal from which it was made, also producing very little smoke. Because of its porosity, relatively high hardness, and higher energy content, coke is used in so-called "blast furnaces" to reduce iron oxide to molten iron, an early step in the production of steel.

There is an increasing concern about the toxic contaminants that are originally contained as minor constituents in coals. Because of the concern about such species getting into the atmosphere, they are often trapped in equipment that is designed to reduce air contamination from coal-burning power plants. They also appear in the coal ash, the noncombustible components in the coal, which are generally stored in open surface ponds or landfills. About 130 million tons of coal ash are produced per year in the United States.

Crude Oil

Petroleum, or crude oil, is also a fossil fuel, similar to coal. But it has the advantage that it is liquid. This makes it much more versatile for a number of applications, and it is more readily transported using vehicles; ships, trucks, and rail, as well as pipelines, and it can be easily stored in tanks. For a number of years it has been less expensive to take crude oil, rather than coal, from the earth, although this disparity has varied with both location and time. As readily extracted natural crude oil supplies are consumed, it is necessary to dig ever deeper, with greater associated costs.

The specific energy of typical crude oil is about 42 kJ kg^{-1}. This is higher than any other fossil fuel. It has become a world, rather than only local, commodity, and is shipped all over the world. Refineries convert it to a variety of products, such as heavy fuel oil, diesel fuel, gasoline, kerosene, etc. Subsequently, significant amounts of these liquid materials are also converted into a variety of solid plastics.

Mankind's use of petrochemical fuels based on crude oil as energy sources is actually only quite recent. Table 1 lists a number of the major crude oil discoveries and their dates. It can be seen that these have all occurred in the last century and a half. Thus, the "oil age" has been just a recent episode in the history of modern civilization.

The Problem of the Depletion of Fossil Fuels

As mentioned earlier, fossil fuels are not infinitely available, and the sources that are found become depleted as their contents are removed. It was long thought that this would not become a problem, for new oil fields would be found to replace those that became depleted. Contrary to this, M. King Hubbert published predictions of future oil production in both the United States [1] and the world [2] that indicated

Table 1 The discovery of oil – locations and dates

Early discoveries	
Near Bakku, on the Caspian Sea	About 1849
Bend, North of Bucharest, Rumania	1857
Oil Springs, Ontario, Canada	1858
Drake well near Titusville, PA, USA	1859
First major oil fields discovered	
Spindletop, near Beaumont, TX, USA	1901
Others in Oklahoma and California, USA	Shortly therafter
Discoveries in the middle east	
Bahrain	1920
Kirkuk in Irak	1927
Gach Saran in Iran	1935
Dammam in Saudi Arabia	1938
Abqaiq in Saudi Arabia	1940
Ghawar in Saudi Arabia	1948–1949
Alaska, USA	
Prudhoe Bay	1968
North Sea Areas	
Forties, UK	1970
Ekofisk, Norway	1971
Brent, UK	1971
South America	
Venezuela	1988

that annual oil production would follow bell-shaped curves, reaching a peak, followed by decline in each case.

His predictions were not taken very seriously for some time, for all seemed to be going well, with oil production growing every year. But then US production began to actually decrease as existing fields began to deplete, and the rate of discovery could not keep up. This is shown in Fig. 2.

These same data are included in Fig. 3, which shows them imposed on the 1956 prediction by Hubbert that US production would peak about 1970. It is seen that the correspondence is remarkable. It is also not surprising that the United States is more and more dependent upon imported oil, not just because of its increasing energy demand, but also the decline in its domestic production.

The mathematical methodology that was developed by Hubbert is discussed in some detail in the book by K.S. Deffeyes [4].

The same question naturally has arisen concerning the world supply of oil, and many people have taken the view that there is no problem in this case, for new sources will always be found. Some, however, including M.R. Simmons [3] and K.S. Deffeyes [4], are convinced that the same peak phenomenon is bound to occur, and that the world production peak has already been reached. Hubbert [2] predicted the world production to peak in 2000 and the Deffeyes estimate [4] was 2005. A number of the major known oil fields are now well beyond their production peaks. This can be seen in Fig. 4.

The expectation that the world production of readily accessible crude oil will reach a peak in the future, if it has not already done so, provides a great incentive for

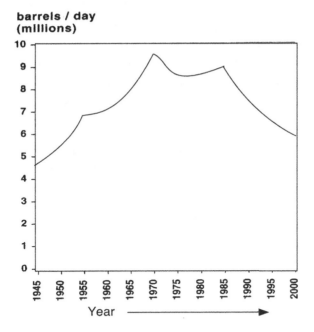

Fig. 2 Annual crude oil production in the United States. Adapted from [3]

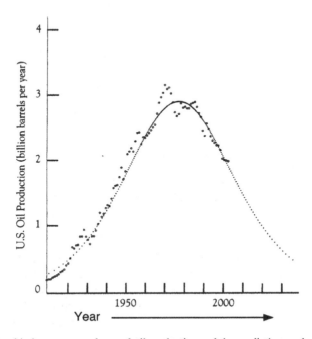

Fig. 3 Relationship between actual annual oil production and the prediction made by Hubbert in 1956. Adapted from [4]

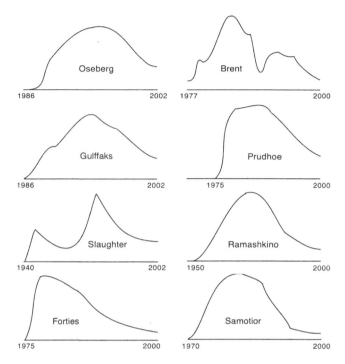

Fig. 4 Time dependence of the relative rate of oil production from eight major oilfields. Adapted from [3]

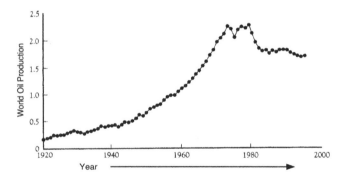

Fig. 5 World production of oil on a per person basis. Adapted from [4]

the development of alternative sources of energy. This is compounded by the fact that the world's population is rapidly expanding. The world's production of oil on a per person basis has already reached a peak and is declining. This is shown in Fig. 5.

The result is the realization that the "oil age" will come to an end and is actually only a relatively short episode in the history of the world – or of modern civilization. This can be represented schematically as shown in Fig. 6.

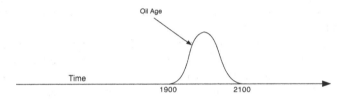

Fig. 6 Period of readily obtainable petrochemical fuels on a long-term time scale

On top of all this, it is reasonable to expect that people in some, or all, of the less developed countries will expect to have access to comparable amounts of energy as those that are currently enjoyed in the developed nations.

Thus, the long-term crude oil supply situation is very bleak. Unlike wood, carbon and petrochemical, fossil fuels must be considered to be nonrenewable resources, for they take millions of years to form. Conservation measures and the search for alternatives to crude oil are imperative.

Other Possible Solid Petrochemical Energy Sources

Coal, and its various grades, was discussed briefly earlier. Other related natural carbon sources are also known, and some are evidently very abundant in selected locations. These include quite large tar sand and oil-containing shale deposits. The pursuit of the possibility of their extraction and use elicits strong, and sometimes politically charged, emotions in some cases.

These are comparatively dilute sources of carbon, and their extraction and conversion into useful fuels are relatively expensive. In addition, they involve the use of other materials, such as natural gas and water, in the conversion process.

Although very large amounts of this family of solid hydrocarbons are known in a number of places, it must be recognized that they are also depletable resources. The time required for their regeneration in nature is immense, so they can also only be counted upon once.

Natural Gas

Natural gas consists primarily of methane, CH_4. It is found associated with liquid fossil fuels and in coal beds. It is also present under the ocean as methane clathrates, as will be mentioned in the next section.

At high pressure underground, natural gas dissolves in oil, and it is often emitted when oil is pumped to the surface. This associated gas is sometimes simply burned at the well sites as "gas flares," for its value is considered to be considerably less than the associated oil. Anyone flying over the Middle East at night can readily identify the oil well areas from the light emitted from these flares. As the price of natural gas has risen, there has been an increasing tendency to capture, rather than simply burning, it.

Oil is generally recovered at depths between 7,500 and 15,000 feet below the surface. The deeper the pressure, the higher the temperature, and the larger molecules in the oil become cracked into smaller units. Below about 15,000 feet the oil is cracked down to molecules that have only one carbon atom, i.e., methane, CH_4. Thus, gas wells are often drilled to depths considerably deeper than those aimed at the extraction of oil.

Early gas wells were actually drilled before the first oil wells. One early successful one was drilled in New York State in 1821. That was 35 years before the famous Drake oil well in Northern Pennsylvania.

Drilling gas wells is generally much faster and less expensive than drilling for oil, and there are currently many more gas wells than oil wells in the United States. Information about natural gas production and available reserves is much more difficult to obtain than that for oil. Thus, there are no predictions available for natural gas comparable to those using the Hubbert method for the predictions about the oil supply.

In addition to methane, natural gas can contain significant quantities of heavier hydrocarbons, such as ethane, propane, butane, and pentane. Beside these related materials, CO_2, nitrogen, hydrogen sulfide, and helium are also often present. A significant effort is often undertaken to separate some of these more valuable components.

Methane-rich gases can also be produced by the anaerobic decay of nonfossil organic matter, such as manure and the waste in landfills. These are typically called *biogas*.

Methane released into the atmosphere is a potent greenhouse gas and is generally considered to be a pollutant. It gradually becomes oxidized to form CO_2 and water, which are also greenhouse gases. The kinetics of this process are such that methane in the atmosphere can be considered to have a half-life of about 7 years.

Natural gas can be used as a fuel for electricity generation in steam turbines and higher temperature gas turbines. It burns cleaner than either petroleum or coal and produces less CO_2 per unit energy obtained. For an equivalent amount of heat, natural gas produces 30% less CO_2 than burning petroleum and 45% less than burning coal. It is therefore generally considered to be the cleanest of the fossil fuels. It can be used for heating and cooking in homes and also as a vehicle fuel instead of gasoline or diesel fuel.

The standard unit of volume of natural gas is 1,000 cubic feet (28.3 m^3) at room temperature and pressure. This gives about one million BTUs, which is approximately 1 GJ of energy. The energy content of 6,000 cubic feet (169.8 m^3) of gas is equivalent to one barrel of crude oil.

As will be discussed in Chap.8, some hydrogen produced from natural gas is also reacted with nitrogen in the high pressure Haber process to produce ammonia, an important feedstock for the production of fertilizers. Some 3–5% of the world natural gas production is used for this purpose.

Natural gas can be readily transported in pipelines. It also can be liquefied, and stored and transported in refrigerated tanks, even across oceans in ships.

Hydrogen

Hydrogen is not actually an energy source, for it is obtained by using energy. Instead, it is generally described as being an energy carrier. Some 9–10 million tons of hydrogen are now produced in the United States per year. It is used for many purposes, as a feedstock and intermediate in chemical production, petroleum refining, and metals treating. Its use to convert heavy petroleum into lighter fractions is called *hydrocracking*. The current applications in transportation and as the fuel in fuel cells represent a very minor fraction of the total.

There are several methods by which hydrogen is currently obtained. The least expensive of these, which accounts for some 95% of the total, is the production of hydrogen by the treatment of natural gas at elevated temperatures. The composition of the natural gas obtained from the earth varies from place to place. It can contain significant amounts of sulfur in some locations. In some cases, it is obtained as a by-product of the extraction of petroleum. As mentioned earlier, the natural gas that is extracted along with the petroleum in the Middle East is often thrown away, for its value may not be sufficient to justify the cost of containing and transporting it.

Hydrogen can also be obtained in other ways. One of the most prominent is by the electrolysis of water, which currently accounts for about 4% of the total. These alternatives will be discussed in Chap. 8.

Methane Clathrates

Another potentially immense source of energy that might be used in the future is related to the presence of large amounts of *methane clathrate* deep in the oceans. This material is also sometimes known as *methane hydrate* or as *methane ice*. It can be described as a crystalline solid H_2O–ice structure within which methane is trapped.

The widespread existence of such materials was first recognized during the 1960s and 1970s, and they are now recognized to be common constituents of the shallow ($<2,000$ m) marine geosphere, where they occur both as deep sedimentary deposits and as outcrops on the ocean floor. They are typically found in the ocean at depths greater than 300 m, and where the bottom water temperature is around $2°C$. They have been found to be widespread along the continental shelves, and are also apparently present in deep lakes, such as Lake Baikal in Siberia.

In addition, methane clathrates have been found trapped in continental rocks of sandstone and siltstone at depths of less than 800 m in cold areas, such as Alaska, Siberia, and Northern Canada.

They are generated as the result of bacterial degradation of organic matter in low oxygen aqueous environments. It is believed that large volumes of methane may occur as bubbles of free gas below the zone of clathrates. They are stable up to about $0°C$, above which they decompose to form liquid water and gaseous methane, CH_4. At higher pressures they can remain stable to higher temperatures.

A typical deposit composition has 1 mol of methane for 5.75 mols of water. As a result, the melting of 1 L of solid clathrate produces 168 L of methane gas at a pressure of one atmosphere.

There are two general methods that have been employed to extract the methane from these clathrates. One is by heating, and the other is by reducing the pressure. The latter evidently requires significantly less energy.

Estimates of the amount of clathrate hydrates on the earth have varied widely and have decreased somewhat with time. Recent values run from 1 to $5 \times 10^{15} m^3$. These values are equivalent to 500–2,500 gigatons of carbon, values smaller than the estimate of 5,000 gigatons of carbon for all other fossil fuel reserves, but substantially greater than current estimates of natural gas (also mostly methane) reserves.

Although methane clathrates are potentially a very important source of fuel, there has evidently been only one commercial development to date, near Norilsk in Russia. Research and development projects are currently underway in both Japan and China.

Since these deposits are very large in some cases, they certainly represent potentially important energy resources. The incentives to develop economical methods to recover the trapped fuel within them are immense.

Chemically Derived Fuels

In addition to the depletable energy sources found in nature, a significant amount of attention is now being given to the production of liquid fuels from renewable sources besides wood.

It has been known for hundreds of years in Asia and Europe that an oil can be produced that can be used to burn in lamps, as a machine lubricant, and also for cooking, by simply pressing rapeseed. More recently, it was found that a modification of this natural oil can be used as a diesel fuel. This plant, called "Raps" in German, has beautiful bright yellow blossoms in the spring. They provide a striking vista in large areas of the landscape in Northern Germany, where the author lived for a number of years. This material is called "Canola" in the United States and Canada.

There are a number of other biological materials that can provide useful fuels. There are two general types of such *agrofuels*. One is to use oilseed crops, such as rapeseed, soy beans, palm oil, and some seeds and nuts that contain high amounts of vegetable oil. Jatropha nuts are up to 40% oil.

The other type includes crops that are high in sugar, e.g., sugar cane, sugar beet and sweet sorghum, or starch, e.g., corn or switch grass (*Panicum virgatum*). These go through a yeast fermentation process to produce ethyl alcohol that can be used as a liquid fuel. Such biofuels currently provide 45% of Brazil's fuel, where all autos run on ethanol.

When the prices of petrochemical fuels are high, the possibility of producing an economically competitive alternative fuel using corn becomes attractive. This has resulted in the diversion of a substantial fraction of the food corn production in that direction. In turn, the costs of related foods to the consumer have escalated and become a political problem.

A further fuel source that is beginning to become economically attractive on a modest scale involves the use of animal fats. Whale oil, obtained from the fat (blubber) of whales, was the first animal oil to become commercially viable for use as a fuel, being used as a lamp oil, and a candle wax. Transesterification (catalytically reacting fats with short-chain aliphatic alcohols) can be used to produce useful liquid fuels from a number of different types of animal waste. One of the attractive features of this process is that it represents a convenient method for disposing of otherwise unusable animal waste from such places as meat processors and high volume chicken restaurants.

Other Alternative Energy Sources

A number of other energy sources are being pursued, and several of these are gradually taking over some of the energy demand burden. Among the prominent alternatives are solar and wind energy technologies. Others involve the use of geothermal sources, collected rainfall, and making use of both tidal variations and currents in the world's oceans. A further approach to the extraction of energy from natural phenomena involves various schemes to harness the power of ocean waves [5].

There is also continued interest in the use of nuclear fission to produce heat, and its use to supply electricity. In this latter case, in addition to cost, there is the persistent, and extremely serious, problem of what to do with the inevitable long-lived radioactive waste. Until the waste problem is solved, this approach will continue to be extremely dangerous.

A significant aspect of most of these alternative energy sources is that they tend to be time-dependent. The availability of water for hydroelectric and other uses varies with the time of the year, as well as the weather. The sun rises and sets, winds come and go, and tidal flows are periodic. This leads to the problem of either matching the time dependence of the energy source with the time dependence of energy needs or the development and use of effective energy storage methods.

As will be described in the text, an extremely important aspect of the use of these time-dependent energy sources, as well as the time dependence of energy needs, is the availability of energy storage mechanisms to help match the supplies with the requirements. On the macroscopic scale, one of the ways that this is currently accomplished is to use the large-scale electrical transmission grid as a buffer. This grid is generally supplied by the use of fossil fuels such as coal or crude oil, although nuclear or hydroelectric sources are also used in some locations. Energy from intermittent sources can be fed into this large system when it is available, reducing the requirements from the normal grid-supplying sources. This can be especially useful if energy from such alternative sources is available at times when the grid's customers' requirements are high, and the additional capacity is quite expensive.

But in addition, the need for energy from the electrical grid generally depends upon the time of day and the day of the week. It also can vary greatly with the weather and time of year. The large-scale energy storage problem of matching supply with need, sometimes called "load leveling" and "peak shaving," will also be dealt with later in this text. There are many other nonsteady-state uses. An example is the use of oil, natural gas, or even electricity, for heating.

An entirely different problem arises when the energy is required for vehicles, or smaller, and likely portable uses, such as computers and telephones. In such cases, energy transfer occurs in one direction, and the storage device is likely charged, either directly or indirectly, from the electrical grid.

Thus, it can be seen that there are many different types of needs for energy storage, requiring many different types of solutions. These are topics that are addressed in this book.

The very serious problem of the depletion of the traditional energy sources and the need to find and develop alternatives has attracted the attention of, and led to action by, governmental bodies worldwide. An important example in the United States was the Energy Independence and Security Act (EISA) that was passed by the Congress and became law in December, 2007 [6]. The intent of this law was to increase energy efficiency and the availability of renewable energy. It included provisions to improve the efficiency of vehicles, appliances, and lighting and to increase the production of biofuels. It also called for accelerated research and development on other energy sources and on related energy storage methods.

Further action in these directions was mandated in the American Recovery and Reinvestment Act of 2009, which became law in February, 2009 [7]. In addition to large increases in funding for energy efficiency and renewable energy, and research and development in energy-related areas, a considerable amount of money was allocated for the development of manufacturing facilities in the United States related to electrochemical energy storage.

It is clear that energy matters have recently attracted an increasing amount of attention worldwide. Energy storage will be an increasingly important component of the overall energy supply picture in the future. This will be particularly critical as the alternative technologies, such as solar and wind sources, where energy production is intermittent, become more widespread.

The various ways in which energy can be stored for later use are discussed in the following chapters of this book.

Acknowledgments The author gladly acknowledges with gratitude the important contributions made to the development of the understanding in this area by his many students and associates in the Solid State Ionics Laboratory of the Department of Materials Science at Stanford University over many years, as well as those in the Center for Solar Energy and Hydrogen Research (ZSW) in Ulm, Germany and in the Faculty of Engineering of the Christian Albrechts University in Kiel, Germany.

References

1. M. King Hubbert, "Nuclear Energy and the Fossil Fuels", American Petroleum institute Drilling and Production Practice Proceedings (1956), p. 5
2. M. King Hubbert, "Energy Resources", in National Research Council, Committee on Resources and Man, *Resources and Man*, W.H. Freeman (1969), p. 196
3. M.R. Simmons, *Twilight in the Desert*: The coming Saudi Oil Shock and the World Economy, John Wiley and Sons, Inc., Hoboken, New Jersey (2005)
4. K.S. Deffeyes, *Beyond Oil*: The View from Hubbert's Peak Hill and Wang, New York (2005)
5. J. Cruz, ed., *Ocean Wave Energy: Current Status and Future Perspectives*, Springer (2008)
6. *Energy Independence and Security Act of 2007*, which became law in the United States in December, 2007
7. *American Recovery and Reinvestment Act of 2009*, which became law in the United States February, 2009

Contents

Chapter 1
Introduction

1.1 Introduction

In addition to the inevitable depletion of the fossil fuels that are now the major sources of energy, and the relatively smaller current alternatives, there is another matter that is very important in considering the effective use of the energy that is available. This is the relationship between the several types of energy supplies and the various uses of energy.

Worldwide energy consumption is between 500 and 600 EJ (5–6 $\times$ 10^{20} J). In terms of consumption rate, this is 15–18 TW (1.5–1.8 $\times$ 10^{13} W). The United States consumes about 25% of the total, although its share of the World's population is about 5%.

A recent estimate of the major United States sources of energy is shown in Table 1.1. The current distribution of energy use, by major category, is indicated in Table 1.2.

These different types of applications have different requirements for access to energy, and different characteristics of its use. One of the important problems with the effective use of available energy supplies is that the schedule of energy use is often not synchronous with its acquisition, even from natural sources. Thus, buffer, or storage systems are necessary.

This requirement for storage mechanisms is highly dependent upon the type of use. Those that use fossil fuels, or their derivatives, for combustion purposes, such as for space heating or internal combustion-powered automobiles, require one type of storage and distribution system. Another, quite different, category involves the various applications that acquire their energy from the large-scale electric power transmission and distribution (T&D) grid. In that case, there are two types of storage systems to consider. One has to do with the electric power grid system itself, and the time dependence of its energy supplies and demands, and the other has to do with storage mechanisms applicable to the various systems and devices that acquire their energy from the grid.

R.A. Huggins, *Energy Storage*,
DOI 10.1007/978-1-4419-1024-0_1, © Springer Science+Business Media, LLC 2010

Table 1.1 Major sources of energy used in the United States in 2007 [1]

Source	Percent of total
Petroleum	39.4
Natural gas	23.3
Coal	22.5
Renewable energies	6.6
Nuclear electric	8.2
Total	100

Table 1.2 Major uses of energy [1]

Type of use	Percent of total energy use
Transportation	28.6
Industrial	21.1
Residential, commercial buildings	10.3
Electric power	40.0
Total	100

There are mechanisms whereby one type of source or storage technology is converted to another. One example is the use of pumped-hydro storage to both take and return electricity to the grid, as described later. Pumped-hydro technology actually operates by a mechanical storage mechanism, based upon the gravitational difference between two different water storage reservoirs. Another is the use of flywheels to reversibly convert mechanical energy into electrical energy.

It will be seen that there are many different types of energy storage technologies or methods. Some of the largest of these are owned and/or operated by energy suppliers. Smaller ones are primarily related to energy users.

1.2 Storage in the Fuel Distribution System

In the simplest case, natural fuels such as wood, coal or crude oil can be stored within the transportation system, in piles, ships and pipelines. For example, crude oil and some of the lighter petrochemical products are stored in tanks in oil depots (sometimes called oil terminal tank farms) in the vicinity of the refineries. These are often near marine tanker harbors or pipelines. This type of transient buffer storage generally has a relatively small capacity, however, and must be supplemented by other methods that can handle much larger amounts of energy.

1.3 Periodic Storage

A number of energy sources do not provide energy at a constant rate, but instead, are intermittent. Sometimes, they have a good measure of periodicity. As an example, some of the biofuels, such as switch grass, sugar cane, corn and oilseeds are only available during part of the year. Likewise, solar, wind and ocean motion energy sources have roughly daily cycles.

The time dependence of the uses of energy often does not correspond to the time variance of such sources. If these were to match perfectly, there would be no need for a storage mechanism. Perfect time-matching is not likely, however. Instead, at least part of the energy is supplied into some type of storage mechanism, from which it is (later) extracted when it is needed.

What is needed is a situation in which the combination of current energy production and stored energy matches the energy and power requirements at all times. This is sometimes called "load management", but it should actually be called "resource management". Overcapacity in either energy sources or storage systems is expensive, and a great deal of attention is given to most effectively and inexpensively meet the needs of energy users.

1.3.1 Long-Term, or Seasonal, Storage

Although there are exceptions, such as the local storage of wood mentioned in the Preface, seasonal storage generally involves very large installations, such as reservoirs and dams that accumulate water primarily during the rainy (or snowy) season of the year. Electric power is produced in hydroelectric facilities by passing water through large turbines. The water collected in such facilities is often used for agricultural purposes, as well as for energy production. Agricultural needs also vary with time during the year.

1.3.2 Daily and Weekly Storage

A number of energy sources produce energy on a daily cycle, related to the periodic characteristics of the sun, tides, and sometimes, wind. It is thus necessary to have mechanisms whereby this energy can be available when needed, and temporarily stored when not needed. But energy from these sources can also vary with the time of the year, and can be significantly affected by changes in the weather, on both long and short time scales. It is common for these sources to be connected to the large electrical transmission grid, providing energy to it when they are in operation. Thus, the grid also acts as a buffer and storage medium.

1.4 The Problem of Load Leveling

The electrical load varies significantly with both the location and the time of day. An example is shown in Fig. 1.1. The major components, industrial, commercial, and residential, require energy at different times during the day. For example,

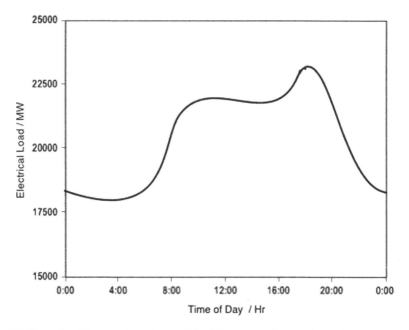

Fig. 1.1 Example of the time dependence of the daily electrical power demand

residential use includes lighting, space heating and air conditioning, and often, electric water heaters and cooking appliances.

The magnitude of the electrical power demand also varies with the time of the year. There is more power used in the winter for heating, and in the summer for air conditioning. The variation with time during the day is also generally greater in the summer than in the winter.

The use of energy also varies with the day of the week in many cases. Whereas it is easy to understand that there is a daily pattern of energy use, the needs are not the same every day of the week because many activities are different on weekends than they are during workdays. This can be seen in Fig. 1.2, which shows a typical pattern of weekly energy use [2].

Whereas both the time-dependence and the magnitudes can vary appreciably with the location, weather, and time of year, these general patterns are almost always present, and pose a serious problem for the electric utility firms that both supply and manage the transmission and distribution (T&D) electric power grid.

The electric utilities can supply this power to the grid from a number of different sources. There are often two or three different technologies, depending upon the load level. The least expensive is the use of coal or oil in large base-load facilities. Thus, the utilities try to cover as much as possible of the need from such sources. However, they are not very flexible, requiring 30–60 min to start up. In addition, utilities typically have a modest amount of *operating reserve*, additional capacity that is available to the system operator within a modest amount of time, for

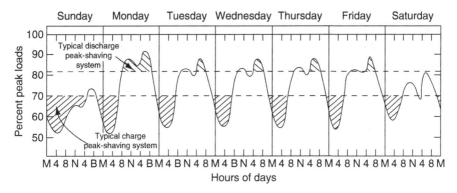

Fig. 1.2 Hourly data on electric load during a full week, showing the possible use of a storage system. After [2]

example, 10 min, to meet the demand if a generator fails or there is another disruption in the supply.

This operating reserve can be divided into two types. One, called *spinning reserve*, is extra generating capacity that can be made available by a relatively simple modification of the operation of the major turbines.

The other is *supplemental*, or *non-spinning*, *reserve*. This label describes capacity that is not currently connected to the system, but that can be brought online after only a short delay. It may involve the use of fast-start generators, or importing power from other interconnected power systems. Generators used for either spinning reserve or supplemental reserve can generally be put into operation in 10 minutes or so.

In addition, there are additional secondary source technologies that are more flexible, but significantly more expensive, and are employed to handle any need for extra capacity. In some cases these involve the use of gas turbine power, similar to the engines that are used to power airplanes.

1.5 Methods That Can Be Used to Reduce the Magnitude of the Variations in Energy Demand

It is self-evident that if the large-scale variations in energy demand can be reduced, the less expensive base load technologies can play a greater role. An obvious solution would be to use *load shifting*, in which some energy needs are shifted from times when the overall demand is large to periods in which other needs are reduced. One way to do this is to use time-of-day pricing, a method that is more common in Europe than in the United States at the present time.

The author lived for a number of years in Germany, and spent some time in a house that used hot water for space heating. A large amount of water was stored in a large insulated vessel that was heated electrically at night, when the power costs were very much lower than during daytime. This hot water was circulated as needed throughout the day.

Another approach is to use energy storage methods to absorb electrical energy when it is available and inexpensive, and to supply it back into the grid system when the demand is higher. This is shown schematically in Fig. 1.3 for the ideal situation, in which the load curve would be entirely flattened.

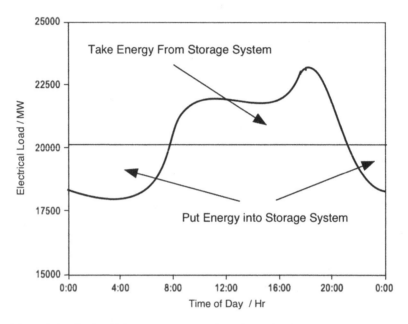

Fig. 1.3 Ideal situation in which energy storage methods flatten the time-dependence of the energy supply requirement

The dominant means of electricity storage for daily load shifting now is the use of pumped hydro facilities. They are also used for spinning reserve and operating reserve applications. They typically produce hundreds of megawatts (MW) for up to 10 h. Another, but not so widespread, approach involves the use of compressed air energy storage.

About 2.5% of the total electric power delivered in the U S is currently cycled through a large-scale storage facility, most commonly pumped hydro. This is different from the practices in Europe and Japan, where about 10 and 15%, respectively, of the delivered power is cycled through such storage facilities. In Japan, this is also a reflection of both higher electricity prices, and a much greater difference between peak and off-peak energy costs.

The development of additional pumped hydro facilities is very limited, due to the scarcity of further cost-effective and environmentally acceptable sites in the United States. Countries with more mountainous terrains have an obvious advantage.

Several additional advanced medium and large-scale energy storage technologies are being developed to help with this problem, including Na/Na$_x$S and flow batteries. These will be discussed later in the text.

1.6 Short-Term Transients

In to these major variations in the energy demand, there are also many short-term transients. This is shown in Fig. 1.4.

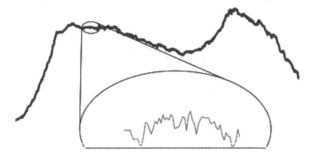

Fig. 1.4 Short time transients superimposed upon the load curve. After [2]

These transients can lead to rotor angle instability, leading to oscillations and an unstable operating condition. Voltage instability can also occur when the load and the associated transmission system require a large amount of reactive, rather than real, power. This can result in a sudden and drastic voltage drop. Short-term (less than 5 min) power outages can also occur. These can be very costly.

A different type of technology is necessary to handle this problem, which is currently handled by making small adjustments in the frequency. However, fast-reaction high power storage mechanisms are ideal for this application.

As will be discussed later, the rapid response characteristics of flywheel systems make it possible to use them to reduce the problem of short-term transients in the load upon the electrical grid. Other options that are being used include Cd/Ni, hydride/Ni and even Pb-acid battery systems. Super-capacitors are also of interest in situations in which their limited energy storage capacity is not limiting. These systems are all discussed in later chapters in the text.

1.7 Portable Applications That Require Energy Storage

In addition to these traditional medium- and large-scale considerations, attention has to be given to the fact that a large number of additional uses of energy, with quite different energy storage requirements, have become very important in recent years. Very obvious examples are the many small-to-medium size electronic devices that one now encounters everywhere. A large fraction of these are portable and thus require batteries for their operation. Their important parameters are quite different from those mentioned above.

An additional area of application for energy storage that has become especially visible in recent years involves vehicles, and hybrid internal combustion – electric vehicles are becoming increasingly numerous. In addition, consideration is again being given to the development of full-electric vehicles. In these cases, the electrical energy storage components are critical.

The important characteristics required by these various types of applications vary widely. As a result, a number of different technologies become important.

1.7.1   Storage Methods for Use with Portable Electronic Devices

Devices such as computers, telephones, music players, camcorders, and personal digital assistants, as well as electronic watches and hearing aids, are now ubiquitous. Almost all of these now are powered by one or another type of electrochemical battery. There are several reasons for this. One is that the need for energy is sporadic, rather than steady. Thus, there is no need to be continually connected to a fixed energy source. Batteries can either be recharged or replaced as needed.

The amount of energy stored per unit weight, the specific energy, is quite high for some types of batteries. This is obviously important for applications in which weight is important. But in some other cases the volume of the energy source, the energy density, is critical. A great deal of progress has been made in these directions in recent years. Then there is the issue of price and, especially, safety. The latter has become more of a problem with the greater the amount of energy stored in small packages. These matters, and some of the important battery technologies, are discussed in some detail in later chapters.

1.7.2   Energy Use and Storage in Vehicles

Vehicle propulsion currently mostly involves the consumption of gasoline or diesel fuel in internal combustion engines. *Hybrid autos* are rapidly being introduced, but still represent only a small fraction of the total. They combine an internal combustion engine with a relatively small high-rate battery that can acquire some of the vehicle's kinetic energy during deceleration. This combination serves to reduce the amount of fuel used, but results in no change in energy source.

A relatively small number of *all-electric* passenger and commercial vehicles are being produced, as well. While most of the attention has been focused on the future of larger vehicles, some 60 million electric scooters and bicycles are already being used in the world. They get their energy from the electric grid, typically by charging overnight. Another variant that is getting increased attention is the category called *plug-in hybrids*. These are different from the normal hybrids in that the on-board battery can be recharged (again, typically overnight) from the electric grid. The charging of the battery component allows these vehicles to travel a limited range on electricity alone. Thus if they do not drive very far, they get their energy from the grid. If they travel farther, the energy input is shared between the electrical grid and a petrochemical fuel.

As the number of plug-in hybrid and all-electric vehicles increases, the overall transportation energy demand will gradually move more toward electricity, and away from liquid fuels. But, in addition, this part of the electric load will be mostly at night, where the other demands are reduced, and thus this will contribute to load leveling.

The energy storage devices used in both hybrid autos and plug-in hybrids are batteries. The desirable characteristics are different in the two cases, however. Hybrids have relatively small batteries, for the amount of energy that must be stored is limited. Instead, the rate of energy absorption during braking can be quite high. Thus these batteries must be able to operate at high rates, or high power. Metal hydride/nickel batteries are now used for this purpose. However, almost all of the auto manufacturers who are involved in the development of these types of vehicles indicate that they expect to use lithium-ion batteries in the future, for they are expected to store considerably more energy per unit weight than the batteries currently used, and also be able to operate at high power.

The all-electric and plug-in hybrid application is different, for in order for the vehicles to have an appreciable range, a large amount of energy must be stored. This requires large, and thus heavy, batteries. And they must be optimized in terms of energy storage, rather than high power. As a result, it is reasonable to expect that this type of vehicles will be designed to employ two different types of batteries, one optimized for energy, and the other for power.

Another obvious factor is cost, especially for the potentially more desirable lithium-ion batteries. This is an area receiving a large amount of research and development attention at the present time, and is discussed in some detail later in this text.

1.8 Hydrogen Propulsion of Vehicles

There has recently been a good deal of interest in the use of hydrogen as a fuel for vehicular propulsion. There are actually two different versions of this topic. One is the direct combustion of hydrogen in internal combustion engines. This can be done, and requires only slight modifications of in current gasoline or diesel engines and injection of gaseous hydrogen at relatively high pressures. The German auto firm BMW has demonstrated such vehicles over a number of years. The hydrogen is stored as a liquid at a low temperature in an insulated tank.

An alternative to the use of liquid hydrogen is to store the required hydrogen in solid metal-hydrogen compounds known as metal hydrides. These are essentially the same materials as the hydrides used in the negative electrodes of the common hydride/nickel batteries. Upon heating, the hydrogen is released so that it can be used in the engine.

The other hydrogen approach, which has now been heavily promoted by the United States government for a number of years, is to use hydrogen-consuming fuel cells for propulsion. But again, the hydrogen must be carried, and stored, in the vehicle.

Because of the metal component, the current metal hydrides are quite heavy, storing only a few percent hydrogen by weight. The United States Department of Energy is supporting a substantial research program aimed at finding other materials that can store at least 6% hydrogen by weight. This target is based upon the assumption that consumers will require a hydrogen-powered vehicle that has approximately the same range as the current internal combustion autos. It is evident that the increasing interest in reduced-range vehicles that is so apparent in connection with those that are now electrically-powered does not yet seem to have migrated to the government research support community.

It is interesting that a firm in Germany is producing military submarines that are powered by hydrogen-consuming fuel cells, and metal hydrides are used to store the necessary hydrogen. The weight of the hydride materials is not a problem in the case of submarines, and some of the materials that are currently known are satisfactory for this purpose.

1.9   Temperature Regulation in Buildings

As indicated in Table 1.2, a significant amount of energy is used in residential and commercial buildings. In addition to lighting, a good measure of this is for the purpose of the regulation of the temperature in living and work spaces. In winter, this involves the supply of heat, and in summer, it also often requires temperature reduction by the use of air conditioning equipment.

In addition to the use of improved thermal insulation, the magnitude of the related energy requirements can be reduced by the use of thermal storage techniques. There are several types of such systems that can be used for this purpose, as will be discussed in Chap. 4.

1.10   Improved Lighting Technologies

It was shown in Table 1.2 that electric power now constitutes about 40% of the total energy use in the United States. Of this, some 22% is consumed by lighting. Thus, over 8% of the total energy use in the United States is due to lighting. On a global basis, about 2,700 TWh, or 19% of total electricity consumption is used for this purpose [3].

Most lighting today involves the use of incandescent bulbs in which tungsten wire is electrically heated to about 3,500 K. The spectrum of the emitted radiation is very broad, covering both the range to which the human eye is sensitive, and also well into the infrared range, where the result is heat. Only about 5% of the electrical energy is converted to visible light, the other 95% is heat. Thus, this is a very inefficient process.

Lighting technology is changing fast, and it is reasonable to expect that its contribution to the total amount of energy use will decrease somewhat in the future. The first step in this process involves greatly increased use of fluorescent lamps, which have light emission efficiencies of about 20–25%, rather than the 5% of incandescent bulbs.

Governmental regulations are being employed in many locations to reduce the use of energy-inefficient incandescent lighting. A law in the State of California requires a minimum standard of 25 lumens/W by 2013, and 60 lumens/W by year 2018. Other states are surely to move in this direction also. In addition to these local initiatives, the US Federal Government effectively banned the use of the common incandescent lights after January, 2014 by passing the Clean Energy Act of 2007.

However, it is now generally thought that the fluorescent light technology will be overtaken by light sources based upon semiconductors, the light-emitting-diodes, or LEDs. Development efforts in this direction have been underway for a long time, with large jumps in their effectiveness [3], and some LED products are now widely used as bicycle lights, in the tail lights of automobiles, and as street lights in cities. The major breakthroughs have involved the development of GaN - based LEDs and the compositions and processes for making them that produce devices with various emitted wavelength ranges [3, 4]. It is now clear that this solid-state type of technology, which is some 15 times as energy efficient as incandescent lights, and which is moving quickly onto the commercial market, will play an even greater role in the future.

The relative energy efficiencies of these three technologies can be seen from the data in Table 1.3.

Table 1.3 Energy efficiency of different lighting technologies

Light source	Lumens(W)
Tungsten incandescent bulb	17.5
Compact fluorescent tubular bulb	85–95
White light-emitting diode	170

1.11 The Structure of This Book

This book is intended to provide a basic understanding of the various mechanisms and technologies that are currently employed for energy storage. An initial chapter introduces relevant basic concepts. This is followed by a group of chapters that describe the most important chemical, mechanical, and electromagnetic methods.

The general principles involved in the various electrochemical technologies that are becoming increasingly important are then introduced, followed by a group of chapters on the most important battery systems. These are followed by a chapter that discusses the storage methods that are most important for the major areas of application.

Much of the discussion will involve both concepts and methods of thermodynamics, mechanics, electrochemistry, and materials science. However, it is not necessary to have a significant amount of prior expertise in such areas, for the background that is necessary will be presented in relevant portions of the text. A number of sources in which further information can be found are included as references [4–11].

References

1. Slight modification of US Energy Information Administration data for the year 2007
2. R. Fernandes. Proc. 9th Intersociety Energy Conf. (1974), p. 413
3. C.J. Humphreys. MRS Bull. *33*, 459 (2008)
4. B. Johnstone. *Brilliant!: Shuji Nakamura and the Revolution in Lighting Technology*. Prometheus Books, New York (2007)
5. J. Jensen and B. Sorenson. *Fundamentals of Energy Storage*. Wiley-Interscience, New York (1984)
6. A. Ter-Gazarian. *Energy Storage for Power Systems*. Peter Peregrinus Ltd., Stevenage, Hertfordshire (1994)
7. *EPRI-DOE Handbook of Energy Storage for Transmission and Distribution Applications*. Electric Power Research Institute, Palo Alto, CA (2003)
8. *Energy Storage for Grid Connected Wind Generation Applications, EPRI-DOE Handbook Supplement*. Electric Power Research Institute, Palo Alto, CA (2004)
9. *Basic Research Needs for Electrical Energy Storage*. Office of Science, Department of Energy, U.S. Government (2007)
10. F.R. Kalhammer, B.M. Kopf, D.H. Swan, V.P. Roan and M.P. Walsh. *Status and Prospects for Zero Emissions Vehicle Technology*. State of California Air Resources Board, Sacramento, CA (2007)
11. R.A. Huggins. *Advanced Batteries: Materials Science Aspects.* Springer, Berlin (2009)

Chapter 2
General Concepts

2.1 Introduction

This book is about energy, and various mechanisms by which it can be stored for use at a later time, for a different purpose, or at a different place.

Energy is the key component of what is called *thermodynamics*. Thermodynamic principles are involved in considerations of the different types of energy and their relation to macroscopic variables such as temperature, pressure, and volume, and chemical and electrical potentials. They are also centrally involved in the transformation of energy between different forms, such as heat and mechanical, electrical, chemical, magnetic, electrostatic, and thermal energy.

It was pointed out in the very influential book by Lewis and Randall [1] that, "aside from the logical and mathematical sciences, there are three great branches of natural science which stand apart by reason of the variety of far-reaching deductions drawn from a small number of primary postulates. They are mechanics, electromagnetics, and thermodynamics". While thermodynamics (from the Greek word *therme*, for heat) is often thought of as being quite esoteric and uninteresting, it can actually be of great practical use, as will be seen later in this text.

2.2 The Mechanical Equivalent of Heat

Thermodynamics originated from the observation that there is a relation between two different forms of energy, heat, and mechanical work.

The first step was the observation by Count Rumford (Benjamin Thompson) in 1798 that the friction of a blunt borer in a cannon caused an increase in the cannon's temperature, and that the increase in temperature was related to the amount of mechanical work done. The quantitative relationship between the amount of mechanical work done on a body and the resultant increase in its temperature was

R.A. Huggins, *Energy Storage*,
DOI 10.1007/978-1-4419-1024-0_2, © Springer Science+Business Media, LLC 2010

determined by James Prescott Joule in the mid nineteenth century. He found that
this relation is

$$1 \, \text{cal} = 4.184 \, \text{J} \tag{2.1}$$

The thermochemical calorie, the unit quantity of heat, is defined as the amount of
heat that must be added to 1 g of water to raise its temperature 1°C. The Joule, a
measure of energy, can be expressed in either electrical or mechanical terms.

$$1 \, \text{J} = 1 \, \text{Ws} = 1 \, \text{VC} \tag{2.2}$$

or

$$1 \, \text{J} = 1 \, \text{Nm} = 1 \, \text{kg} \, \text{m}^2 \, \text{s}^{-2} \tag{2.3}$$

The amount of heat required to raise the temperature of a material 1°C is called
its *heat capacity*, or its *specific heat*. In the latter case, the amount of heat per unit
weight, the dimensions are J kg^{-1} K^{-1}.

2.3   The First Law of Thermodynamics – Conservation of Energy

In a closed system, energy cannot either be created or destroyed. It can only be
converted from one type to another type. This is called the *first law of thermodynamics*, or the *law of the conservation of energy*. It can be expressed as

$$\Delta U = q + w, \tag{2.4}$$

where U is the *internal energy* of a material or a system, assuming that it is not in
motion, and therefore has no kinetic energy, q is heat absorbed by the system, and
w is work done on the system by external forces. In the case of a simple solid, U can
be thought of as the sum of the energy of all of its interatomic bonds. It does not
have an absolute value, but is always compared to some reference value.

2.4   Enthalpy

Another important quantity is the *enthalpy H*, which is sometimes called the *heat
content*. The name came from the Greek *enthalpein*, to warm. This is defined by (2.5)

$$H = U + pv, \tag{2.5}$$

where p is the applied pressure and v the volume. The product pv is generally quite
small for solids in the conditions that will be met in this text.

If a system (e.g., a material) undergoes a change in state, such as melting or a chemical reaction, there will be a change in enthalpy ΔH. A positive value of ΔH means that heat is absorbed, and the reaction is described as *endothermic*. On the other hand, if heat is evolved, ΔH is negative, the internal energy is reduced, and the reaction is called *exothermic*. The latter is always the case for spontaneous processes. This change in heat content when a reaction takes place is called the *latent heat* of the reaction.

H, U, and the *pv* product all have the dimensions of energy, kJ mol^{-1}. Their values are compared to a reference *standard state*. For pressure, the standard state is 1 bar, or 0.1 MPa. The conventional reference temperature is 298.15 K.

The standard state that is generally used for all solid substances is their chemically pure forms at a pressure of 1 bar and a specified temperature. This is indicated by the use of the index "^{0}". At a temperature of 298.15 K and a pressure of 1 bar, the value of H^0 is zero for all pure materials.

2.5 Entropy

A further important quantity in discussions of thermodynamics is the *entropy* (from the Greek word *trope*, a deviation or change). Entropy, S, is a measure of disorder, or randomness. What this means will become evident in the examples to be discussed below.

2.5.1 Thermal Entropy

At a temperature of absolute zero (0 K), the structure of a solid material is "frozen". When the temperature is increased by adding heat, or thermal energy, its constituent particles begin to vibrate in place, acquiring local kinetic energy, similar to the energy in a swinging pendulum or vibrating spring. The magnitude of this vibrational, or motional, kinetic energy, or heat, is proportional to the temperature. The proportional factor is called the *thermal entropy*, S_{th}. This can be simply written as

$$q = TS_{th}. \tag{2.6}$$

This can be rearranged to define the thermal entropy, the randomness of the positions of the vibrating particles at any time.

$$S_{th} = q/T. \tag{2.7}$$

It can be seen that the product TS_{th} also has the dimension of energy.

One can find more sophisticated discussions of the origins and physical meaning of thermal entropy in a number of places, such as [1, 2], but such greater depth is not necessary here.

2.5.2 Configurational Entropy

A different type of randomness also has to be considered in materials systems. In addition to the vibratory motion of the fundamental particles present, there can also be some degree of disorder in the arrangement of the particles, that is, atoms and electrons, in a material. This is sometimes called the *configurational*, or *structural*, *entropy*, S_{conf}. It is a measure of the uniformity or regularity of the internal structure, or the arrangement of the particles in a crystal structure, and also has the dimension of energy/T.

The magnitude of this type of entropy changes when there is a change in crystal structure of a solid, the solid melts to become a liquid, or a chemical reaction takes place in which the entropies of the reactants and the products are different.

The total entropy, indicated here simply by the symbol S, is the sum of the thermal entropy and the configurational entropy.

$$S = S_{th} + S_{conf}. \tag{2.8}$$

2.6 The Energy Available to Do Work

The driving force for reactions or other changes to take place is always a reduction in energy. This is relatively easy to understand in the case of mechanical systems, where the total energy is generally divided into two types, potential energy and kinetic energy. It may seem to be a bit more complicated in chemical systems, however.

Some time ago, the driving force for a chemical reaction was called the *affinity*. For example, if A and B tend to react to form a product AB, the amount of energy released would be the *affinity* of that reaction. But since the work of J. Willard Gibbs [3, 4] it has been recognized that the driving force for any process, and also therefore the maximum amount of work that can be obtained from it, is a change in the quantity G, where

$$G = H - TS. \tag{2.9}$$

The name now generally given to G is the *Gibbs free energy*, although it is sometimes called the *Gibbs energy*. In parts of Europe, it is called the *free enthalpy*. It is also equivalent to the *exergy*, a term introduced in 1956 by Zoran Rant [5], that is used in some branches of engineering.

It is obvious that the energy that is available to be used to do either mechanical or electrical work is less than the heat (total energy) H present by the amount of energy tied up in the entropy, TS.

The Gibbs free energy G will appear many times in this text, for it plays a significant role in many applications. Changes or differences in G, not in H or U, constitute the driving forces for many processes and reactions.

2.7   The Temperature Dependence of G, H and S

The temperature dependence of the Gibbs free energy, G, the enthalpy H, and the total entropy S for a simple metal, pure aluminum, are shown in Fig. 2.1.

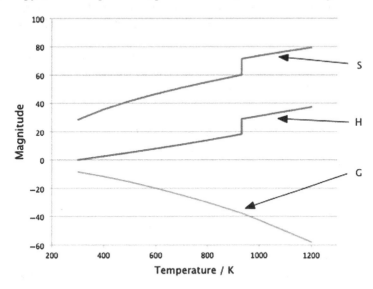

Fig. 2.1 Thermodynamic data for pure aluminum. Dimensions are J K^{-1} mol for S, and kJ mol^{-1} for H and G

It can be seen that there is a discontinuity in both the enthalpy H and the entropy S, but not the Gibbs free energy G, at the melting point 933.45 K. The entropy increases by 11.48 J K^{-1} mol^{-1}, for the liquid aluminum structure has a greater value of configurational entropy than solid aluminum. There is also a corresponding discontinuity in the enthalpy of 10.70 kJ mol^{-1} at that temperature since the process of melting is endothermic, requiring the input of heat at the melting point to convert aluminum from solid to liquid.

The difference between the gradual changes in both H and S with temperature and their discontinuities when melting or other phase changes occur will be discussed further in Chaps. 3 and 4, which deal with the storage of heat. In one case, one considers changes in *sensible heat*, and in the other, the important quantity is the *latent heat* of the reaction.

2.8   Irreversible and Reversible Storage Modes

There are two general types of energy storage to be considered. In one type, energy that is already present, or stored, is available to be used once. In the other, energy can be both used and replaced, that is, the energy is stored reversibly.

2.9   The Carnot Limitation

Considerations of the thermodynamics of energy systems often include discussions of the so-called *Carnot limitation*. It is applicable to energy conversion processes that involve heat engines, and results in the desire to have such devices operate at as high a temperature as possible, producing products at as low a temperature as possible.

The maximum efficiency that can be obtained from any process that involves the input of energy at one (higher) temperature, and its exhaustion at another (lower) temperature is given by

$$\varepsilon_{max} = \frac{T_H - T_L}{T_H} = 1 - \frac{T_L}{T_H}. \tag{2.10}$$

There is also a theoretical limit to the maximum power that can be obtained from such a heat engine [6]:

$$\varepsilon_{max\,power} = 1 - \sqrt{\frac{T_L}{T_H}}. \tag{2.11}$$

Practical power plant efficiencies are now quite close to this limit.

This Carnot limitation is not applicable to other types of energy conversion systems that do not involve temperature changes. As a result, isothermal energy conversion systems can have significantly greater efficiencies. This is a great inherent advantage of the use of fuel cells and batteries, for example.

2.10   Energy Quality

In many applications, not only is the amount of energy important, but also its *quality*. Energy in certain forms can be much more useful than energy in other forms. In cases involving the storage and use of thermal energy, the temperature is important.

As mentioned earlier, the amount of useful energy, that can do work, G is related to the total energy or heat content, the enthalpy H, by

$$G = H - TS. \tag{2.9}$$

The energy quality can be defined as the ratio G/H,

$$\frac{G}{H} = \frac{\Delta T}{T} = 1 - \frac{T_0}{T}. \tag{2.12}$$

where T is the temperature in question, and T_0 is the ambient temperature. The energy quality is thus greater the larger the value of T.

In electrochemical systems the quality of stored energy is dependent upon the voltage. High voltage energy is more useful than low voltage energy. This will be discussed in Chap. 9.

References

1. G.N. Lewis and M. Randall. *"Thermodynamics"*, revised by K.S. Pitzer and L. Brewer. McGraw-Hill Book Co., New York (1961)
2. L. Pauling. *General Chemistry*. Dover, New York (1970)
3. J.W. Gibbs. Trans. Conn. Acad. Arts. Sci. *2*, 309, 382 (1873)
4. J.W. Gibbs. Trans. Conn. Acad. Arts. Sci. *3*, 108, 343 (1875)
5. Z. Rant. Forschung auf dem Gebiete des Ingenieurwesens *22*, 36–37 (1956)
6. M. Rubin, B. Andresen and R. Berry. In *Beyond the Energy Crisis*, ed. by R. Fazzolare and C. Smith. Pergamon Press, New York (1981)

Chapter 3
Thermal Energy Storage

3.1 Introduction

It was mentioned in Chap. 1 that a significant portion of the total energy use is for
lighting and temperature control in living and working spaces. Energy use for
lighting purposes comprises between 20 and 50% of the energy used in homes,
and varies appreciably with both location and time of the year, of course. This is
expected to decrease substantially as the result of the use of fluorescent and light-
emitting-diode (LED) devices in the future.

Residential and commercial heating and cooling needs are currently mostly
taken care of by the use of gas or electric heating and electrically powered air
conditioning. This energy requirement can also be reduced appreciably by various
measures. Among these is the reduction of heat transfer to and from the environment by the use of better insulation. Another is to make use of thermal energy
storage systems. Thermal energy storage is also used in large-scale power plants,
although that will not be included here. Discussions of these matters can be found in
several places [1–4].

Household hot water systems in the United States typically have a gas- or
electrically heated water heater and storage tank, from which water is distributed
using pipes throughout the dwelling. There is thus heat loss from the distribution
piping as well as through the insulation on the storage tank. A different system is
used in parts of Europe, where energy is more expensive. This involves local
heating of water where, and only when it is needed. Thus, there are essentially no
storage losses.

In addition to the storage of heat to maintain something at a high temperature,
thermal energy storage methods, the main topic of this chapter, are sometimes used
to maintain cold temperatures for the storage of food or for other chiller applications. Information about this type of application can be found in [3].

There are two general types of thermal storage mechanisms. As will be seen
below, one is based upon the use of the *sensible heat* in various solid and/or liquid
materials. The other involves the *latent heat* of phase change reactions.

R.A. Huggins, *Energy Storage*,
DOI 10.1007/978-1-4419-1024-0_3, © Springer Science+Business Media, LLC 2010

3.2   Sensible Heat

Energy can be added to a material by simply heating it to a higher temperature. The energy that is involved in changing its temperature is called "sensible heat", and its amount is simply the product of the specific heat and the temperature change.

This sensible heat can be transferred to another, cooler material, or to the environment, by radiation, convection, or conduction. Thus, this is a method for storing energy in the form of heat, and transferring it again. A simple example is the traditional procedure to using a hot rock or a hot water bottle to pre-warm a bed before going to sleep.

This type of energy storage has also been used to control the temperature in living or working spaces. In some cases, the amount of storage material can be quite large, so that there is the obvious concern about its cost. This results in the use of relatively simple and inexpensive materials. Data on the thermal capacity of some examples are presented in Table 3.1.

Table 3.1 Thermal properties of some common materials

Material	Density (kg m^{-3})	Specific heat (J kg^{-1} K^{-1})	Volumetric thermal capacity (10^6 J m^{-3} K^{-1})
Clay	1,458	879	1.28
Brick	1,800	837	1.51
Sandstone	2,200	712	1.57
Wood	700	2,390	1.67
Concrete	2,000	880	1.76
Glass	2,710	837	2.27
Aluminum	2,710	896	2.43
Steel	7,840	465	3.68
Magnetite	5,177	752	3.69
Water	988	4,182	4.17

The amount of heat q that can be transferred from a given mass of material at one temperature to another at a lower temperature is given by

$$q = \rho C_p V \Delta T, \tag{3.1}$$

where ρ is the density, C_p the specific heat at constant pressure, and V the volume of the storage material. ΔT is the temperature difference.

Each of the materials in Table 3.1 has some advantages and some disadvantages. The specific heat of water is more than twice that of most of the other materials, but it is only useful over a limited temperature range (5–95°C). Thus, it appears attractive for systems involved in the control of living space heating and cooling. An example of this was mentioned in Chap. 1.

On the other hand, some of the inexpensive bulk solids can be used over a wider range of temperature, and they can be more compact due to their higher densities. An additional factor that may be important in some cases is the thermal conductivity of the storage material, for that influences the rate at which heat can be either absorbed or extracted.

No matter how well the system is insulated, there are always some losses in this method.

Thus, the insulation or thermal isolation of the storage material can be quite important, particularly if the storage period is substantial.

The rate of heat loss to the surroundings is proportional to the surface area, and also to the temperature difference, but the total amount of thermal storage is proportional to the volume of any storage container. Therefore, it is more effective to use large vessels with shapes that are not far from spherical.

3.3 Latent Heat

A different mechanism for the storage of energy involves *phase transitions* with no change in the chemical composition. It was shown in Fig. 2.2 of Chap. 2 that there is a jump in the value of the entropy, and a corresponding change in the enthalpy, or heat content, but not of the Gibbs free energy, at the melting point of aluminum, a simple metal. This is characteristic of phase transitions in elements and compounds that melt *congruently*, that is, the solid and liquid phases have the same chemical composition. There are also phase transitions in which both phases are solids. As an example, there is a transition between the alpha and beta phases in titanium at 882°C. The alpha, lower temperature, phase has a hexagonal crystal structure, whereas the beta phase has a body-centered cubic crystal structure. Because they have different crystal structures they have different values of configurational entropy, and there is a corresponding change in the enthalpy.

In these cases, latent heat is absorbed or supplied at a constant temperature, rather than over a range of temperature, as it is with sensible heat. Isothermal latent heat systems are generally physically much smaller than sensible heat systems of comparable capacity.

A further simple example is water. At low temperatures it is solid, at intermediate temperatures it becomes a liquid, and at high temperatures it converts to a gas. Thus, it can undergo two phase transitions, with associated changes in entropy and enthalpy. The Gibbs free energy, or chemical potential, is continuous, for the two phases are in equilibrium with each other at the transition temperature.

From

$$\Delta G = \Delta H - T\Delta S = 0, \tag{3.2}$$

at that temperature, the change in heat content ΔH at the transition temperature is equal to $T\Delta S$.

The slope of the temperature dependence of the Gibbs free energy is proportional to the negative value of the entropy, which is different in the different phases, for they have different structures. This is shown schematically in Fig. 3.1.

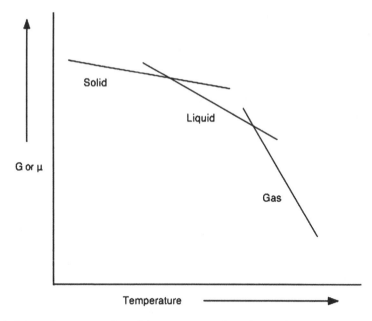

Fig. 3.1 Schematic representation of the temperature dependence of the Gibbs free energy of water

3.3.1 Inorganic Phase Change Materials

There are a number of other materials that also undergo congruent phase transitions. Data on some simple inorganic materials are given in Table 3.2. There are also many cases in which liquids convert isothermally into two or more phases with different compositions by means of *eutectic reactions* upon cooling, with accompanying heat evolution. Upon heating, heat is absorbed by the reverse reactions. Data on a large number of molten salt eutectic systems can be found in [5].

Table 3.2 Heat of fusion data for several simple inorganic materials

Phase	Melting point (°C)	Heat of fusion (MJ kg^{-1})
NH_4NO_3	170	0.12
$NaNO_3$	307	0.13
NaOH	318	0.15
$Ca(NO_3)_2$	561	0.12
LiCl	614	0.31
$FeCl_2$	670	0.34
$MgCl_2$	708	0.45
KCl	776	0.34
NaCl	801	0.50

In addition to the phase changes involved in the conversion of solids into liquids, there are a number of cases in which there is a phase change in the solid state. These involve changes in the crystal structure, and thus changes in the entropy and the heat content at temperatures below their melting points. These materials can be used to store thermal energy at the temperatures of their solid state phase transitions. Some examples are given in Table 3.3.

Table 3.3 Data on solid state phase transition and melting entropy for a number of materials

Material	Transition temperature (°C)	Melting temperature (°C)	Transition entropy (J mol^{-1} K^{-1})	Melting entropy (J mol^{-1} K^{-1})
FeS	138	1,190	4.05	21.51
AgI	148	558	14.61	11.33
Ag$_2$S	177	837	8.86	7.01
Na$_2$SO$_4$	247	884	12.5	18.2
Ag$_2$SO$_4$	427	660	26.66	19.19
Li$_2$SO$_4$	577	860	29.2	7.9
LiNaSO$_4$	518	615	31.2	Small

There are several features of the data in this table that deserve comment. The total entropy change, and thus the heat absorbed, between the low temperature solid state and the molten state is divided between that which occurs at the solid state phase transition and that which occurs upon melting. In general, the crystal structures below and above solid state transformations are not very different. As a result, the entropy change that is involved is not very large, and most of the total entropy change occurs at the melting point. An example of this type that is included in Table 3.3 is FeS.

But there are some materials in which the higher temperature solid phase has a crystal structure in which one of the atomic species has an unusual vibrational amplitude or inter-site mobility. When that phase is stable it has a high value of entropy. Thus the entropy change in its formation is large. Two examples of this are shown in Table 3.3, AgI and Ag$_2$S. Silver ions have an unusually high vibrational amplitude and mobility in the higher temperature phases of both of these materials. AgI is called a *fast ionic conductor*, and Ag$_2$S is a *fast mixed conductor*. The difference is that Ag$_2$S is an electronic, as well as an ionic, conductor, whereas AgI is an electronic insulator.

Na$_2$SO$_4$ and Ag$_2$SO$_4$ are two additional examples of this type of behavior. In these cases the sodium and silver ions are relatively mobile, and the SO$_4$ tetrahedral groups are essentially static parts of the crystal structure.

However, there are some other materials in which the type of structural disorder that leads to large values of entropy is quite different. Whereas atomic or ionic motion in most solids, including AgI and Ag$_2$S, involves species either jumping into adjacent crystallographic vacancies or between dilutely occupied interstitial positions, some materials have been found in which atomic motion occurs in coordinated groups. That is, groups of atoms move together. An example of this type of atomic motion is found in a group of lithium sulfates. It was found that these materials can also have unusually high values of ionic conductivity, different from that of the sodium and silver analogs [6, 7]. The structural reason for this is that there is rotation of SO_4 groups within the "static" crystal structure that assists in the longer-range transport of the lithium ions. This was first described as a *cogwheel mechanism*, but is now generally referred to as a *paddlewheel mechanism* [8]. These materials also exhibit a high degree of mechanical plasticity, and are sometimes also called *plastic crystals*. Reviews of this topic can be found in [9, 10].

It can be seen in Table 3.3 that the entropy changes related to the solid state structural changes in the paddlewheel materials are significantly greater than the entropy changes upon melting. They are thus particularly attractive as heat storage components if the temperature is appropriate. Materials that have large values of latent heat related to solid state phase transitions are discussed in [11].

3.3.2 Organic Phase Change Materials

In addition to these inorganic materials, it is also possible to take advantage of analogous behavior in organic materials. Data for some simple organic materials with large values of the heat of fusion are shown in Table 3.4, and the thermal properties of fatty acids and two simple aromatic materials are presented in Tables 3.5 and 3.6.

Table 3.4 Data relating to some organic phase change materials

Material	Melting temperature (°C)	Heat of fusion (kJ kg^{-1})
Paraffin wax	64	173.6
Polyglycol E 400	8	99.6
Polyglycol E 600	22	127.2
Polyglycol E 6000	66	190.0

Table 3.5 Data relating to some fatty acids used as phase change materials

Material	Melting temperature (°C)	Heat of fusion (kJ kg^{-1})
Stearic acid	69	202.5
Palmitic acid	64	185.4
Capric acid	32	152.7
Caprylic acid	16	148.5

Table 3.6 Some aromatics used as phase change materials

Material	Melting temperature (°C)	Heats of fusion (kJ kg^{-1})
Biphenyl	71	19.2
Napthalene	80	147.7

3.4   Quasi-Latent Heat

Chemical reactions generally result in the generation or absorption of heat, similar to the thermal effects related to phase transitions in materials in which there are no changes in chemical composition. The thermal effects related to chemical reactions are often described in terms of *quasi-latent heat*. They will be discussed in Chap. 4.

3.5   Heat Pumps

Although they are not involved in the storage of heat, the topic of this chapter, it seems desirable to briefly discuss *heat pumps*. They can be used to move heat from one location to another. There are two general types of such systems. One involves the input of mechanical work to move low temperature heat from one location to another location at a higher temperature.

In the second type, which is sometimes called a chemical heat pump, the important feature is the temperature dependence of a chemical reaction. An example of this method is a system involving metal hydride materials in two locations, one that is at a high temperature in which there is heat input, and one that is cooler. Heat is moved by the transport of a chemical species, such as hydrogen, from one location to the other.

References

1. G. Beghi, ed. *Energy Storage and Transportation*. D. Reidel, Boston (1981)
2. J. Jensen and B. Sorenson. *Fundamentals of Energy Storage*. Wiley-Interscience, New York (1984)
3. I. Dincer and M.A. Rosen, eds. *Thermal Energy Storage*. Wiley, New York (2002)
4. A. Ter-Gazarian. *Energy Storage for Power Systems*. Peter Peregrinus, Stevenage (1994)
5. Physical Properties Data Compilations Relevant to Energy Storage. 1. Molten Salts: Eutectic Data, NSRDS-NBS 61, Part 1 (1978)
6. T. Foerland and J. Krogh-Moe. Acta Chem. Scand. *11*, 565 (1957)
7. T. Foerland and J. Krogh-Moe. Acta Cryst. *11*, 224 (1958)
8. A. Kvist and A. Bengtzelius. In *Fast Ion Transport in Solids*, ed. by W. van Gool. North Holland, Amsterdam (1973), p. 193
9. A. Lunden and J.O. Thomas. In *High Conductivity Solid Ionic Conductors*, ed. by T. Takahashi. World Scientific, Singapore (1989), p. 45
10. A. Lunden. In *Fast Ion Transport in Solids*, ed. by B. Scrosati, A. Magistris, C.M. Mari and G. Mariotto. Kluwer Academic Publishers, Dordrecht (1992), p. 181
11. K. Schroeder and C.-A. Sjöblom. High Temp. High Press. *12*, 327 (1980)

Chapter 4
Reversible Chemical Reactions

4.1 Introduction

In the discussion of thermal energy storage in Chap. 3, it was pointed out that energy can be stored in both the *sensible heat* that is related to changes in the temperature of materials and their heat capacities, and the *latent heat* involved in isothermal phase transitions. A common example of such an isothermal phase transition with a significant amount of stored heat is the melting and freezing of water. In such cases, there are no changes in chemical composition. The chemical species below and above the phase transition are the same. Only their physical state is different. Such reactions are said to be *congruent*.

Another type of energy storage involves reversible chemical reactions in which there is a change in the chemical species present. In many cases, such reactions can also be reversible. An example of this would be the reaction of hydrogen and oxygen to form water, and its reverse, the decomposition of water into hydrogen and oxygen. The energy, or heat, involved in this reversible reaction is the heat of reaction, and is sometimes called *quasi-latent heat*.

As was mentioned in Chap. 3, the amounts of heat involved in latent heat reactions can be much larger than those that are involved in sensible heat. As a result they are typically used in situations in which it is desired to maintain the temperature of a system at, or near, a constant value. As will be seen in this chapter, there are a number of materials in which chemical reactions can take place that can exhibit both *enhanced sensible heat* and *quasi-latent heat*.

4.2 Types of Non-congruent Chemical Reactions

In addition to the *congruent* reactions that have already been discussed, there are a number of different types of *chemical reactions* in which one or more materials react to produce products with different compositions. Reactions in which the reactants and products have different compositions are *non-congruent*.

R.A. Huggins, *Energy Storage*,
DOI 10.1007/978-1-4419-1024-0_4, © Springer Science+Business Media, LLC 2010

In some cases, the result of the reaction is a change in the chemical composition of one or more of the phases present by the addition of another species to it, or deletion of a species from it. Such reactions are called by the general label *insertion reactions*.

There are also a number of important chemical reactions in which some phases grow or shrink, or new ones form and others disappear. The result is that the microstructure of the material gets significantly changed, or reconstituted. Such reactions, of which there are several types, are called *reconstitution reactions*.

In order to see how to make use of them for energy storage, it is important to understand the major types of *reaction mechanisms*, as well as the *driving forces* that tend to cause such reactions to occur.

4.2.1 Insertion Reactions

Changes in the chemical composition of materials can be the result of the *insertion* of guest species into normally unoccupied crystallographic sites in the crystal structure of an existing stable *host material*. The opposite can also occur, in which atoms are *deleted* from crystallographic sites within the host material.

Such reactions, in which the composition of an existing phase is changed by the incorporation of guest species, can also be thought of as a *solution* of the guest into the host material. Therefore, such processes are also sometimes called *solid solution reactions*. In the particular case of the insertion of species into materials with layer-type crystal structures, insertion reactions are sometimes called *intercalation reactions*.

Although the chemical composition of the host phase initially present can be changed substantially, this type of reaction does not result in a change in the identity, the basic crystal structure, or amounts of the phases in the micro-structure.

However, in most cases, the addition of interstitial species into previously unoccupied locations in the structure, or their deletion, causes a change in volume. This involves mechanical stresses, and mechanical energy. The mechanical energy related to the insertion and extraction of interstitial species plays a significant role in the *hysteresis*, and thus energy loss, observed in a number of reversible battery electrode reactions.

Generally, the incorporation of such guest species occurs *topotactically*. This means that the guest species tend to be present at specific (low energy) locations inside the crystal structure of the host species.

A simple reaction of this type might be the reaction of an amount x of species A with a phase BX to produce the *solid solution* product A_xBX. This can be written as

$$xA + BX = A_xBX. \tag{4.1}$$

4.2.2 Formation Reactions

A different type of reaction results in the formation of a new phase. A simple example can be the reaction of species A with species B to form a new phase AB. This is a *formation reaction*, and can be represented simply by (4.2)

$$A + B = AB. \tag{4.2}$$

Since this modifies the microstructure, this is an example of one type of *reconstitution reaction*.

There are many examples of this type of formation reaction. There can also be subsequent additional formation reactions whereby other phases can be formed by further reaction of the product of an original reaction.

The *driving force* for the simple reaction in (4.2) is the difference in the values of the *standard Gibbs free energy of formation* of the products, only AB in this case, and the standard Gibbs free energies of formation of the reactants, A and B.

$$\Delta G_r^0 = \sum \Delta G_f^0 (\text{products}) - \sum \Delta G_f^0 (\text{reactants}). \tag{4.3}$$

The standard Gibbs free energy of formation of all elements is zero, so if A and B are simple elements, the value of the standard Gibbs free energy change, ΔG_r^0, that results per mol of this reaction is simply the standard Gibbs free energy of formation per mol of AB, $\Delta G_f^0 (AB)$. That is:

$$\Delta G_r^0 = \Delta G_f^0 (AB). \tag{4.4}$$

Values of the standard Gibbs free energy of formation for many materials can be found in a number of sources, for example, [1]. These values change with the temperature, due to the change in the product of the temperature and the entropy related to the reaction, as was discussed in Chap. 2.

In addition to the change in Gibbs free energy, there is a change in enthalpy, and thus the stored heat, related to such reactions. This is the difference between the enthalpies of the products and the reactants

$$\Delta H_r^0 = \sum \Delta H_f^0 (\text{products}) - \sum \Delta H_f^0 (\text{reactants}). \tag{4.5}$$

Values of the enthalpy of most materials are relatively temperature-insensitive, except when phase changes are taking place that involve structural changes, and therefore changes in the entropy.

4.2.3   Decomposition Reactions

A number of materials can undergo *decomposition reactions* in which one phase is converted into two related phases. This is the reverse of *formation reactions*, and can be represented simply as

$$AB = A + B. \tag{4.6}$$

There are two general types of these reactions. In one case, one phase decomposes into two other phases when it is cooled, and in the other, a single phase decomposes into two other phases when it is heated. There are specific names for these reactions. If a single liquid phase is converted into two solid phases upon cooling, it is called a *eutectic reaction*. On the other hand, if the high temperature single phase is also solid, rather than being a liquid, this is a *eutectoid reaction*.

In the case of a single solid phase decomposing into a solid phase and a liquid phase when it is heated, the mechanism is called a *peritectic reaction*. Likewise, if all phases are solids, what happens is called a *peritectoid reaction*.

These types of reactions will be discussed in more detail later.

4.2.4   Displacement Reactions

Another type of *reconstitution reaction* involves a *displacement* process that can be simply represented as

$$A + BX = AX + B, \tag{4.7}$$

in which species A displaces species B in the simple binary phase BX, to form AX instead. A new phase, consisting of elemental B, will be formed in addition. This will tend to occur if phase AX has a greater stability, that is, has a more negative value of $\Delta G_f{}^0$, than the phase BX. An example of this type is

$$Li + Cu_2O = Li_2O + Cu, \tag{4.8}$$

in which the reaction of lithium with Cu_2O results in the formation of two new phases, Li_2O and elemental copper. In this case,

$$\Delta G_r{}^0 = \Delta G_f{}^0(Li_2O) - \Delta G_f{}^0(Cu_2O), \tag{4.9}$$

since the standard Gibbs free energy of formation of all elements is zero.

The thermal effect of such a reaction is therefore

$$\Delta H_r{}^0 = \Delta H_f{}^0(Li_2O) - \Delta H_f{}^0(Cu_2O). \tag{4.10}$$

It is also possible to have a *displacement reaction* occur by the replacement of one interstitial species by another inside a stable host material. In this case, only one additional phase is formed, the material that is displaced. The term *extrusion* is sometimes used in the technical literature to describe this type of process.

In some cases, the new element or phase that is formed by such an *interstitial displacement process* is *crystalline*, whereas in other cases, it can be *amorphous*.

4.3   Phase Diagrams

Phase diagrams are *thinking tools* that are useful to help understand these various types of reactions, and associated phenomena. They are graphical representations that indicate the phases and their compositions that are present in a materials system under equilibrium conditions, and were often called *constitution diagrams* in the past. Thus, it is reasonable that reactions in which there is a change in the identity or amounts of the phases present are designated as *reconstitution reactions*.

Phase diagrams are widely used in materials science, and are especially useful in understanding the relationships between thermal treatments, microstructures and properties of metals, alloys and other solids.

4.3.1   The Gibbs Phase Rule

The *Phase Rule* was proposed by J. Willard Gibbs, a physicist, chemist, and mathematician, in the 1870s. Gibbs is often considered to have been one of the greatest American scientists. He is given credit for much of the theoretical foundation of chemical thermodynamics and physical chemistry. He introduced the concepts of chemical potential, free energy, made great contributions to statistical mechanics, and also invented vector analysis.

What is now generally called the *Gibbs phase rule* is an especially useful *thinking tool* when considering reactions within and between materials. It will appear a number of times in this text.

For present purposes, it can be written as

$$F = C - P + 2, \qquad (4.11)$$

in which C is the *number of components* (e.g., chemical elements), and P is the *number of phases present* in this materials system in a given experiment.

A phase is formally described as a distinct and homogeneous form of matter separated by its surface from other forms. That may not be very useful. But consider a mixture of salt and pepper. They are different phases, for it is possible to physically separate them as black and white particles. Likewise, water and ice can be identified and separated, even though they have the same chemical composition, H_2O. Therefore, they are separate phases.

The quantity F may also be difficult to understand. It is the *number of degrees of freedom*; that means the number of *intensive thermodynamic parameters* that must be specified in order to *define the system* and *all of its associated properties*.

Intensive parameters have values that are independent of the amount of material present. For this purpose, the most useful thermodynamic parameters are the temperature, the overall pressure, and either the chemical potential or the chemical composition of each of the phases present. In electrochemical systems, the electric potential can also be an important intensive parameter.

4.3.2   Binary Phase Diagrams

As mentioned earlier, *phase diagrams* are figures that graphically represent the equilibrium state of a chemical system. There are various types of phase diagrams, but in the most common case they are 2D plots that indicate the temperature and compositional conditions for the stability of various phases and their compositions under equilibrium conditions.

A *binary phase diagram* is a 2D plot of temperature versus the overall composition of materials (*alloys*) composed of two different components (elements). It shows the temperature-composition conditions for the stability and composition ranges of the various phases that can form in a given system, and is commonly used in materials science. It will be seen that there are temperature-composition regions in which only a single phase is stable, and regions in which two phases are stable.

A very simple binary system is shown in Fig. 4.1. In this case, it is assumed that the two components (elements) A and B are completely miscible, that is, they can dissolve in each other over the complete range of composition, from pure element A to pure element B, both when they are liquids and when they are solids. Thus, one can speak of a liquid solution and a solid solution in different regions of temperature-composition space. Since the elements have different melting points, the temperature above which their liquid solution is stable (called the *liquidus*) will vary with composition across the diagram. The temperature below which the material is completely solid (called the *solidus*) is also composition-dependent. At any composition, there is a range of temperature between the solidus and the liquidus within which two phases are present, a liquid solution and a solid solution.

Likewise, at any temperature between the melting points of pure A and pure B, the liquid and solid phases that are in equilibrium with each other have different compositions, corresponding to the two compositional limits of the two-phase region in the middle part of this type of phase diagram. This is shown schematically in Fig. 4.2.

The compositions of the solid and liquid phases that are in equilibrium with each other at a particular temperature of interest can be directly read off the composition scale at the bottom of the diagram. It is also obvious that these two compositions will both be different for other temperatures, due to the slopes of the solidus and

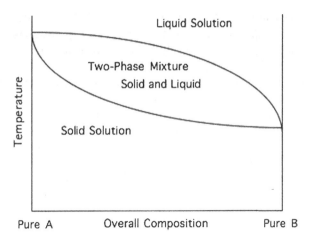

Fig. 4.1 Schematic phase diagram for binary system with complete miscibility in both the liquid and solid phases

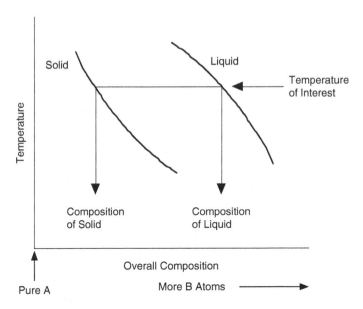

Fig. 4.2 Compositions of liquid and solid phases in equilibrium with each other at a particular temperature between the melting points of the two elements

liquidus curves. At a fixed temperature, the relative amounts, but not the compositions, of the two phases depend upon the overall composition.

The overall composition is, of course, not limited to the range within the two-phase mixture region of the phase diagram. Consider an isothermal experiment, and suppose that the overall composition were to start far to the left at the temperature of interest in the case illustrated in Fig. 4.2, perhaps as far as pure A. Now if more and more atoms of element B are continually added, a solid solution of B atoms in A will be formed whose composition will gradually rise. This can continue over a relatively wide compositional range until the overall composition reaches the solidus line that signifies the 1-phase/2-phase border.

Further addition of B atoms causes the overall composition to keep changing, of course. However, the composition of the solid solution cannot become indefinitely B-rich. Instead, when the overall composition arrives at the *solidus* line, some liquid phase begins to form. Its composition is different from that of the solid solution, being determined by the composition limit of the liquid solution at that temperature, the *liquidus* line. Thus, the microstructure contains two phases with quite different local compositions. Further changes in the overall composition at this temperature result in the formation of more and more of the fixed-composition liquid solution phase at the expense of the fixed-composition solid solution phase. By the time that the overall composition reaches that of the liquid solution (the liquidus line), there is no more of the solid phase left. Further addition of B atoms then causes the composition of the liquid solution to gradually become more and more B-rich.

This same type of behavior also occurs if all of the phases in the relevant part of a phase diagram are solids. The same rules apply. In any two-phase region at a fixed temperature, the compositions of the two end phases are constant, and variations of the overall composition are accomplished by changes in the amounts of the two fixed-composition phases. It is thus obvious that 1-phase regions are always separated by two-phase regions in such binary (two-component) phase diagrams at a constant temperature.

It can readily be seen that this behavior is consistent with the Gibbs phase rule, $F = C–P + 2$. For a binary, 2-component system, $C = 2$, A and B. In single-phase regions, $P = 1$. Therefore, $F = 3$. This means that three intensive parameters must be specified to determine such a system, and all of its properties. Thus, if the pressure is 1 atm, the specification of the composition and the temperature, two more intensive variables, determine all the properties in one-phase regions of this part of the phase diagram.

On the other hand, in two-phase region, that is, where a mixture of a solid solution and a liquid solution are both present, $C = 2$, and $P = 2$. Therefore, $F = 2$. This means that if the pressure is fixed, such as at 1 atm, only one other parameter must be specified to determine all the properties. This parameter might be the temperature, for example. As shown in Fig. 4.2, that will automatically determine the compositions of each of the two phases. Alternatively, if the composition of the solid solution were specified, that can only occur at one temperature, and with only one composition of the liquid solution.

4.3.3   The Lever Rule

The relation between the overall composition and the amounts of each of the phases present in a two-phase region of a binary phase diagram can be found by use of a simple mechanical analog, and is called the *lever rule*.

If two different masses M_1 and M_2 are hung on a bar that is supported by a fulcrum, the location of the fulcrum can be adjusted so that the bar will be in balance. This is shown in Fig. 4.3. The condition for balance is that the ratio of the lengths L_2 and L_1 is equal to the ratio of the masses M_1 and M_2. That is,

$$M_1/M_2 = L_2/L_1. \tag{4.12}$$

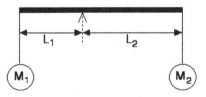

Fig. 4.3 Mechanical lever analog

Analogously, the amounts of the two phases in a two-phase region can be found from the lengths L_1 and L_2 on the composition scale. The ratio of the amounts of phases 1 and 2 is related to the ratio of the deviations of their compositions L_2 and L_1 from the overall composition on the composition scale. This is illustrated in Fig. 4.4, and can be expressed as

$$Q_1/Q_2 = L_2/L_1, \tag{4.13}$$

in which Q_1 and Q_2 represent the amounts of phases 1 and 2.

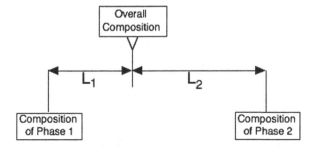

Fig. 4.4 Application of the lever rule to compositions in a two-phase region in a binary phase diagram

4.3.4 Three-Phase Reactions in Binary Systems

It was mentioned in Sect. 4.2.3 that a number of materials undergo decomposition reactions, that is, peritectic, eutectic, or their all-solid analogs, peritectoid and eutectoid, reactions upon changes in the temperature. These reactions all must involve three phases, A, B, and AB, or more generally, α, β and γ.

Considering again the Gibbs phase rule, in a binary system where $C = 2$, and $P = 3$, F must be 1. At a fixed pressure of 1 atm, no more parameters can be varied. What this means is that three phases can only be in equilibrium under one set of conditions. The compositions of each of the phases present can only have a single value, and the temperature must also be fixed.

Thus when three phase reactions take place at a fixed pressure in binary systems, all of the phases must be present and in equilibrium with each other at a specific temperature, and with particular compositions. Two will touch the isothermal composition line at the ends, and the third will do so at a single composition between them. Both above and below the unique peritectic temperature only one-phase and two-phase regions are possible.

In peritectic reactions one of the phases at the ends must be a liquid, whereas the other two phases are solids. This situation can be understood by the consideration of the simple binary system containing a peritectic reaction shown in Fig. 4.5. In this case, the terminal (end) phases on the A and B component sides of the phase diagram are labeled α and β.

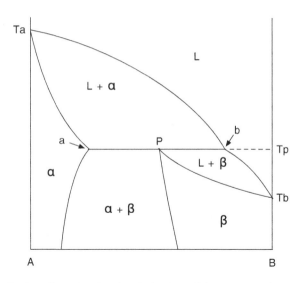

Fig. 4.5 Generalized configuration showing the isothermal arrangement of phases in a peritectic reaction

The melting points of components A and B are indicated by T_A and T_B, whereas the temperature of the peritectic reaction is T_P. At this temperature phase α has a composition a, phase β has a composition P, and the liquid a composition b.

The situation at the peritectic temperature can also be represented schematically as shown in Fig. 4.6.

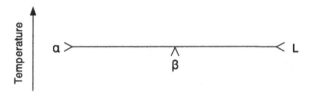

Fig. 4.6 Simple schematic representation of a peritectic reaction

4.3.5 Examples of Materials Systems with Peritectic Reactions

An example of a group of materials that undergo peritectic reactions is the family of salt hydrates. Upon heating, they form two product phases, a saturated aqueous solution plus a solid salt phase, at a specific transition temperature. This type of reaction could be represented as shown in Fig. 4.7. The transition temperature is the peritectic temperature.

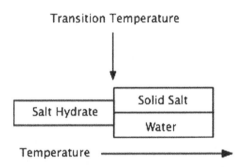

Fig. 4.7 Schematic representation of a salt hydrate decomposition reaction

The phase diagram for this type of system, in which the salt forms a monohydrate, is shown in Fig. 4.8.

Some examples of materials that can undergo this type of structural change at modest temperatures are included in Table 4.1. It can be seen that they can store a rather large amount of heat per unit volume.

The properties of some salt ammoniates that also undergo congruent reactions are presented in Table 4.2.

4.3.6 Binary Systems That Contain Eutectic Reactions

The discussion thus far has been about materials systems containing peritectic reactions. There are a number of cases in which eutectic reactions are present, in which a liquid phase decomposes to form two solid phases, α and β upon cooling. This can be represented schematically as shown in Fig. 4.9.

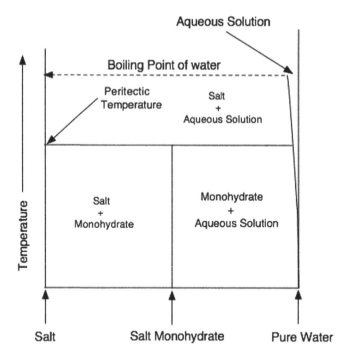

Fig. 4.8 Phase diagram of a salt–water system in which a salt monohydrate forms at low temperatures

Table 4.1 Data on several salt hydrates

Hydrate	Temperature (°C)	Heat of fusion (kJ kg^{-1})	Latent heat density (MJ m^{-3})
$CaCl_2 \cdot 6H_2O$	29	190.8	281
$Na_2SO_4 \cdot 10H_2O$	32		342
$Na_2CO_3 \cdot 10H_2O$	33		360
$CaBr_2 \cdot 6H_2O$	34	115.5	
$Na_2HPO_4 \cdot 12H_2O$	35		205
$Zn(NO_3)_2 \cdot 6H_2O$	36	146.9	
$Na_2HPO_4 \cdot 7H_2O$	48		302
$Na_2S_2O_3 \cdot 5H_2O$	48		346
$Ba(OH)_2 \cdot 8H_2O$	78	265.7	655
$Mg(NO_3)_2 \cdot 6H_2O$	89	162.8	162.8
$MgCl_2 \cdot 6H_2O$	117	168.6	168.6

A real example of this is the lead–tin metallurgical system, whose phase diagram is shown in Fig. 4.10. In this case, the liquid phase reacts to form the terminal lead and tin solid solutions at the eutectic temperature, 185°C.

The discussion thus far has assumed that the overall composition of the liquid is the same as the eutectic composition. This is, of course, not necessary.

Table 4.2 Data on several salt ammoniates

Reaction	Dissociation temperature at 1 bar (°C)	ΔH (kJ mol^{-1})
$CaCl_2 \cdot 8NH_3 = CaCl_2 \cdot 4NH_3 + 4NH_3$	27	184
$BaBr_2 \cdot 4NH_3 = BaBr_2 \cdot 2NH_3 + 2NH_3$	40	86
$LiCl \cdot 3NH_3 = LiCl \cdot 2NH_3 + NH_3$	56	46
$BaBr_2 \cdot 2NH_3 = BaBr_2 \cdot NH_3 + NH_3$	66	46
$MnCl_2 \cdot 6NH_3 = MnCl_2 \cdot 2NH_3 + 4NH_3$	87	200
$CuSO_4 \cdot 5NH_3 = CuSO_4 \cdot 4NH_3 + NH_3$	100	60

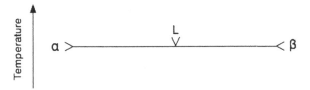

Fig. 4.9 Simple schematic representation of a eutectic reaction

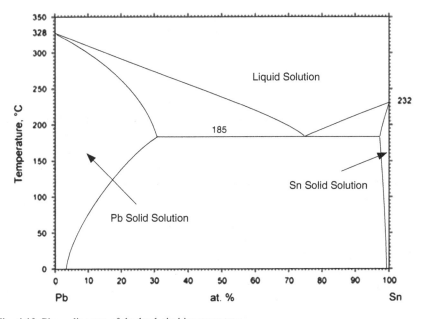

Fig. 4.10 Phase diagram of the lead–tin binary system

If the overall composition does not coincide with the *eutectic composition*, or the *peritectic composition* in the case of phase diagrams containing peritectic reactions, the overall compositions traverses a two-phase field during cooling (or heating). This is illustrated in Fig. 4.11 for a eutectic system.

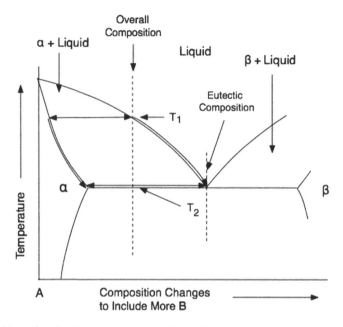

Fig. 4.11 Illustration showing how the compositions of both the solid phase α and the remaining liquid change with temperature above the eutectic temperature

In this case, a material with an overall composition on the A-rich side of the eutectic composition is cooled from a temperature at which it is entirely liquid. When the temperature reaches T_1, the phase α begins to form from the liquid. The composition of the α phase that initially forms is indicated by the left end of the arrow at that temperature. As the temperature decreases further, the compositions of both the solid α phase and the liquid continuously change, both incorporating more B atoms, as shown by the curved arrows along the phase boundaries. The compositional changes of both phase α and the remaining liquid in this temperature region are the result of a chemical reaction in which

$$\text{Liquid}_1 = \alpha + \text{Liquid}_2, \tag{4.14}$$

where the composition of α has a lower B content, and $Liquid_2$ has a higher B content, than $Liquid_1$. As with all chemical reactions, there are associated changes in Gibbs free energy and enthalpy, that is, heat. The implications of this are discussed in the following section.

4.4 Thermal Effects Related to Liquid and Solid Reactions

An experimental method that is often used to study phase transformations of materials is what is commonly called *thermal analysis*. In a simple example, the temperature is raised to a high value, the material is allowed to cool, and the temperature recorded as a function of time.

Changes in the slope of the resulting temperature-time *cooling curve* give information about the type of reactions that are taking place, and over what temperature intervals they occur. This is another example of a useful *thinking tool*. If the rate of extraction of heat is maintained constant, it is also possible to get quantitative thermodynamic information in this way.

As discussed in Chap. 3, the amount of heat given off during the cooling of a given mass of material is proportional to the product of the heat capacity and the temperature change. This can be expressed as

$$\int dq = \rho V C_P \int dT, \tag{4.15}$$

where q is the heat evolved, ρ the density, V the volume, C_P the heat capacity at constant volume, and T the temperature, if no chemical changes take place.

But when a chemical change also takes place as the temperature is changed in a cooling reaction, an additional heat term must be added to this relation. If the temperature-dependent reaction causes the evolution of heat, the *effective heat capacity* will be greater, with the result that the material will cool at a slower rate. This is illustrated schematically in Fig. 4.12 for a material with an overall composition shown in Fig. 4.11. The cooling curve has a relatively steep slope

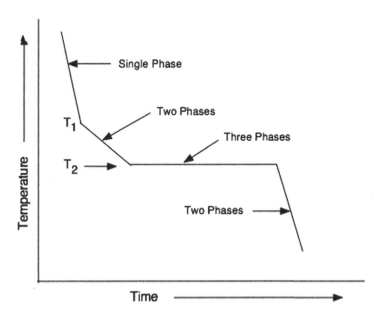

Fig. 4.12 Schematic temperature-time cooling curve showing changes in slope during a constant rate of heat extraction experiment with a material with the overall composition indicated in Fig. 4.11

when the material is entirely liquid. When temperature T_1 is reached upon cooling, the reaction indicated in (4.14) begins to take place, causing the generation of heat. Thus, the effective heat capacity is increased, and the rate of cooling, the slope of the cooling curve, decreases. When temperature T_2, the eutectic temperature, is reached, the composition of the remaining liquid is Liquid$_{eut}$, the eutectic composition, and there is no further cooling as the eutectic reaction

$$\text{Liquid}_{eut} = \alpha + \beta, \tag{4.16}$$

takes place. When that reaction has come to completion, and there is no more liquid left, the slope of the cooling curve again increases.

The characteristics of the cooling curve will change as the overall composition is varied. This is shown schematically in Figs. 4.13 and 4.14 for three different values of the overall composition.

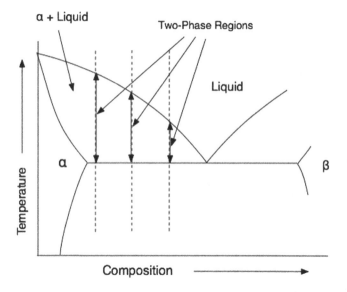

Fig. 4.13 Illustration of the temperature ranges over which the two-phase (α + Liquid) structure is formed for three different overall compositions

The thermal behavior of materials with this type of behavior can be used for temperature control or maintenance purposes. By changing the overall composition of the material being used, it is possible to vary the relative effects of the isothermal latent heat reaction and the enhanced sensible heat reaction that takes place over a range of temperature.

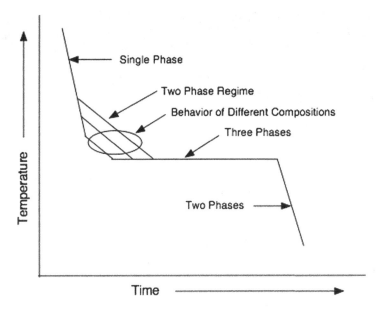

Fig. 4.14 Changes in the cooling curve as the initial overall composition is varied

4.5 Thermal Effects Related to Reversible Gas Phase Reactions

There are a number of simple gas phase reactions of the type

$$A + B = C + D. \qquad (4.17)$$

The relation between the Gibbs free energy, the enthalpy, and the temperature

$$G = H - TS \qquad (4.18)$$

was introduced in Chap. 2. Changes in G constitute the driving forces for all reactions, whereas changes in H indicate changes in heat content.

As has been mentioned earlier, the change in the Gibbs free energy resulting from a reaction of the type in (4.17) under standard conditions, ΔG_r^0, is the difference between the standard Gibbs free energies of formation of the products and of the reactants. Likewise, the change in enthalpy resulting from such a reaction, ΔG_r^0, is the difference between the standard enthalpies of formation of the products and the reactants.

Reactions such as (4.17) can generally take place over a substantial range of temperatures. The driving forces for such reactions can sometimes be reversed by changes in the temperature, depending upon the magnitudes of the entropy values of the species involved on both sides of the reaction, according to (4.18).

On the other hand, the magnitude of the heat generated or absorbed, determined by the enthalpies of the reactants and products, is relatively temperature-independent. However, it goes in the opposite direction if the reaction is reversed.

The first example of the use of this phenomenon to store and transmit heat involved the reaction of methane with steam, generally described as the *steam reforming of natural gas* [2, 3]. It is discussed further in Chap. 8, for it is commonly used to produce hydrogen from natural gas.

$$CH_4 + H_2O = CO + 3H_2. \tag{4.19}$$

The temperature dependence of the direction of this reaction can be understood by the calculation of the standard Gibbs free energy change $\Delta G_r{}^0$ from information about the standard Gibbs free energies of formation of the species involved

$$\Delta G_r{}^0 = \Delta G_f{}^0(CO) + 3\Delta G_f{}^0(H_2) - \Delta G_f{}^0(CH_4) - \Delta G_f{}^0(H_2O). \tag{4.20}$$

Values of the standard Gibbs free energy of formation of the species in this reaction for three different temperatures are given in Table 4.3.

Table 4.3 Temperature dependence of the standard Gibbs free energies of formation of species in reaction (4.20)

Species	$\Delta G_f{}^0$ (400 K) (kJ mol^{-1})	$\Delta G_f{}^0$ (800 K) (kJ mol^{-1})	$\Delta G_f{}^0$ (1,200 K) (kJ mol^{-1})
CO	−146.4	−182.5	−217.8
H_2O	−224.0	−203.6	−181.6
CH_4	−42.0	−2.1	+41.6
H_2	0	0	0

From these data, it is possible to obtain the standard Gibbs free energy of reaction as a function of temperature. The results are listed in Table 4.4, and plotted in Fig. 4.15. It is seen that this reaction will go forward at temperatures above about 900 K, and in the reverse direction at lower temperatures.

Table 4.4 Temperature dependence of standard Gibbs free energy of reaction (4.20)

Temperature (K)	$\Delta G_r{}^0$ (kJ mol^{-1})
400	+119.6
800	+23.2
1,200	−77.8

The thermal effects of this reaction can be obtained from information on the standard enthalpies of the species involved. These data are shown in Table 4.5. From these data, the temperature dependence of the standard enthalpy change of the reaction can be obtained. The results are listed in Table 4.6, and plotted in Fig. 4.16. As expected, there is not very much variation of the heat effect with temperature.

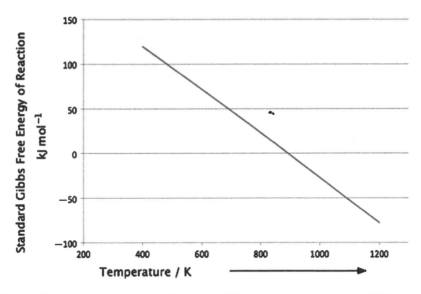

Fig. 4.15 Temperature dependence of the standard Gibbs free energy of reaction (4.20)

Table 4.5 Temperature dependence of the standard enthalpies of species in reaction (4.20)

Species	ΔH_f^0 (400 K) (kJ mol^{-1})	ΔH_f^0 (800 K) (kJ mol^{-1})	ΔH_f^0 (1,200 K) (kJ mol^{-1})
CO	−110.1	−110.9	−113.2
H$_2$O	−242.8	−246.4	−249.0
CH$_4$	−78.0	−87.3	−91.5
H$_2$	0	0	0

Table 4.6 Temperature dependence of standard enthalpy of reaction (4.20)

Temperature (K)	ΔH_r^0 (kJ mol^{-1})
400	+210.7
800	+222.8
1,200	+227.3

But the important point is that the reaction tends to go forward, and is endothermic, at high temperatures, and in the reverse, exothermic, direction at low temperatures.

The temperature-dependent reversal of this reaction can be used, for it can be driven at high temperatures at one location, where it is endothermic, and the product gases sent through pipes to a distant location, where the reaction is caused to react exothermically in the reverse direction, giving off heat. The same thing could be done in a single location, of course, using heat when it is available, and supplying it when it is needed.

A number of other reactions in the carbon–hydrogen–oxygen system can also be used for this purpose. Some of these are listed in Table 4.5.

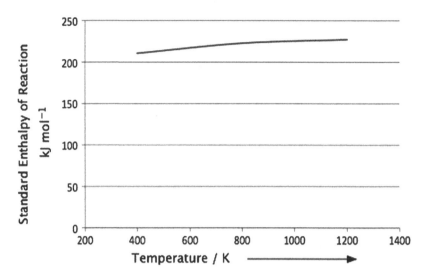

Fig. 4.16 Temperature dependence of the standard enthalpy of reaction (4.20)

Table 4.7 Reversible chemical reactions in the C–H–O system

System	Enthalpy of reaction, ΔH_r^0 (kJ mol^{-1})	Temperature range (K)
$C_{10}H_{18} = C_{10}H_8 + 5H_2$	314	450–700
$C_7H_{14} = C_7H_8 + 3H_2$	213	450–700
$C_6H_{12} = C_6H_6 + 3H_2$	207	500–700
$CH_4 + 2H_2O = CO_2 + 4H_2$	165	500–700
$CH_4 + CO_2 = 2CO + 2H_2$	247	700–1,200
$CH_4 + H_2O = CO + 3H_2$	225	700–1,200

References

1. I. Barin. *Thermochemical Data of Pure Substances*, 3rd Edition. VCH, Wienheim (1995)
2. R. Schulten, C.B. Van der Decken, K. Kugler and H. Barnert. Proc. British Nuclear Energy Soc. (1974)
3. R. Harth, J. Range and U. Boltendahl. In G. Beghi, ed. *Energy Storage and Transportation*. D. Riedel, Dordrecht (1981), p. 358

Chapter 5
Energy Storage in Organic Fuels

5.1 Introduction

The Preface included a discussion of the several types of natural materials that can be obtained from the Earth and used as fuels. The major ones are wood and the several fossil fuels, including the various types of coals, crude oil, and natural gas. The fossil fuels, which now play such a major role in the energy supply, will surely gradually become less important as they become depleted.

There are also the organic materials that can be considered to be renewable, the agrofuels, that contain significant amounts of vegetable oil, and the crops that are high in sugar or starch. These are often discussed in terms of *biomass* and *bioenergy*.

While these materials are basically fuels, and therefore energy carriers, they can also be thought of as energy storage media, for they accumulate energy as they grow that can be utilized in the future. But, in addition, some of them have other characteristics, such as being nutrients, or serving as raw materials for a number of industries.

Another aspect of biomass is that this category should also include living materials, including animals. There is, of course, a crossover between these categories, for animals consume agrofuels, and also contribute energy in the form of food. They can also provide mechanical energy.

5.2 Storage of Energy in Living Biomass

It has been estimated that the energy storage densities of living plants and animals are 10–30 MJ kg^{-1} of dry weight [1]. The question is how this energy can be usefully acquired. In the case of dry wood and straw, this energy can be converted to heat by burning. But most organic material contains a significant amount of water, and the energy that has to be furnished to dry it can exceed the amount that is obtained by its burning.

R.A. Huggins, *Energy Storage*,
DOI 10.1007/978-1-4419-1024-0_5, © Springer Science+Business Media, LLC 2010

There are other methods by which the energy contained in biomass can be retrieved, however. One of these is the consumption of biomass by a wide variety of living organisms as food without having to dry it.

The amount of energy stored in living biomass on the Earth is roughly 1.5×10^{22} J, and its average living residence time is about 3.5 years. There is a difference between living biomass in the oceans and on the land. Growth is generally more rapid in the oceans, mostly in the form of phytoplankton, which has a very short lifetime, of the order of a few weeks. On land, the biomass growth rate is much lower than in the oceans, but the average residence time is longer.

For biomass to be considered as a renewable energy storage mechanism, the rate of growth must be at least as rapid as the rate of the extraction of the energy by harvesting. This can be influenced in various ways, such as by the use of fertilizers and/or artificial irrigation or lighting. Growth typically involves the consumption of CO_2, which is also generally considered positive for other (environmental) reasons. It has been shown that increasing the concentration of CO_2 in the air can enhance the rate of growth of many plants, in some cases up to a factor of two [2]. This is done in commercial greenhouses in some places. Pest control can also be important.

The rapid growth of plants can be very impressive. The efficiency of the use of solar radiation to produce stored energy by a number of plant types is shown in Table 5.1 [2]. The efficiency is defined as the energy content of the crop divided by the accumulated incident solar energy for a given land surface area.

Table 5.1 Examples of high yield plant species

Plant	Short-term efficiency	Annual average efficiency	Annual yield $(kg\ m^{-2}\ year^{-1})$
Sugarcane		0.028	11.2
Napier grass	0.024		
Sorghum	0.032	0.009	3.6
Corn	0.032		
Alfalfa	0.014	0.007	2.9
Sugar beet	0.019	0.008	3.3
Chlorella	0.017		
Eucalyptus		0.013	5.4

These data are general averages, and actual yields can be increased by the use of fertilizers, well-designed irrigation, etc. The matter of irrigation is a critical world-wide problem, for the needs are growing more rapidly, partly because of the greater increase in human population, than sources are being developed.

As mentioned earlier, the situation is somewhat different for aquatic plants. Open ocean photosynthesis is dominated by phytoplankton. The limiting factor determining production is the availability of nutrients, particularly nitrogen and phosphate. Where these are readily available, for example, from agricultural runoff, the growth in nearby waters can be remarkable. This is not always desirable, however.

The author saw this happen in Clear Lake in Northern California, where there was evidently a lot of agricultural fertilizer runoff. He was there to participate in small sailboat racing. The water was so full of green algae marine growth that it was essentially opaque. It was like sailing through thick pea soup. This problem has evidently been subsequently alleviated.

The energy production in the North Atlantic Ocean is about 0.05 W m^{-2} [3]. But in coastal areas, the productivity typically rises to about 0.5 W m^{-2}. For coral reefs and areas where currents bring nutrient-rich water, such as along the Peru coast, the conversion efficiency can be 2–3% of the incident solar energy. Under ideal conditions, with an artificial nutrient supply, energy conversion efficiencies can reach 4% [4].

Despite these attractive data, the annual marine plant harvest is only about one million metric tons. Most of this is consumed as food in Japan and Korea. Most algae are rich in protein, and it would seem to make sense to increase their use in food. Unfortunately, pollution interferes with this in many areas.

5.3 Storage via Animals

Domesticated animals are fed mostly by plant material that has grown by accepting energy from the sun. Animals store energy acquired from the consumption of plants. This energy can be delivered in the form of food, both to humans, and in some cases, to other animals.

Roughly two-thirds of the total harvested plant material is fed to animals, yet it is estimated [1] that animal production constitutes only 14% of the total food production. This difference between the 33% input and the 14% output shows that the use of plant food to raise animals is actually not a very efficient use of the energy in the plants. There are, however, other arguments why this is done. These involve concerns about the desired distribution of meat, vegetable, and milk products in the human diet. Thus, energy efficiency is not always the dominant concern.

In the past, animals also have contributed mechanical energy to a number of important applications, such as transportation and the tilling of soil to assist agriculture. In the more affluent countries, these applications have been mostly taken over by mechanical alternatives, such as tractors, which consume fossil fuels.

There are also some other ways in which the stored energy in animals is utilized. In the Black Forest region of Germany, one notices that the older buildings were always constructed so that the animals were kept underneath the living quarters of the farming families. This was done to utilize the heat from the animals to increase the temperature of the living quarters in the winter.

Jensen [1] provided some interesting figures on several aspects of energy consumption, storage, and output related to domestic animals. He estimated that there are some 1.5×10^9 domesticated large animals in the world (cattle, horses, yaks, buffalo, donkeys, camels, etc.). About 400 million of these may be contributing mechanical work, at an average power level of 375 W per animal for some 6 h a day. If the average food intake is 600 W per animal, the efficiency of the conversion of the food energy to mechanical energy would be 16%. In many applications, however, only a fraction of the animals' mechanical power is actually used for a useful purpose. There is no useful energy output, although there is energy consumption, when a horse or cow merely walks around.

5.4   Hard Biomass

The growth of wood and other hard biomass by the absorption of solar energy, and its consumption, after at least some level of storage, by oxidation to provide heat for cooking, space heating, and other purposes is important in many relatively less highly developed societies. But much of this heat is actually wasted. For example, less than 10% of the heat combustion actually reaches the contents of a cooking pot. But, in addition, some of the energy is given off as heat to the surroundings. As a result, up to 50% of the heat energy can be actually used, although this depends to a considerable extent upon the design and operation of the stove. Some of the cast iron stoves used in Europe were actually quite good at converting and storing the heat of combustion.

5.5   Synthetic Liquid Fuels

A number of fuels can be made by the modification of species found in nature. The most interesting of these are methanol (CH_3OH), ethanol (C_2H_5OH), methane (CH_4), ammonia (NH_3), and methylcyclohexane (C_7H_{14}).

The production of methanol from coal involves the *steam reforming reaction*, followed by further reaction of CO with hydrogen

$$CO + 2H_2 = CH_3OH. \tag{5.1}$$

Alternatively, it is possible to react CO_2 with hydrogen to form methanol and water

$$CO_2 + 3H_2 = CH_3OH + H_2O. \tag{5.2}$$

Ethanol can be produced from CO_2 and H_2 in a similar way as methanol by upgrading the hydrogen content.

5.6   Gaseous Fuels Stored as Liquids

Some liquid fuels can readily be stored in small high-pressure tanks. This is often done in cases where a fuel is used for small scale heating or cooking, for example, in recreational vehicles, boats, or when camping. The fuel is kept as a liquid under pressure, but as soon as the pressure is reduced by opening a valve, it emits as a gas.

The approximate pressure at ambient temperature can be calculated from the boiling point at 1 atm pressure, T_{bp}, by the use of the ideal gas equation

$$PV = nRT,$$ (5.3)

where V is the volume, n is the number of moles and R the gas constant. So for a fixed volume and amount of material, the pressure at ambient temperature, assumed to be 298 K, is given by

$$P_{298} = 298/T_{bp}.$$ (5.4)

Some common examples are shown in Table 5.2.

Table 5.2 Data on materials commonly used for small scale heating

Material	Boiling temperature (K)	Pressure at 298 K (atm)
Butane	272.5	1.1
Propane	231	1.3
Methane	111	2.7

5.7 The Energy Content of Various Materials Used as Fuels

To put some of these data in perspective, the specific energies of a number of materials that are used as fuels are shown in Table 5.3. It is interesting to compare these numbers with the specific energy of the well-known explosive TNT, which is only 3.6 MJ kg^{-1}, which, when converted to electrical units, is about 1 kWh.

Table 5.3 Specific energy of various materials used as fuels

Fuel	Specific energy (MJ kg^{-1})
Crude oil	42
Coal	32
Dry wood	15
Hydrogen gas	120
Methanol	21
Ethanol	28
Propane	47
Butane	46
Gasoline	44
Diesel fuel	43

References

1. J. Jensen and B. Sorensen. *Fundamentals of Energy Storage.* Wiley-Interscience, New York (1984)
2. J. Bassham. Science *197*, 630 (1977)
3. E. Odum. *Ecology.* Holt-Reinhardt and Winston, New York (1972)
4. B. Sorensen. *Renewable Energy.* Academic Press, London (1979)

Chapter 6
Mechanical Energy Storage

6.1 Introduction

There are two basic types of energy storage that result from the application of forces upon material systems. One of these involves changes in potential energy, and the other involves changes in the motion of mass, and thus kinetic energy. This chapter will focus upon the major types of potential energy and kinetic energy storage. It will be seen that it is possible to translate between these two types of energy, as well as to convert these energies to heat or work.

6.2 Potential Energy Storage

Potential energy always involves the imposition of forces upon materials systems, and the energy stored is the integral of the force times the distance over which it operates. Thus

$$\text{Energy} = \int (\text{force})(\text{distance}). \qquad (6.1)$$

Consider the application of a tensile stress upon a solid rod, causing it to elongate. This is illustrated simply in Fig. 6.1.

The stress σ is the force per unit cross-sectional area, and the resultant fractional change in length $\Delta x/x_0$ is the strain ε.

In metals, the strain is proportional to the force, and this can be represented as a stress/strain diagram, as shown in Fig. 6.2. The proportionality constant is the Young's modulus Y, and this linear relation is called "Hooke's Law."

R.A. Huggins, *Energy Storage*,
DOI 10.1007/978-1-4419-1024-0_6, © Springer Science+Business Media, LLC 2010

Fig. 6.1 Simple example of the elongation of a solid rod as the result of an applied tensile force upon its ends

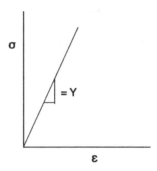

Fig. 6.2 Schematic stress/strain diagram for an elastic metal

If this mechanical deformation is elastic, the work W that is done on the spring is the area under the stress/strain curve. This is obviously proportional to the magnitude of the applied stress. That is

$$W = \frac{1}{2}\sigma\varepsilon = \frac{1}{2}\gamma\varepsilon^2. \tag{6.2}$$

If this mechanical process is reversible without any losses, the work is equal to the amount of stored energy in this simple system.

In metals and ceramics, Young's modulus is a constant up to a critical value of the stress, called the yield point. This is because the interatomic forces in such materials are linear at small displacements. At higher values of stress, however, there can be plastic (nonreversible) deformation, and then, ultimately, fracture.

In polymers and rubbers, Young's modulus can vary with the value of the strain, due to the action of different physical processes in their microstructures.

An example of a stress/strain curve for a common rubber is shown schematically in Fig. 6.3.

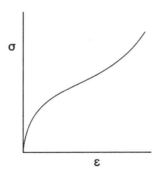

Fig. 6.3 Schematic stress/strain curve for rubber

The deformation of a metallic spring in a mechanical clock, and the use of stretched rubber bands to power model airplanes are simple examples of this type of stored mechanical potential energy.

6.3   Energy Storage in Pressurized Gas

Everyone who has had to pump up a bicycle tire knows that process requires work, and that the required force becomes greater as the pressure increases. If there is a leak, or the valve is opened, the gas stored in the tire is released. This is a simple example of the storage of energy in a gas.

It is possible to store energy by making use of the elastic properties of gases in a manner similar to that of the elastic properties of solids.

This can be readily understood by the consideration of the ideal gas law, or the equation of state of an ideal gas, that can be written as

$$PV = nRT, \qquad (6.3)$$

where P is the absolute pressure of the gas, V its volume, n the number of moles, R the gas constant, and T the absolute temperature. The value of R is 8.314 J mol^{-1} K^{-1}, or 0.082 L atm K^{-1} mol^{-1}. Using this latter value, the volume of a mole of gas can be readily found to be 22.4 L at 273 K or 0°C.

For a constant volume, such as that of a bicycle tire, the pressure is proportional to the amount of gas (air), n, that has been pumped into it. This can be simply represented as shown in Fig. 6.4, which is seen to be directly analogous to Fig. 6.2.

Gases can be compressed and stored in simple mechanical tanks, so long as the pressure is not so large as to cause mechanical damage. This is one of the ways in which the hydrogen used as the fuel in the fuel cells that are being developed for the

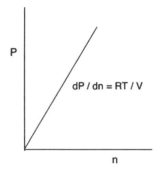

Fig. 6.4 Elastic behavior of gas under pressure

propulsion of vehicles is contained. Such tanks have been traditionally made of high strength metals, but carbon fiber composite materials, which can have greater strength and lighter weight, are becoming more attractive for use at high pressures.

The tanks that are used to store gaseous hydrogen to power the fuel cells in the automobiles currently under development can operate up to a pressure of 10^4 psi. One of these autos carries a total of 8 kg of hydrogen in its tanks.

Another alternative for this purpose is to store the hydrogen in solid metal hydride materials. That topic will not be discussed at this point, however. It appears in Chap. 8.

It is also possible to store gases under elevated pressure in underground cavities, if they are gas tight. This is the case for some large salt caverns, depleted oil wells, or underground aquifers.

Since this process is generally close to adiabatic if these are done rapidly, heat is given off during compression, and there is cooling during expansion. Some sort of heat transfer system must be employed to take care of this problem. This can be a serious consideration. For example, compression to a pressure of 70 atm can produce a temperature of about 1,000 K. This can cause the overall efficiency of a pressure storage system to be significantly reduced.

6.4 Potential Energy Storage Using Gravity

Instead of depending upon the elastic properties of solids or gases, there are energy production and storage methods that are based upon gravitational forces.

One example that is familiar to many people is a type of clock that is driven by the gravitational force on a mass, or "weight". Some of these are called "grandfather clocks", others are "cuckoo clocks". These types of clocks evolved from the realization by Galileo in the early 1600s that the period of the swing of a pendulum is independent of its amplitude, and that this phenomenon might be used for timekeeping. This led to the invention of the pendulum clock by Christiaan Huygens in 1656, which was shortly followed by the invention of the anchor escapement mechanism by Robert Hooke in 1657.

"Cuckoo clocks" driven by weights have been produced in the Black Forest area of Germany since the middle of the 1700s. They typically include a moving bird, and a small bellow is used to make the bird call sound. These have been popular tourist items for some time.

These pendulum-regulated clocks are driven by the action of gravity upon a weight, instead of a metallic spring. This mechanism is illustrated schematically in Fig. 6.5.

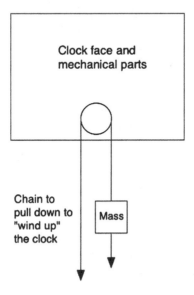

Fig. 6.5 Schematic illustration of the mechanism used to provide energy to pendulum-regulated clocks

In these cases, the potential energy involves the attractive force between two bodies, W_{pot}, where

$$W_{pot} = -G\frac{Mm}{r},\qquad(6.4)$$

G is the gravitational constant (6.67×10^{-11} m^3 kg^{-1} s^{-2}), M is the mass of the earth (5.98×10^{24} kg), m is the mass being moved, and r is the distance between their centers. Thus

$$\Delta W_{pot} = -GMm\left(\frac{1}{r+\Delta r} - \frac{1}{r}\right) = mg\Delta r,\qquad(6.5)$$

where

$$g = GMr^{-2} = 9.81 \text{ ms}^{-2}.\qquad(6.6)$$

6.5 Hydroelectric Power

Water is evaporated from the Earth's surface by solar energy as part of the global climate cycle. This evaporation is partly from the land masses, but primarily from the world's oceans. The moisture rises and condenses to form clouds in the sky, which are transported by the global air circulation. This moisture can later precipitate in the form of rain or snow, sometimes at higher elevations. The water from the rain and snow that fall at high altitudes can be stored in reservoirs, from which it can be run through turbines to lower elevations, producing electricity.

This "hydroelectric power" is a major source of electrical energy in a number of countries, including Switzerland and Norway. It is also an important component of the total energy picture in parts of the United States.

One of its major advantages is that the flow through the turbines can be turned on and off in response to the current need. This is not instantaneous, however, for there is a startup time for the turbines of the order of a few minutes.

There are some disadvantages to this method of energy acquisition and storage as well, for the large reservoirs and their related water collection areas can require a considerable amount of real estate, and this can become a political problem. In addition, the dams, themselves, can be quite expensive.

The use of the gravitation force on collected water to produce energy can also involve much smaller scale facilities. Years ago there were many small water-wheels that were driven by falling water to provide either mechanical or electrical output. These were generally located on smaller waterways, and often used modest mill-ponds to store the water before it was fed onto the water mills.

Another form of hydroelectric energy production takes advantage of the tidal rise and fall of the ocean surface as the result of the gravitational forces of the moon and sun, coupled with the effects of the Earth's rotation. Variations in the wind can also have a temporary effect.

The gravitational effect of the moon would theoretically cause the surface of the ocean to rise about 54 cm at its highest point, if it were to have a uniform depth, if there were no land masses, and if the earth were not rotating. The gravitational effect of the sun is somewhat smaller, theoretically producing an amplitude of about 25 cm under comparable conditions. Tides rise and fall with a cycle time near 12 h, with their magnitudes dependent upon the relative positions of the sun and the moon. Tides with the greatest amplitudes occur when the sun and moon are in line, and are called "spring tides." Those with the smallest amplitudes are "neap tides".

Tidal amplitudes vary greatly from place to place as well, depending upon variations in local ocean depth as well as the nearby underwater land mass topography. In some locations the difference between low and high tide is quite large. Examples include some areas near the English Channel, along the coast of New Zealand, and in the Bay of Fundy and Ungava Bay in Eastern Canada.

Programs have been initiated to make use of the large variations in water level at several locations by the construction of "impoundment ponds" that admit and release seawater through turbines. The power that is generated in this way is, of course, periodic, related to the timing of the tides.

6.6   Pumped-Hydro Storage

A modification of hydroelectric storage and power is called "pumped-hydro" storage. The general configuration in this case involves water storage facilities at two different elevations. They can either be natural or artificially constructed, and can have a wide range of sizes. They could include underground caverns, old mine shafts, volumes formerly occupied by oil, or newly excavated volume.

Water can be run through turbines from the upper one to the lower one, producing electricity, as in simple hydroelectric power systems. But then water can be pumped back up to the storage area at the higher elevation, effectively recharging the system. In some cases, this involves the use of two-way turbines. This can make sense if the price of electricity varies significantly at different times of the day or the week. This type of energy storage can be especially useful in connection with daily peak shaving and load leveling, as well as weekly and seasonal variations in the energy demand.

This scheme is illustrated schematically in Fig. 6.6.

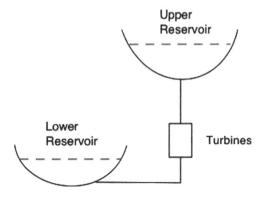

Fig. 6.6 Schematic illustration of a pumped-hydro system

Whereas the efficiency of large-scale water-driven turbines can be quite high, even over 95%, the efficiency of the dual cycle reversible storage system typically is about 80%. There are other losses, of course, such as water evaporation from one or both of the reservoirs, leakage around the turbine, and losses due to friction of the moving water.

There are many such pumped-hydro storage systems in the world. Some of them are listed in Table 6.1.

Table 6.1 Examples of large pumped hydro systems

Country	Name	Capacity (MW)
Argentina	Rio Grande-Cerro Pelado	750
Australia	Tumut Three	1,500
Austria	Malta-Haupstufe	730
Bulgaria	PAVEC Chaira	864
China	Guangzhou	2,400
France	Montezic	920
Germany	Goldisthal	1,060
	Markersbach	1,050
India	Purulia	900
Iran	Siah Bisheh	1,140
Italy	Chiotas	1,184
Japan	Kannagawa	2,700
Russia	Zagorsk	1,320
Switzerland	Lac des Dix	2,099
Taiwan	Mingtan	1,620
United Kingdom	Dinorwig, Wales	1,728
United States	Castaic Dam	1,566
	Pyramid Lake	1,495
	Mount Elbert	1,212
	Northfield Mountain	1,080
	Ludington	1,872
	Mt. Hope	2,000
	Blenheim-Gilboa	1,200
	Raccoon Mountain	1,530
	Bath County	2,710

6.7 Use of the Kinetic Energy in Moving Water

It is possible to extract power from moving water by the immersion of a water-driven propeller or turbine. This could be done in flowing rivers, where the flow is relatively constant with time. It can also be done in locations in which there are significant tidal currents. In this case, however, the current and, therefore the power available, are periodic, with rather substantial periods between.

Whereas this may appear to be rather simple, some practical matters require attention. These include seawater corrosion and the growth of barnacles and other biological species on underwater surfaces. Some recent designs involve retractable propellers that can be periodically cleaned.

6.8 Kinetic Energy in Mechanical Systems

In addition to potential energy, it is also possible to store kinetic energy. This is energy that is related to the motion of mass. This will be discussed in two parts, linear motion of a mass, and rotational motion of mass. More information on this topic can be found in [1].

6.8.1 Linear Kinetic Energy

The kinetic energy E_{kin} related to a body in linear motion can be written as

$$E_{kin} = \frac{1}{2}mv^2, \tag{6.7}$$

where m is its mass, and v its linear velocity.

A simple hammer is an example of this principle. The kinetic energy in its moving mass is utilized to drive nails, as well as for other purposes.

A somewhat more exotic example involves hybrid automobiles. In some of the current hybrid internal combustion/electric vehicles, an electric motor/generator in the drive train drives the wheels. It is fed from a high-rate battery, or sometimes an *ultracapacitor*, or a *supercapacitor* that is recharged as needed by an efficient internal combustion engine.

When the vehicle is slowed down, or braked, the motor/generator operates in reverse, and some of the vehicle's kinetic energy is converted to electrical energy that is fed back into the battery. This energy can then be used subsequently for propulsion. The amount of this recovered energy is typically about 10% of the basic propulsion energy in urban driving. This system is shown schematically in Fig. 6.7.

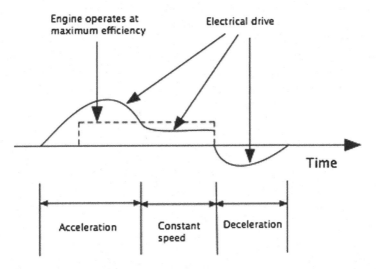

Fig. 6.7 Schematic representation of the interaction between the internal combustion engine and the battery in a hybrid vehicle

A variant on this model is the "plug-in hybrid," in which the net energy consumed from the battery is replaced by the use of an electrical recharger when the vehicle is not being used. This might involve connection to the electrical system of a home overnight, for example.

6.8.2 *Rotational Kinetic Energy*

Kinetic energy is also present when a body is rotated. In this case,

$$E_{\text{kin}} = I\omega^2, \tag{6.8}$$

where I is the moment of inertia and ω is the angular velocity.

Flywheels can be used for the purpose of storing kinetic energy, and there are ready methods whereby this mechanical energy can be converted to and from electrical energy. Current flywheels can store up to about 125 Wh kg^{-1} of energy, and can have capacities above 2 kWh.

The moment of inertia of a rotating body can be expressed as

$$I = \int \rho(x) r^2 \text{d}x, \tag{6.9}$$

where $\rho(x)$ is the mass distribution, and r is the distance from the center of rotation.

Thus, the magnitude of the kinetic energy in a rotational system is increased if the rotational velocity ω and I are large. The latter can be achieved by having a large mass at a large value of r. Rotational velocities can be very large, typically 20,000–100,000 rpm.

Aerodynamic drag can be substantial in high velocity flywheels. As a result, they are typically operated under vacuum conditions. It is also general practice to use magnetic bearings.

Consideration must be given to the strength of the material from which the flywheel is constructed, for it must be able to withstand the centrifugal force. This can be represented schematically as shown in Fig. 6.8.

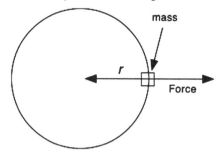

Fig. 6.8 Schematic representation of the centrifugal force operating on a mass in rotation

The centrifugal force is given by

$$\text{Force} = (\text{mass})(\text{acceleration}) = mr\omega^2. \tag{6.10}$$

Flywheels can have a variety of shapes. One optimization strategy is to use a disk design in which the stress is the same everywhere. If the material with which the flywheel is constructed is uniform, this results in a shape such as that illustrated schematically in Figs. 6.9 and 6.10.

The local thickness b is given by

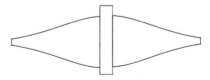

Fig. 6.9 Schematic drawing of the cross-section of a disk upon a central shaft in which the centrifugal stress in the disk is the same everywhere

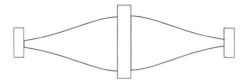

Fig. 6.10 Constant stress disk with constant thickness outer rim

$$b = b_0 \exp\left[(\text{constant})(r^2)\right]. \tag{6.11}$$

The radial strain causes tangential strain, and thus tangential stress. Thus if such a flywheel were to be reinforced by the use of high strength carbon fibers, this is done by placement of the fibers around the outside, so that they carry the tangential force.

In this case, the specific energy, the kinetic energy per unit mass, in the flywheel is given by

$$\frac{E_{\text{kin}}}{\text{mass}} = \frac{\sigma_{\text{max}}}{\rho} K_m, \tag{6.12}$$

where σ_{max} is the maximum allowable stress, ρ is the material density, and K_m the shape factor. Note that the ratio σ_{max}/ρ leads to a strength-to-weight ratio, not just the material's strength.

Some values of the shape factor are given in Table 6.2.

Table 6.2 Values of the shape factor for several simple disk shapes

Shape	K_m
Brush shape	0.33
Flat disk	0.6
Constant stress disk	1.0
Thin rim only	0.5
Thin rim on constant stress disk	0.6–1.0

Because of the advantage of operation at as high a stress as possible, there is a tendency to construct high performance flywheels of fiber reinforced materials, either Kevlar or carbon. These materials are, however, not isotropic, so that simple disk shapes are not practical. Instead, constant stress central shapes with high stress rims are often used. By using such materials it is possible to achieve energy storage values as high as 200 kJ kg^{-1} in modern flywheels. A range of flywheel types, and their characteristics are shown in Table 6.3.

Table 6.3 Examples of flywheel characteristics

Object	K, shape factor	Mass (kg)	Diameter (m)	Angular velocity (rpm)	Energy stored	Energy stored (kWh)
Bicycle wheel	1	1	0.7	150	15 J	4×10^{-7}
Flintstone's stone wheel	0.5	245	0.5	200	1,680 J	4.7×10^{-4}
Train wheel, 60 km h^{-1}	0.5	942	1	318	65,000 J	1.8×10^{-2}
Large truck wheel, 18 mph	0.5	1,000	2	79	17,000 J	4.8×10^{-3}
Train braking flywheel	0.5	3,000	0.5	8,000	33 MJ	9.1
Electrical power backup flywheel	0.5	600	0.5	30,000	92 MJ	26

It should be pointed out that flywheels can be dangerous. If they get out of balance or begin to come apart, their parts can become very high velocity projectiles. Thus, it is safer to construct them of many small pieces. This is the reason for the circular brush shape concept. They are generally housed in very robust steel containment, and large units are placed underground for safety reasons.

Flywheels also store energy in the form of mechanical strain potential energy – like springs – due to the forces upon them. The magnitude of this potential energy is typically small, for example, 5%, compared to their kinetic energy, however.

Another consideration in the use of flywheels is the rate at which energy can be added or deleted, that is, their power. The maximum power that can be applied or extracted is determined by the mechanical properties of the central shaft. The maximum torque τ_{max} that can be withstood can be expressed as

$$\tau_{max} = \frac{2}{3}\pi\sigma_s R_0{}^3, \tag{6.13}$$

where R_0 is the shaft radius and σ_s is the maximum shear strength of the shaft material.

The torque involved in a change in the rotational velocity of the flywheel

$$\tau = I\frac{d\omega}{dt}, \tag{6.14}$$

for a disk of radius R and thickness T,

$$\tau = \frac{\pi}{2}\rho TR^4 \frac{d\omega}{dt}. \tag{6.15}$$

The maximum possible acceleration can be obtained by setting this value of τ equal to the value of τ_{max} calculated above.

$$\frac{d\omega}{dt} = \frac{4}{3}\left(\frac{\sigma_s}{\rho RT}\right)\left(\frac{R_0}{R}\right)^3. \tag{6.16}$$

The power is the rate of change of the kinetic energy. So the maximum power is

$$P_{max} = \frac{\pi}{2}\rho \omega TR^4 \frac{d\omega_{max}}{dt} = \frac{\pi}{3}R_0^3 \sigma_{max}, \tag{6.17}$$

for the case in which the strength of material from which the shaft is made is the same as the strength of the flywheel material.

But the maximum rotational velocity of a flywheel is related to the strength of its material by

$$\omega_{max} = \frac{1}{R}\left(\frac{2\sigma_{max}}{\rho}\right)^{1/2}. \tag{6.18}$$

To see the magnitudes involved, consider a flywheel with a weight of 4.54 kg, $R = 17$ cm, $T = R_0 = R/10$, $\sigma_{max} = 1.5 \times 10^6$ lb in^{-2} and $\rho = 3.0$.

If ω_{max} is 1.6×10^6 rad/s, the maximum power, P_{max} is 5×10^8 W, or 10^6 W kg^{-1}.

These are very large numbers. Thus flywheels are very good at handling high power, and therefore energy transients.

For comparison, the power per unit weight of a typical battery might be of the order of 100 W kg^{-1}.

6.9 Internal Structural Energy Storage

It is also possible to introduce a different type of mechanical energy into solid materials by plastically deforming them such that changes occur in their microstructure. These can involve changes in the concentrations or distributions of dislocations or crystallographic point defects that form during mechanical deformation or irradiation. In some extreme cases, such as heavy forging at elevated temperatures, this can become quite evident, for it can be seen from color changes that the extensive mechanical deformation causes the internal temperature to rise. Although at least some of this energy can be recovered upon annealing, this increase of the internal energy in solids is not readily reversible, and is therefore not of interest for the types of applications that are generally considered in this text.

References

1. G. Genta. *Kinetic Energy Storage*. Butterworths, London (1985)

Chapter 7
Electromagnetic Energy Storage

7.1 Introduction

Several of the prior chapters in this text have shown that there is a wide range of energy storage needs with widely different time periods. Some involve seasonal, weekly, or daily cycles, and others require energy intermittently, sometimes over much shorter time periods. A variety of different technologies are employed to meet these requirements.

This chapter deals with two general mechanisms by which electrical energy can be stored. One involves capacitors, in which energy can be stored by the separation of negative and positive electrical charges. The other involves the relationship between electrical and magnetic phenomena.

It will be seen that both of these mechanisms are most applicable to situations in which there is a requirement for the storage of modest amounts of energy under very transient conditions, for relatively short times and sometimes at high rates. Such applications, therefore, emphasize fast kinetics and high power, rather than the amount of energy that can be stored. A very long cycle life is also generally very important. It will be seen later that the amount of energy that can be stored by such methods, however, is generally much less than can be stored by chemical and electrochemical methods.

The range of both current and potential future requirements and applications is very broad. In addition to the reduction in short-term transients in the large electrical power distribution grid system mentioned in Chap. 1, typical examples that are now highly visible include digital communication devices that require pulses in the millisecond range, implanted medical devices that require pulses with characteristic times of the order of seconds, and hybrid vehicle traction applications, where the high power demand can extend from seconds up to minutes, and the ability to absorb large currents upon braking is also important.

There are two general approaches to the solution of these types of requirements. One involves the use of electrical devices and systems in which energy is stored in materials and configurations that exhibit capacitor-like characteristics. The other involves the storage of energy in electromagnets. These are discussed in the following sections.

R.A. Huggins, *Energy Storage*,
DOI 10.1007/978-1-4419-1024-0_7, © Springer Science+Business Media, LLC 2010

7.2   Energy Storage in Capacitors

Energy can be reversibly stored in materials within electric fields and in the vicinity of interfaces in devices called capacitors. There are two general types of such devices, and they can have a wide range of values of the important practical parameters, the amount of energy that can be stored, and the rate at which it can be absorbed and released.

7.2.1   Energy in a Parallel Plate Capacitor

In a parallel plate capacitor, such as that shown schematically in Fig. 7.1, a dielectric material sits between two metallic plates. If an electrical potential difference is applied between the plates, there will be an electric field across the material located between them that causes local displacements of the negative and positive charges within it.

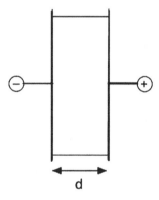

Fig. 7.1 Schematic view of a simple parallel plate capacitor

The energy stored W_C in a material between capacitor plates of area A separated by a distance d, is given by

$$W_C = \frac{1}{2}\varepsilon A \frac{V^2}{d} \tag{7.1}$$

where ε is the permittivity of the material between the plates, and V is the applied voltage.

The capacity C of this configuration is given by

$$C = \varepsilon \frac{A}{d} \tag{7.2}$$

and

$$\varepsilon = \varepsilon_r \varepsilon_0 \tag{7.3}$$

where ε_r is the relative permittivity of the material, and ε_0 is the permittivity of a vacuum, 8.854×10^{-12} F/m. The permittivity was sometimes called the dielectric constant in the past.

Values of the relative permittivity of several materials are shown in Table 7.1.

Table 7.1 Values of the relative permittivity of some materials at ambient temperature

Material	ε_r
Nylon	2.1
Teflon	2.6
Bakelite	4.9
Soft glass	6–7
Distilled water	80
High permeability oxides	$10–15 \times 10^3$

By substituting (7.2) and (7.3) into (7.1), the energy stored in the capacitor can be written as

$$W_C = \frac{1}{2} C V^2 \tag{7.4}$$

Since the charge stored in a capacitor, Q, is related to its capacitance and the voltage by

$$Q = CV \tag{7.5}$$

the stored energy can also be expressed as

$$W_C = \frac{1}{2} Q V \tag{7.6}$$

It is interesting that the amount of the energy stored in such a capacitor is inversely proportional to the volume of the dielectric material between the plates of the capacitor. This is counter-intuitive, for we normally think that if a material carries energy, the greater the amount of the material, the greater amount of energy that it contains.

7.3 Electrochemical Charge Storage Mechanisms

The behavior of such a parallel plate configuration becomes very different if the space between the electrodes contains a material that has the properties of a liquid, or even a solid, electrolyte. In such a case, the charge is stored in the interfacial region where the electronically conducting material meets the ionically conducting material. The physical separation between the positive and the balancing negative charges is then very small. This can be thought of as equivalent to a parallel plate capacitor with a thickness of the order of interatomic distances, and results in much greater amounts of charge storage per unit area.

Devices with this type of local structure are called electrochemical capacitors, and there are two general types. One involves the storage of charge in the electrical double-layer at or near the electrolyte/electronic material interface, and such devices are called *ultracapacitors*. The other type makes use of the transient additional reversible absorption of atomic species into the crystal structure of the solid electronically conducting electrode, and such devices are called *supercapacitors*. Both of these mechanisms can lead to much larger values of capacitance than capacitors with dielectric materials between their plates discussed above.

7.3.1   Electrostatic Energy Storage in the Electrical Double-Layer in the Vicinity of an Electrolyte/Electrode Interface

As mentioned above, the interface between a chemically inert electronic conductor electrode and an adjacent electrolyte with mobile ionic charges can function as a simple capacitor with a very small distance separating two parallel plates.

The amount of charge that can be stored in such a configuration is generally of the order of 15–40 $\mu F/cm^2$ of interface [1–3]. Thus, efforts are made to optimize the amount of interface in the device microstructure. Techniques have been devised to produce various types of carbon, as well as some other electronically conducting, but chemically inert, materials in very highly dispersed form, leading to very large interfacial areas. Typical values of specific capacitance, in Farads per gram, of a number of double-layer electrode materials are included in Table 7.2.

Table 7.2 Characteristics of some double-layer electrode materials

Electrode material	Specific capacitance(F/g)
Graphite paper	0.13
Carbon cloth	35
Aerogel carbon	30–40
Cellulose-based foamed carbon	70–180

The potential difference that can be applied is limited by the decomposition voltage of the electrolyte, which is 1.23 V for aqueous electrolytes, but can be up to 4–5 V for some organic solvent electrolytes. The result is that the specific energies of aqueous electrolyte systems of this type are generally in the range of 1–1.5 W/kg, whereas those that use organic solvent electrolytes can be 7–10 W/kg.

A device in which this is the dominant charge storage mechanism will behave like a pure capacitor in series with its internal resistance. Its time constant is equal to the product of the capacitance and the series resistance. Thus, it is important to keep the resistance as low as possible if rapid response is desired. Organic electrolytes characteristically have much lower values of ionic conductivity, and thus provide greater resistance, and longer time constants than aqueous electrolytes. The conductivity of acids, such as H_2SO_4, is somewhat greater than that of aqueous bases. Furthermore, the larger the capacitance, the greater the time constant, the slower the device, and the lower the power level.

Another important feature of devices that operate by the double-layer mechanism is that the amount of charge stored is a linear function of the voltage according to the Equation (7.5). The voltage therefore falls linearly with the amount of charge extracted. Thus, voltage-dependent applications can only utilize a fraction of the total energy stored in such systems. The power supplied to resistive applications is proportional to the square of the instantaneous voltage, so this can be an important limitation.

Although there is some confusion in the literature about terminology, devices of this type that have been developed with large values of such capacitance have generally been called either EDLC (electrical double layer capacitive) devices, or *ultracapacitors* [1–3]. The label *supercapacitors* is now becoming more and more common, and will be discussed later.

Ultracapacitor devices utilizing the storage of charge in the electrochemical double layer have been developed and produced in large numbers in Japan for a considerable period of time [4]. These are primarily used for semiconductor memory backup purposes, as well as for several types of small actuators.

7.3.2 Underpotential Faradaic Two-Dimensional Adsorption on the Surface of a Solid Electrode

Due to the characteristics of the electrolyte/electrode surface structure and its related thermodynamics, it is often found that modest amounts of *Faradaic electrodeposition* can occur at potentials somewhat removed from those needed for the bulk deposition of a new phase. This results in the occupation of specific crystallographic sites on the surface of the solid electrode. This mechanism typically results in only partial surface coverage, and thus the production of an *adsorption pseudo-capacitance* of some 200–400 $\mu F/cm^2$ of interfacial area [2]. This is significantly larger than the amount of charge stored per unit area in the electrochemical double layer. However, materials with which this mechanism can be effectively used are rare, and it is not common.

7.3.3 Faradaic Deposition That Results in the Three-Dimensional Absorption of the Electroactive Species into the Bulk Solid Electrode Material by an Insertion Reaction

As will be discussed in later chapters, many materials are now known in which atoms from the electrolyte can move into the surface of a solid electrode material.

These are called *solid solution electrodes*. In such cases, the *electroactive species* diffuses into and out of the interior of the crystal structure of the solid electrode as its potential is changed. Since the amount of energy stored is proportional to the amount of the electroactive species that can be absorbed by the electrode, this bulk storage mechanism can lead to much higher values of energy storage per unit volume of electrode structure than any surface-related process. Because it makes no sense to express this bulk phenomenon in terms of the capacitance per unit interfacial area, for it depends upon how far into the solid the material from the electrolyte penetrates, values of capacitance are generally given as *Farads per gram* in this case.

This mechanism was called *redox pseudo-capacitance* by the Conway group [1–3]. They also first started the use of the term *supercapacitors* to describe devices utilizing this type of charge storage. In this way, solid solution bulk storage *supercapacitors* can be distinguished from double-layer storage *ultracapacitors* in which atoms from the electrolyte do not enter the solid electrode material.

The bulk storage *supercapacitor* mechanism is utilized in the devices that are most interesting for energy-sensitive pulse applications. Since the kinetic behavior of such devices is related to the electrolyte/electrode area, it is important that they also have very fine large surface area microstructures.

During investigations of the *dimensionally stable electrodes* that are used as positive electrodes in the *chlor-alkali process,* it was noticed that RuO_2 seemed to behave as though its interface with the electrolyte had an unusually large capacitance. This swiftly led to several investigations of the capacitive behavior of such materials [5–8].

The possibility of the development of RuO_2-type materials as commercial capacitors began in Canada around 1975, the key players being D.R. Craig [9, 10] and B.E. Conway. This soon evolved into a proprietary development program at the Continental Group, Inc. laboratory in California, which subsequently was taken over by Pinnacle Research Institute. The products that were initially developed and manufactured were all oriented toward the military market. This orientation is changing now, and activities are being undertaken by several firms to produce this type of product for the civilian market. Activities in this area have also been initiated more recently in Europe, including work at Daimler-Benz (Dornier) and Siemens in Germany, and at Thompson-CSF in France.

It was originally thought that charge storage in these materials is due to *redox reactions* at or very near the interface between the electrolyte and RuO_2. Careful measurements showed that the capacitance is large, and proportional to the surface area [11].

It is now known, however, that the charge is stored by hydrogen insertion into the bulk of the RuO_2, and the capacitive behavior is not just on the surface [12]. At relatively short times, the depth of hydrogen diffusion will be limited, and the amount of charge stored in the bulk will not reach its ultimate saturation value. Measurements of the variation of the potential with the hydrogen content have also shown that the electrochemical titration curve is quite steep, limiting the depth of penetration. This helps to lead to apparent capacitor-like behavior.

Experiments [6, 13] showed that the chemical diffusion coefficient of hydrogen in bulk crystalline RuO_2 is about 5×10^{-14} cm^2/s. Thus, the penetration into the bulk crystalline solid is rather shallow at the relatively high frequencies typically used in capacitor experiments. It was shown in later work [14] that the apparent hydrogen solubility is considerably higher in amorphous RuO_2 than in crystalline material. Experiments on hydrated RuO_2 [12, 15, 16] demonstrated that it has a substantially larger charge storage capacity than anhydrous RuO_2.

It has also been found [16] that the amount of charge that can be stored in hydrated RuO_2 is independent of the surface area, but proportional to the total mass. This is shown in Fig. 7.2. Over 1 hydrogen atom can be reversibly inserted into the structure per Ru atom [12]. The *coulometric titration* curve is shown in Fig. 7.3.

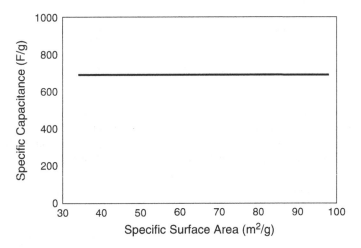

Fig. 7.2 Apparent capacitance of RuO_2 hydrate as a function of surface area. After [16]

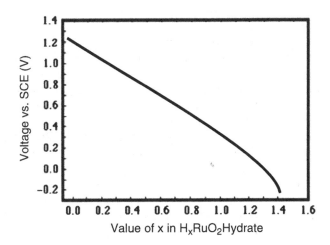

Fig. 7.3 Dependence of the potential of RuO_2 hydrate upon the amount of inserted hydrogen. After [12]

Crystalline RuO_2 is a very good electronic conductor. Its electronic resistivity is about 10^{-5} Ω cm. This is about a factor of 100 lower than that of bulk carbons. The hydrated material, on the other hand, has a considerably higher resistivity, and it has been found to be advantageous to add some carbon to the microstructure in order to reduce the electronic resistance in that case.

Several other materials that are *electrochromic*, that is, change color as charge is inserted or deleted, show similar pseudo-capacitive behavior, such as NiOOH and IrO_2, and thus could be used in supercapacitors. This clearly indicates the insertion of species into the bulk crystal structure.

Typical values of specific capacitance of a number of insertion reaction electrode materials are included in Table 7.3.

Table 7.3 Characteristics of some insertion reaction electrode materials

Electrode material	Specific capacitance (F/g)
Polymers (e.g. polyaniline)	400–500
RuO_2	380
RuO_2 hydrate	760

The insertion of guest species into the host crystal structure in such insertion reactions generally results in some change in the volume. This can lead to morphological changes and a reduction in capacity upon cycling. The volume change is generally roughly proportional to the concentration of the guest species. As a result, it is often found that the magnitude of this degradation depends upon the depth of the charge–discharge cycles.

7.3.4   Faradaically – Driven Reconstitution Reactions

The electrodes in many battery systems undergo *reconstitution reactions*, in which different phases form and others are consumed. In accordance with the *Gibbs Phase Rule*, this often results in an open-circuit electrode potential that is independent of the state of charge. As discussed elsewhere in this text, the amount of charge storage is determined by the characteristics of the related phase diagram and can be quite large. Some reactions of this type can also have relatively rapid kinetics. However, there is a potential difficulty in the use of this type of reaction in applications that require many repeatable cycles, for they generally involve microstructural changes that are not entirely reversible. Thus the possibility of a cycle-life limitation must be kept in mind.

A special strategy whereby this microstructural irreversibility may be avoided or reduced in certain cases has been proposed [17]. This involves the use of an all-solid electrode in which a mixed-conducting solid matrix phase with a very high chemical diffusion coefficient surrounds small particles of the reactant phases.

7.4   Comparative Magnitudes of Energy Storage

The maximum amount of energy that can be stored in any device is the integral of its voltage-charge product, and cannot exceed the product of its maximum voltage and the maximum amount of charge it can store. On this basis, we can make a simple comparison can be made between these different types of energy storage mechanisms.

The results are shown schematically in Fig. 7.4, in which the relationship between the potential and the amount of charge delivered is plotted for three different types of systems, a double-layer electrode, an insertion reaction electrode, and a reconstitution reaction electrode. Electrodes that involve two-dimensional *Faradaic underpotential* deposition are not included, as they do not constitute a practical alternative.

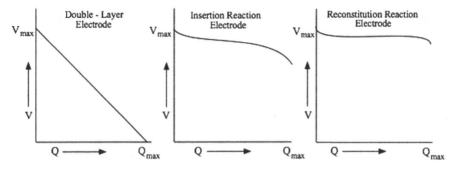

Fig. 7.4 Comparison of the variation of the potential with the amount of charge extracted for different types of energy storage mechanisms

In the case of a true capacitor, the amount of charge stored is a linear function of the applied voltage. Thus, as shown on the left side of Fig. 7.4, the voltage falls off linearly with the amount of charge delivered.

A single-phase solid solution insertion-reaction type of electrode characteristically has a potential-charge relation of the type shown in the middle. The thermodynamic basis for this shape, in which the potential is composition-dependent, and thus state of charge-dependent, was discussed earlier in Chap. 6.

The characteristic behavior of a reconstitution-reaction electrode system is shown on the right side. In this case, it is assumed that the temperature and pressure are fixed, and that the number of components is equal to the number of phases, so that from a thermodynamic point of view there are no degrees of freedom. This means that all of the intensive variables, including the electrode potential, are independent of the overall composition, and thus independent of the amount of charge delivered. Thus the discharge characteristic consists of a voltage plateau.

As mentioned above, the maximum amount of energy that is available in each case is the area under the V/Q curve. This is indicated in Fig. 7.5 for the three cases

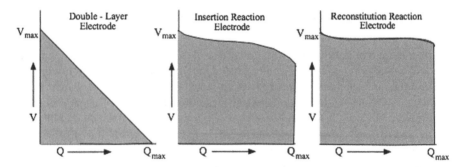

Fig. 7.5 The amount of energy available for materials with different types of storage mechanisms, indicated by the area under their curves

of interest. It is seen that the maximum amount of energy that can be stored in an electrode that behaves as a capacitor is $1/2$ $(V_{max})(Q_{max})$. The actual amount of available energy will, of course, depend upon the power level, due to unavoidable losses, such as that due to the inevitable internal resistance of the system.

7.5 Importance of the Quality of the Stored Energy

As mentioned earlier, the *quality* of heat is a commonly used concept in engineering thermodynamics. High temperature heat is generally much more useful than low temperature heat. Thus, in considering a practical thermal system, one has to consider both the amount of heat and its quality (the temperature at which it is available).

One can consider an analogous situation in the application of energy storage devices and systems. In such cases, in addition to the total amount of energy that can be stored, one should also consider the voltage at which it is available. Thus, it is useful to consider the quality of the stored electrical energy.

If this factor is taken into account, an additional difference between systems that utilize electrodes that operate by these three different types of mechanisms can be seen. This is indicated in Table 7.4, in which the amount of higher value energy available in the different cases is compared. In that case, only a simple distinction is used. Energy at a potential above $V_{max}/2$ is considered to be high value energy.

Table 7.4 Maximum amount of high value energy available

Type of electrode	High value energy (%)
	Where (V > $V_{max}/2$)
Double layer electrode	37.5
Insertion reaction electrode	About 80
Reconstitution reaction electrode	About 90

Thus, there are a number of parameters that determine important properties of a transient storage system. These are listed in Table 7.5.

Table 7.5 Parameters that determine the values of maximum potential, maximum charge, and maximum energy stored

Type of electrode	V_{max} *determined by*
Double-layer electrode	The electrolyte stability window
Insertion reaction electrode	Thermodynamics of guest-host phase
Reconstitution reaction electrode	Thermodynamics of polyphase reactions
Type of electrode	Q_{max} *determined by*
Double-layer electrode	Electrode microstructure, electrolyte
Insertion reaction electrode	Mass of electrode, thermodynamics
Reconstitution reaction electrode	Mass of electrode, thermodynamics

7.6 Transient Behavior of a Capacitor

In addition to the question of the amount of energy that can be stored in a capacitor, consideration must also be given to the rate at which it can be obtained.

There is always some series resistance connected to a capacitor. This can be indicated in the case of a simple parallel plate capacitor and *ultracapacitors* in the simple equivalent circuit of Fig. 7.6. If one or both of the electrodes undergo some insertion reaction, as is the case in *supercapacitors*, the kinetics become more complicated. The kinetic behavior in that case can be treated by the use of Laplace transform techniques, and will be discussed later in this chapter.

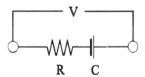

Fig. 7.6 Simple equivalent circuit of a capacitor and its associated series resistance

If such a capacitor is charged, and then shorted, which sets the external voltage V to zero, the voltage across the capacitor will be balanced by the voltage across the resistor, for

$$V = V_R + V_C \tag{7.7}$$

As the capacitor discharges, its voltage decreases, as does the voltage across the resistor and the current through it.

The instantaneous current $i_{(t)}$ through the resistor decays exponentially with time according to

$$i_{(t)} - i_0 \exp\left(\frac{-t}{RC}\right) \tag{7.8}$$

where i_0 is the initial current at the start of discharge.

The product RC is called the *time constant*, τ. It is a useful parameter in understanding the rate at which energy can be obtained from capacitor-based systems.

Taking the logarithm of both sides of (7.8),

$$\ln\left(\frac{i}{i_0}\right) = -\frac{t}{RC} \tag{7.9}$$

Thus, the time t has the value of the time constant RC when

$$\left(\frac{i}{i_0}\right) = \exp(-1) \tag{7.10}$$

The value of $\exp(-1)$ is approximately 0.3679.

In the case of a series arrangement of a resistor R and an inductance L, the time constant is given by

$$\tau = \frac{R}{L} \tag{7.11}$$

The variation of the current through the resistor R with time in a capacitor system is shown in Fig. 7.7.

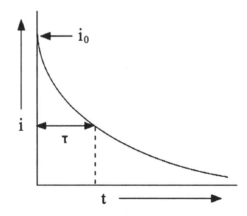

Fig. 7.7 Time-dependent current decay in a capacitor system

The rate at which energy can be stored – or supplied – is determined by the time constant. Therefore, it is desirable that the series resistance R is as small as possible. But it is also obvious that if the capacitance C is large, the time constant will be large.

During operation some of the applied power goes into heating the resistor

$$\text{Power} = i_{(t)}V_R = I_{(t)}^2 R \qquad (7.12)$$

The maximum power is determined by the value of V^2/R, and can be very large for a short time, and values up to 10^9 W/m^3 can be attained in some designs.

$$P = I_0^2 R \left(\frac{V}{R}\right)^2 R = \frac{V^2}{R} \qquad (7.13)$$

7.7 Modeling Transient Behavior of Electrochemical Systems Containing Capacitive Components Using LaPlace Transforms

7.7.1 Introduction

The quantitative understanding of the application of capacitive, or other, components in actual devices requires the knowledge of the relationship between component properties and system behavior, for systems typically involve more than one component. As an example, in addition to the electrode impedances, there are almost always resistive, and/or capacitive, impedances present, both relating to internal phenomena and to external factors.

It is often very useful to utilize *equivalent electrical circuits* whose electrical behavior is analogous to the behavior of physical systems as *thinking tools* to obtain insight into the important parameters and their inter-relationships. This allows the use of the methods that have been developed in electrical engineering for circuit analysis to evaluate the overall behavior of interdependent physical phenomena.

A useful way to do this is based upon the simple concept of a relation between a *driving force* and the *response* of a device or system to it. This relation can be written very generally as:

Driving Function = (Transfer Function) * (Response Function)

In an electrochemical system, the driving function represents the current or voltage demands imposed by the application, and the response function is the output of the electrochemical system in response to these demands. The key element of this approach is the determination of the (time-dependent) transfer function of the device or system, for that determines the relationship between application demand and system output.

7.7.2 Use of LaPlace Transform Techniques

The general method that has been developed for electrical device and circuit analysis involves the use of *LaPlace transform* techniques. There are several basic steps in this analysis. They involve:

1. The determination of the transfer function of the individual equivalent circuit components
2. The calculation of the transfer function of the total system
3. The introduction of the driving function determined by the application, and finally
4. The calculation of the system (energy source) output.

Some readers of this chapter may not be familiar with LaPlace transform methods. But they can be readily understood by the use of an analogy. Consider the use of logarithms to multiply two numbers, for example, A and B. The general procedure is to find the logarithms (transforms) of both A and B, to add them together, and then to use antilogarithms to reconvert the sum of the logarithms (transforms) into a normal number.

This method has been applied to some simple electrochemical situations, including the influence of the presence of series resistance upon the rate of charge accumulation in an insertion reaction electrode [18] and the electrical response of electrochemical capacitors [11].

The calculation of the transient electrical response of an insertion, or solid-solution, electrode involves the solution of the diffusion equation for boundary conditions that are appropriate to the particular form of applied signal. In addition, the relation between the concentration of the electroactive mobile species and their activity is necessary. This approach has been utilized to determine the kinetic properties of individual materials by employing current and/or voltage steps or pulses.

But in real electrochemical systems or devices one has to consider the presence of other components and phenomena, that is, other circuit elements. As a simple example, there is always an electrolyte, and thus a series resistance, present, and the behavior of the electrolyte/electrode interface may also have to be considered. Thus, the simple solution of Fick's diffusion laws for the electrochemical behavior of the electrode alone may not be satisfactory.

Examples of the LaPlace transforms of several common functions are given in Table 7.6.

Table 7.6 Examples of LaPlace transforms

Function	LaPlace transform
General impedance function	$Z(p) = E(p)/I(p)$
Fick's second law	$pC - c(t = 0) = D\frac{d^2C}{dx^2}$
Current step d(t)	$I(p) = 1$
Potential vs. time	$E(p) = V(dE/dy)$
Impedance of insertion reaction electrode	$Z(p) = Q/Da$

Where $Q = \frac{V(dE/dy)}{nFs}$, $a = (p/D)^{1/2}$, dE/dy = slope of coulometric titration curve, y = composition parameter, n = stoichiometric coefficient, F = Faraday constant, s = surface area, p = complex frequency variable, x = positional coordinate, V = molar volume, $q(t)$ = charge accumulated in electrode, $i(t)$ = instantaneous current, $F(t)$ = instantaneous electrode potential.

7.7.3   Simple Examples

To illustrate this method, the response of an insertion electrode under both a step in potential and a step in current, as well as a system consisting of a simple series arrangement of a resistance and an insertion reaction electrode that has a diffusional impedance will be described.

a. Upon the imposition of a step in potential F_0, the time dependence of the current $i(t)$ is given by

$$i(t) = \frac{F_0}{Q} \left(\frac{D}{pt} \right)^{1/2} \tag{7.14}$$

b. The time dependence of the charge accumulated (or produced) $q(t)$ is

$$q(t) = \frac{2F_0}{Q} \left(\frac{t}{p} \right)^{1/2} \tag{7.15}$$

c. For the case of a step in current i_0, the time dependence of the electrode potential $F(t)$ is given by

$$F(t) = 2Q \left(\frac{t}{pD} \right)^{1/2} \tag{7.16}$$

d. The time dependence of the current after the imposition of a step potential of F_0 for the more complicated case of a resistance in series with an insertion reaction electrode is found to be

$$i(t) = (F_0/R) \exp \left[\left(\frac{Q}{R} \right)^2 t \right] erfc \left[\frac{Qt^{1/2}}{R} \right] \tag{7.17}$$

e. The charge accumulated (or produced) in the case of this series combination is found to be

$$q(t) = \frac{F_0 R}{Q^{1/2}} \left[\exp \left[\left(\frac{Q}{R} \right)^2 t \right] erfc \left[\frac{Qt^{1/2}}{R} \right] - 1 \right] + \frac{2F_0}{Q} \left(\frac{t}{p} \right)^{1/2} \tag{7.18}$$

The influence of the value of the series resistance can readily be seen in Fig. 7.8 [18].

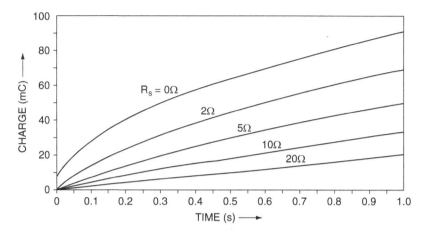

Fig. 7.8 The charge accumulated in an insertion reaction electrode as a function of time for various values of series resistance. After [18]

The parameters used in the calculation illustrated in Fig. 7.8 are:

$$Q^{-1} = 6.33\,\Omega/\mathrm{s}^{1/2}$$

Derived from $D = 10^{-8}$ cm^2/s, $V_m = 30$ cm^3/mol, $dE/dy = -2$ V, $s = 1$ cm^2 and $n = 1$.

The applied voltage step was 0.5 V.

Examination of the solutions for the behavior of single components under the first two sets of conditions obtained by this method shows that they are equivalent to those for the analytical solution of the diffusion equation under equivalent experimental conditions.

The impedance of an insertion reaction electrode alone under an AC driving force has also been described in [19].

This LaPlace transform procedure is thus an alternative to the more conventional analytical approach. But the real value of the LaPlace transform approach becomes clearer, however, under more complex conditions, such as when more than one component is present, and in which the normal procedures become quite cumbersome.

7.8 Energy Storage in Magnetic Systems

The energy storage capability of electromagnets can be much greater than that of capacitors of comparable size. Especially interesting is the possibility of the use of superconductor alloys to carry current in such devices. But before that is discussed, it is necessary to consider the basic aspects of energy storage in magnetic systems.

7.8.1 Energy in a Material in a Magnetic Field

It was shown earlier in this chapter that the energy stored in a parallel plate capacitor with spacing d and area A when a voltage V is applied across it can be written as

$$W_C = \frac{1}{2}\varepsilon A \frac{V^2}{d} = \frac{1}{2}CV^2 \qquad (7.19)$$

ε is the *permittivity*, a measure of the polarization of the material between the plates by the electric field, and C the *capacitance*.

Energy can also be stored in magnetic materials and systems. The analogous relation is

$$W_M = \frac{1}{2}\mu H^2 \qquad (7.20)$$

where H is the intensity of a *magnetic field*, and μ is the *permeability*, a constant that is dependent upon the material within the field, analogous to the *permittivity*. The magnetic field H is sometimes called the *magnetizing field*, or the *magnetizing force*.

There is also a relation equivalent to that in (7.3):

$$\mu = \mu_r \mu_0 \qquad (7.21)$$

where μ_r is called the *relative permeability* of the material present in the magnetic field, and μ_0 the *permeability of vacuum*, 1.257×10^{-6} Henries per meter.

When a material is placed in a magnetic field, an internal magnetic field will be induced within it whose magnitude depends upon the material's permeability μ. This internal *induced magnetic field*, B, which is sometimes called the *magnetic induction*, or the *magnetic flux density*, is thus related to the external field H by

$$B = \mu H \qquad (7.22)$$

Equation (7.20) can therefore be rewritten in terms of the *induced magnetic field* B inside the material instead of the external field H as

$$W_M = \frac{1}{2\mu}B^2 = \frac{1}{2}BH \qquad (7.23)$$

Equation (7.22) can be written to show the separate influence of the external field and the internal properties of the material as

$$B = \mu_0 H + \mu_0 M \qquad (7.24)$$

M is called the *magnetization*, and $\mu_0 M$ is the additional induced magnetic field due to the properties of the solid. The *magnetization* can also be expressed as

$$M = \frac{(\mu - \mu_0)H}{\mu_0} = \frac{\mu H}{\mu_0} - 1 = \mu_r H - 1 \tag{7.25}$$

The magnetic properties of the solid can also be expressed in terms of the *susceptibility*, X, which is dimensionless, where

$$X = \frac{M}{H} = \frac{\mu - \mu_0}{\mu_0} = \frac{\mu}{\mu_0} - 1 = \mu_r - 1 \tag{7.26}$$

and

$$B = \mu_0(H + M) \tag{7.27}$$

One way to generate a magnetic field H is to pass current through a nearby electrical conductor. In the case of a wire shaped into a spiral, or helix, the value of the H field inside it is

$$H = 4\pi n I \tag{7.28}$$

where n is the number of turns per unit length of the spiral, and I is the magnitude of the current. The direction of this field is parallel to the length of the spiral. This is shown schematically in Fig. 7.9.

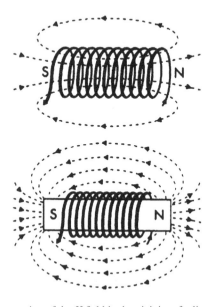

Fig. 7.9 Schematic representation of the H field in the vicinity of a linear helix. It is seen that the presence of the material with a high value of susceptibility amplifies the external H field

The H field is continuous through the material, as well as in its external environment, whereas the B field is only within the solid.

Since magnetic units and their dimensions may not be as familiar to many people as electrostatic units, a list of some of them is given in Table 7.7.

Table 7.7 Magnetic quantities, units and dimensions

Quantity	Unit	Symbol	Dimensions
Magnetic field	Henry	H	A/m
Magnetic induction	Tesla	B	$Wb/m^2 = V\ s/A\ m^2$
Permeability		μ	V s/A m
Energy product		BH	kJ/m^3
Magnetization per unit volume		M_v	A/m
Magnetization per unit mass		M_m	A m^2/kg
Magnetic flux	Weber	Wb	V s
Inductance	Henry	L	$Wb/A = V\ s/A$

The unit of permeability in the SI system is the Henry. One Henry has the value of 1 Weber per ampere, and is the *inductance* that produces one volt when the current in a circuit varies at a uniform rate of 1 amp/s.

It was shown in (7.23) that the energy in an electromagnet is proportional to the product of B and H. These quantities are related to each other by the value of the magnetization M, or the relative permeability, μ_r.

There are two general classes of magnetic materials, one group is generally described as *soft magnetic materials*, and the other is called *hard magnetic materials*. Samples of the latter are also often named *permanent magnets*. Their characteristics are shown schematically in Fig. 7.10.

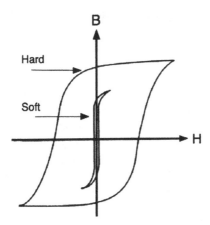

Fig. 7.10 Schematic B–H curves for soft and hard magnetic materials

It can be seen that these are very different. The major difference is the large area in the case of the hard material. If such a material is magnetized by increasing H, and then H is reduced to zero, B, and thus the magnetization, remains at a high value. In order to reduce B, or the magnetization to zero, a large negative value of H is required. Thus, the material tends to remain magnetized, that is, "hard". The magnitude of this reversed H field that is necessary to demagnetize this material is called the *coercivity*. Thus, such a material stores a lot of magnetic energy, but it is very difficult to get it out. The area inside the B–H curve, a measure of hysteresis, represents the energy lost each time the magnetization is reversed. Data on several hard magnetic materials are included in Table 7.8.

Table 7.8 B–H products of several hard magnetic materials

Material	(BH)max/Wb A/m^3
Alnico	36,000
Platinum–Cobalt	70,000
Samarium–Cobalt	120,000

It can be seen that the area within the curve for the soft magnetic material in Fig. 7.10 is much smaller than that for the hard magnetic material. It is soft, rather than hard, magnetic materials that are used to reversibly store energy in electromagnetic systems under transient conditions. For this purpose, the energy loss due to hysteresis (the area inside the B–H curve) should be as small as possible. Data on the magnetic properties of several types of soft magnetic materials are shown in Table 7.9.

Table 7.9 Data on some soft magnetic materials

Material	Composition/weight percent	Relative permeability	Resistivity/Ω m
Commercial cast iron	99.95 Fe	150	1×10^{-7}
Oriented silicon iron alloy	97 Fe, 3 Si	1,400	4.7×10^{-7}
Permalloy	55 Fe, 45 Ni	2,500	4.5×10^{-7}
Supermalloy	79 Ni, 15 Fe, 5 Mo	75,000	6×10^{-7}
Ferroxcube A	48 $MnFe_2O_4$, 52 $ZnFe_2O_4$	1,400	2,000
Ferroxcube B	64 $ZnFe_2O_4$, 36 $NiFe_2O_4$	650	107

Because they have a large influence over efficiency, a considerable amount of work has been done over many years to optimize soft magnetic materials for their different uses. An important step forward was the development of iron–silicon alloys with relatively high electronic resistivity, thus reducing the hysteresis losses due to induced *eddy currents*. Processing these materials so that they develop a preferred crystallographic orientation also increases their permeability appreciably, as does annealing them in moist hydrogen to reduce the amount of carbon impurity present. These alloys are now commonly used as insulated laminated sheets in transformers for the transmission of moderate to large amount of electrical power.

Another class of soft magnetic materials includes the nickel–iron *permalloy* type of alloys. These materials have very large values of permeability, and are generally used in very low power applications that require large changes in magnetization with relatively small applied fields.

A third class of soft magnetic materials includes transition metal oxide ceramics called ferrites. Because they have very high values of electronic resistivity, and thus have no appreciable eddy current loss, they can be used in electronic equipment at very high frequencies.

According to (7.23), the energy stored in a magnetic material is one half the product of B and H, and therefore the area under a plot of B vs. H. From (7.22), the slope of such a curve is the material's permeability. This is shown in Fig. 7.11 for two soft magnetic materials with different values of relative permeability in the same H field.

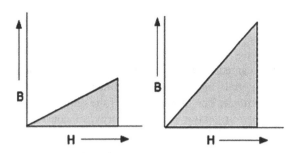

Fig. 7.11 Areas indicating the amount of stored energy for two materials with different values of relative permeability up to the same value of H field

In this case, it is assumed that there is very little hysteresis, so that the data measured when the H field is increased are essentially the same as those when the H field is decreased.

This is analogous to the energy under a stress/strain curve in mechanical materials systems, and the energy stored under the voltage/composition curve in electrode materials in electrochemical systems.

7.8.2 Energy Storage in Superconducting Magnetic Systems

The magnetic energy of materials in external H fields is dependent upon the intensity of that field. If the H field is produced by current passing through a surrounding spiral conductor, its magnitude is proportional to the current according to (7.28). It is obvious that high currents are desirable if one wants to store large amounts of energy.

However, the passage of current through a metal wire causes Joule heating according to (7.29).

$$\text{Heat} = I^2 R \qquad (7.29)$$

Thus, it is desirable to consider the use of a superconducting material that has essentially no resistance to carry the current. Such systems are generally designed with the high permeability soft magnetic material within a superconductor coil in the shape of a toroid.

Energy can be fed into such a system by use of a DC power supply. Once the current is established in the superconductor, the power supply can be disconnected. The energy is then stored in the magnetic material inside the superconducting coil, where it can be maintained as long as desired without the need for further input.

The transmission of energy to and from the DC superconductor electromagnetic storage system requires special high power AC/DC conversion rectifier, inverter, and control systems. This power conditioning system causes a 2–3% energy loss in each direction.

An additional feature that must be taken into account is the generation of large mechanical forces acting on the materials by the large magnetic fields present. The mitigation of this can add considerably to the cost of the whole system.

Superconductor materials have to be maintained below a so-called material-specific *critical temperature*. The maintenance of the required low temperature by the use of a cryostatic refrigerator requires energy, of course. There is another complication in that, superconducting materials lose their superconductive property if the value of the surrounding H field is above a critical value, called the *critical field*. Since the field is caused by the current in the superconductor, another way of looking at this limitation is in terms of a *critical current*, rather than a *critical field*.

Such systems are generally used for short-term energy storage, such as improving the power quality and stability of the transmission distribution system, where the rapid response and high short-term power are a distinct advantage.

There is a serious potential danger if either the temperature or the field becomes too high, so that the material is no longer superconducting. Its resistance then becomes "normal." This can result in very large, and dangerous, amounts of Joule heating. Safety considerations have meant that plans for superconducting energy storage devices of any appreciable magnitude generally involve their being placed in caverns deep underground.

7.8.3 *Superconductive Materials*

The phenomenon of superconductivity was discovered in 1911 by H. Kammerlingh Onnes [20]. He found that the electrical resistance of solid Hg disappeared below about 4 K. Research and development activities aimed at finding materials that remain superconducting to higher temperatures were pursued over many years. Several groups of metals and alloys with higher critical temperatures were gradually found. Intermetallic compounds containing niobium were shown to have the most attractive properties. The gradual improvement in the superconducting transition temperature resulting from the development of different alloys is shown in Fig. 7.12.

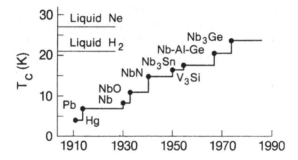

Fig. 7.12 The time-dependent increase in the superconducting transition temperature resulting from the development of new materials. After [23]

During the 1970s and 1980s, it was generally concluded from both experimental and theoretical work that the maximum possible temperature for superconductivity was about 24 K. Then, in 1986, J.G. Bednorz and K.A. Müller showed that superconductivity could remain present in a ceramic oxide material, lanthanum barium yttrium oxide, up to 35 K [21]. A discussion of their work leading up to this discovery can be found in [22].

This unleashed a frantic effort in many laboratories to confirm this surprising result, and investigate the possibility that superconductivity could be extended to even higher temperatures in other non-metallic materials. An important step was the discovery that the phase yttrium barium copper oxide remains superconducting up to 93 K [23]. This discovery was particularly important, for this temperature is above the boiling point of relatively inexpensive liquid nitrogen, 77 K. The details of the synthesis of this material have a large influence upon its properties. Particularly important is the oxygen content [24].

Another quite different material that has recently been found to be superconducting is Mg diboride, MgB_2 [25]. Although its critical temperature is only 39 K, it may become important for some applications, for it is relatively inexpensive and easy to synthesize.

Data on the critical properties of some of the important superconductive materials are included in Table 7.10.

Table 7.10 Values of the critical temperature and critical field of superconducting materials

Material	Critical temperature (K)	Critical field (Tesla)
Nb–Ti	10	15
Nb_3Al	18	?
Nb_3Ge	23.2	37
Nb_3Sn	18.3	30
NbTi	10	15
MgB_2	39	74
$YBa_2Cu_3O_7$	92	?

All of the higher temperature superconductors are inherently brittle. It has proved to be difficult to find a way to make these materials into the long lengths necessary to make magnetic coils. As a result, the wires used in electromagnets are now typically made from the alloys that have lower critical temperatures. Nb–Ti alloys, which have a ductile BCC crystal structure, can be formed into wires and made into coils and are commonly used for this purpose. Fibers of this material are imbedded in an aluminum or copper matrix for structural purposes.

However, they cannot withstand as high a magnetic field intensity as the more expensive Nb_3Sn, which is harder to use. Nb_3Sn has the A15 crystal structure that will be discussed in Chap. 8. It is very brittle, and cannot be drawn into a wire shape. To overcome this problem, a composite microstructure containing ductile precursor phases is used, with separate alloys of Nb, Cu, and Sn. After the material is made into wire and worked into the final shape, it is heat treated, during which the Sn reacts with Nb to form the Nb_3Sn phase. Because of its high critical field, this material is preferred for the production of high power magnets and electrical machinery.

The compounds MgB_2 and $YBa_2Cu_3O_7$, are also brittle, and cannot readily be formed into wires and other shapes. It is expected that this problem can be circumvented, however, by forming them in situ within other materials that can be readily shaped, and a considerable amount of development effort is aimed in this direction.

References

1. B. E. Conway, J. Electrochem. Soc. *138*, 1539 (1991)
2. B.E. Conway, *Proceedings of Symposium on New Sealed Rechargeable Batteries and Supercapacitors*, B.M. Barnett, E. Dowgiallo, G. Halpert, Y. Matsuda and Z. Takeharas, Eds., The Electrochemical Society (1993), p. 15
3. B.E. Conway, *Electrochemical Supercapacitors: Scientific Fundamentals and Technological Applications*, Plenum Press, New York (1999)
4. A. Nishino, *Proceedings of Symposium on New Sealed Rechargeable Batteries and Supercapacitors*, B.M. Barnett, E. Dowgiallo, G. Halpert, Y. Matsuda and Z. Takehara, Eds. The Electrochemical Society (1993), p. 1
5. L.D. Burke, O.J. Murphy, J.F. O'Neill and S. Venkatesan, J. Chem. Soc. Faraday Trans. *1*, 73 (1977)
6. T. Arikado, C. Iwakura and H. Tamura, Electrochem. Acta *22*, 513 (1977)
7. D. Michell, D.A.J. Rand and R. Woods, J. Electroanal. Chem. *89*, 11 (1978)
8. S. Trasatti and G. Lodi, in *Electrodes of Conductive Metallic Oxides-Part A.*, ed. by S. Trasatti, Elsevier, Amsterdam (1980), p. 301
9. D.R. Craig, Canadian Patent 1,196,683 (1985)
10. D.R. Craig, "Electric Energy Storage Devices", European Patent Application 82,109,061.0, submitted 30 September, 1982, Publication No. 0 078 404, 11 May 1983
11. I.D. Raistrick and R.J. Sherman, *Proceedings of Symposium on Electrode Materials and Processes for Energy Conversion and Storage*, S. Srinivasan, S. Wagner, and H. Wroblowa, Eds. Electrochem. Soc. (1987), p. 582
12. T.R. Jow and J.P. Zheng, J. Electrochem. Soc. *145*, 49 (1998)
13. J.E. Weston and B.C.H. Steele, J. Appl. Electrochem. *10*, 49 (1980)

14. J.P. Zheng, T.R. Jow, Q.X. Jia and X.D. Wu, J. Electrochem. Soc. *143*, 1068 (1996)
15. J.P. Zheng and T.R. Jow, J. Electrochem. Soc. *142*, L6 (1995)
16. J.P. Zheng, P.J. Cygan and T.R. Jow J. Electrochem. Soc. *142*, 2699 (1995)
17. B.A. Boukamp, G.C. Lesh and R.A. Huggins, J. Electrochem. Soc. *128*, 725 (1981)
18. I.D. Raistrick and R.A. Huggins, Solid State Ionics *7*, 213 (1982)
19. C. Ho, I.D. Raistrick and R.A. Huggins, J. Electrochem. Soc. *127*, 343 (1980)
20. H. Kammerlingh Onnes, Commun. Phys. Lab. Univ. Leiden, *12*, 120 (1911)
21. J.G. Bednorz and K.A. Müller, Z. Phys. B *64*, 189 (1986)
22. J.G. Bednorz and K.A. Müller, Rev. Mod. Phys. *60*, 585 (1988)
23. M.K. Wu, J.R. Ashburn, C.J. Torng, P.H. Hor, R.L. Meng, L. Gao, Z.J. Huang, Y.Q. Wang, and C.W. Chu, Phys. Rev. Lett. *58,* 908 (1987)
24. R. Beyers, E.M. Engler, B.T. Ahn, T.M. Gür and R.A. Huggins, in *High-Temperature Super-conductors,* ed. by M.B. Brodsky, R.C. Dynes, K. Kitazawa and H.L. Tuller, vol. 99, Materials Research Society (1988), p. 77
25. J. Nagamatsu, N. Nakagawa, T. Muranaka, Y. Zenitani and J. Akimitsu, Nature *410,* 63 (2001)

Chapter 8
Hydrogen Storage

8.1 Introduction

Hydrogen is an important energy carrier, and when used as a fuel, can be considered as an alternate to the major fossil fuels, coal, crude oil, and natural gas, and their derivatives. It has the potential to be a clean, reliable, and affordable energy source, and has the major advantage that the product of its combustion with oxygen is water, rather than CO and CO_2, which contain carbon and are considered greenhouse gases. It is expected to play a major role in future energy systems.

It has been shown that hydrogen can be used directly in internal reciprocating combustion engines, requiring relatively minor modifications, if it is raised to a moderately high pressure, as well as in turbines and process heaters.

It can also be used in hydrogen/oxygen fuel cells to directly produce electricity. Again, the only product is water. The energy efficiency of fuel cells can be as high as 60%. Fossil fuel systems, on the other hand, are typically about 34% efficient. When high temperature fuel cells are used, it is possible to obtain electricity and also to use the heat generated in the fuel cell, related to its inefficiency, for heating purposes. This is called *cogeneration*, and it is possible to obtain total energy efficiencies up to 80% in this manner.

Electrically powered vehicles have the advantage that electric motors can have energy efficiencies of about 90%, whereas typical internal combustion engines are about 25% efficient. On the other hand, fuel cells now cost about 100 times as much as equivalent internal combustion engines of comparable power. One can, however, expect some reduction in cost with mass production and further development of fuel cells.

Because of these attractive features, a number of people have long advocated the concept of a simple *Hydrogen Economy*, in which hydrogen is used as the major fuel.

It reacts with oxygen, either by combustion or in fuel cells, to give energy, and the only product is water. It can be regenerated directly from water by electrolysis.

R.A. Huggins, *Energy Storage*,
DOI 10.1007/978-1-4419-1024-0_8, © Springer Science+Business Media, LLC 2010

This is a closed chemical cycle. No chemical compounds are created or destroyed, but there is a net flow of energy.

This concept is illustrated schematically in Fig. 8.1. It is important when considering this concept that a full hydrogen economy is not necessary. Even partial implementation would be desirable as a way to reduce environmental pollution problems.

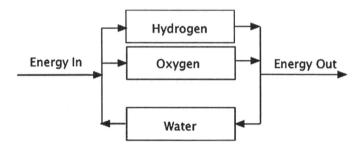

Fig. 8.1 The general concept of the ideal hydrogen economy

The major use of hydrogen at the present time, however, is not as an energy carrier, but as a reactant in a number of important large-scale chemical processes.

8.2 The Production of Hydrogen

As discussed in the Preface, the major, and least expensive, way to obtain hydrogen is to extract it from natural gas, which is primarily methane. The most common method for the conversion of methane to hydrogen involves the use of the *steam reforming*, followed by the *water–gas shift reaction*. It now provides some 95% of all the hydrogen produced in the United States. This process, as well as several other methods for making hydrogen, will be briefly discussed in the following sections of this chapter.

8.2.1 The Steam Reforming Process

The first step in this procedure is the elimination of impurities, such as sulfur, from the methane-rich natural gas. The methane is then reacted with steam at a relatively high temperature, using nickel oxide as a catalyst. This process is called *steam reforming*, and was already discussed in Chap. 4. It can be written as

$$CH_4 + H_2O = CO + 3H_2 \tag{8.1}$$

This can be followed by a second step in which air is added to convert any residual methane that did not react during the steam reforming.

$$2CH_4 + O_2 = 2CO + 4H_2 \tag{8.2}$$

This is then followed by the *water–gas shift reaction* at a somewhat lower temperature that produces more hydrogen from the CO and steam

$$CO + H_2O = CO_2 + H_2 \tag{8.3}$$

As discussed already, the driving force for any reaction is the standard Gibbs free energy change, ΔG_r^0 that occurs as the result. This is the difference between the sum of the standard Gibbs free energies of formation of the products and the sum of the standard Gibbs free energies of formation of the reactants. In this case, this can be expressed as

$$\Delta G_r^0 = \Delta G_f^0(CO) + 3\Delta G_f^0(H_2) - \Delta G_f^0(CH_4) - \Delta G_f^0(H_2O) \tag{8.4}$$

Values of the standard Gibbs free energy of formation of the relevant species for three different temperatures are given in Table 8.1.

Table 8.1 Temperature dependence of the standard Gibbs free energies of formation of species in (8.1)

Species	ΔG_f^0 (400 K) (kJ mol^{-1})	ΔG_f^0 (800 K) (kJ mol^{-1})	ΔG_f^0 (1,200 K) (kJ mol^{-1})
CO	−146.4	−182.5	−217.8
H$_2$O	−224.0	−203.6	−181.6
CO$_2$	−394.6	−395.5	−396.0
CH$_4$	−42.0	−2.1	+41.6

From these data, it is possible to obtain the standard Gibbs free energy of relevant reactions in this system at those temperatures. These are shown in Table 8.2, and the results for (8.1) and (8.3) are plotted in Fig. 8.2.

Table 8.2 Standard Gibbs free energies of reaction at several temperatures

Reaction	ΔG_r^0 (400 K) (kJ mol^{-1})	ΔG_r^0 (800 K) (kJ mol^{-1})	ΔG_r^0 (1,200 K) (kJ mol^{-1})
$CH_4 + H_2O = CO + 3 H_2$	+119.6	+23.2	−77.8
$2 CH_4 + O_2 = 2 CO + 4 H_2$	−208.8	−360.8	−518.8
$CO + H_2O = CO_2 + H_2$	−24.2	−9.4	+3.4
$CH_4 + 2 O_2 = CO_2 + 2 H_2O$	−800.6	−800.6	−800.8
$C + H_2O = CO + H_2$	+77.6	+21.1	−36.2

It can be seen that the steam reforming reaction will only go forward if the temperature is *above* about 900 K. Likewise, the subsequent water–gas shift reaction will only proceed if the temperature is *below* about 1,025 K.

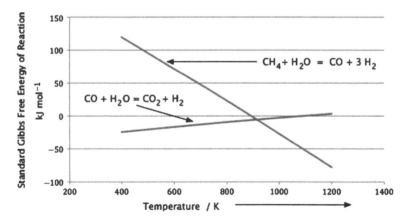

Fig. 8.2 Temperature dependence of the standard Gibbs free energy of reactions involved in the extraction of hydrogen from methane

8.2.2 The Reaction of Steam with Carbon

A number of countries, especially the United States and China, have large amounts of coal that they can use as an energy source. It is possible to react steam with solid carbon instead of with methane. In this case, the product is *syngas*, a mixture of CO and hydrogen. This reaction is included in Table 8.2. The resultant CO can then be reacted with steam in the water–gas shift reaction, just as is done in the case of steam reforming of methane.

The temperature dependence of the standard Gibbs free energies of these two reactions is shown in Fig. 8.3. As was the case with the reaction of steam with methane, the standard Gibbs free energy of the reaction of steam with carbon also decreases as the temperature increases. It can be seen that this reaction will only go forward if the temperature is above about 940 K. Again, the water–gas shift reaction proceeds only at lower temperatures.

The power plants that generate hydrogen from coal by this two-step process have overall efficiencies of about 34%. But if they capture the effluent CO_2 from the water–gas reaction, the efficiencies can rise to above 40%.

The thermal behavior of these various reactions can also be determined from data on the standard enthalpies of the species in these reactions. Such data for three temperatures are included in Table 8.3.

From these data for the various species, the enthalpy (heat) effects of these reactions can be calculated. The results are included in Table 8.4, and plotted in Fig. 8.4.

It is seen that the water–gas reaction is endothermic, whereas the other reactions are both exothermic.

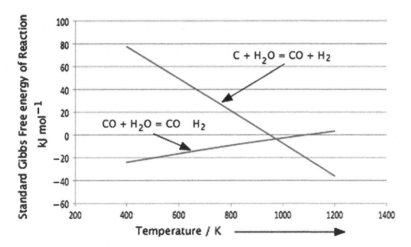

Fig. 8.3 Temperature dependence of the standard Gibbs free energies of reactions involved in reaction of steam with carbon

Table 8.3 Temperature dependence of the standard enthalpies of relevant species

Species	ΔH_f^0 (400 K) (kJ mol^{-1})	ΔH_f^0 (800 K) (kJ mol^{-1})	ΔH_f^0 (1,200 K) (kJ mol^{-1})
CO	−110.1	−110.9	−113.2
H_2O	−242.8	−246.4	−249.0
CO_2	−393.6	−394.2	−395.0
CH_4	−78.0	−87.3	−91.5

Table 8.4 Standard enthalpies of reactions

Reaction	ΔH_r^0 (400 K) (kJ mol^{-1})	ΔH_r^0 (800 K) (kJ mol^{-1})	ΔH_r^0 (1,200 K) (kJ mol^{-1})
$C + H_2O = CO + H_2$	+132.7	+135.5	+135.8
$CO + H_2O = CO_2 + H_2$	−40.7	−36.9	−32.8

If the hydrogen is to be used in a low temperature fuel cell, the gas mixture resulting from the water–gas reaction also generally undergoes a further step, called *methanation*, in which the remaining CO is converted back into methane, which is recycled. This is necessary because CO poisons the platinum catalysts that are typically used in such fuel cells.

It should be noted that all of these gas phase reactions produce products that consist of mixtures of gases. The separation of hydrogen from the other gas components must also be done, and there is a need for the development of better selective membranes for this purpose.

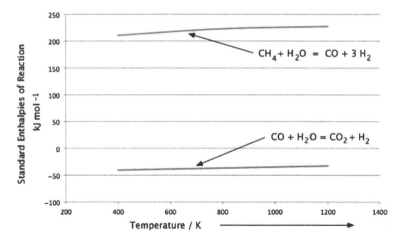

Fig. 8.4 Temperature dependence of the standard enthalpies of reactions

8.2.3 Electrolytic Production of Hydrogen

The second major method for the production of hydrogen involves the electrolysis of water by imposing a voltage between two electrodes within it that exceeds its thermodynamic stability range. The result is the evolution of hydrogen gas at the negative electrode, and oxygen gas at the positive electrode. Both of these gases have a significant commercial value. It is easy to design systems to collect them separately.

Relatively pure hydrogen can be produced by the electrolysis of water, and this method appears to be attractive in the long run. There is an enormous amount of water on the earth, with the potential to supply an almost limitless supply of hydrogen. About 4% of the hydrogen currently used in the world is produced by electrolysis.

The problem with this apparently attractive scenario is that the electrolytic decomposition of water to produce hydrogen is currently quite expensive. Thus, it is only an attractive large-scale option where the cost of electricity is relatively low.

An advantage of water electrolysis to produce hydrogen is that instead of requiring large central facilities, there can be distributed generation using smaller units. They can be located at places near where the hydrogen will be used in order to reduce transportation costs.

The equilibrium (zero current) cell voltage E_{eq} required to decompose water can be found from the value of the standard Gibbs free energy of the formation of water, ΔG_f^0, a thermodynamic quantity. Data of this type for many materials over a range of temperatures can be found in [1].

$$E = -\Delta G_f^0/2F \qquad\qquad (8.5)$$

F is the Faraday constant, or 96,500 Joules per volt equivalent. The value 2 is present because the formation of water from hydrogen and oxygen involves two electronic charges (two equivalents).

$$H_2 + \frac{1}{2}O_2 = H_2O \tag{8.6}$$

The value of ΔG_f^0 is -237.1 kJ/mol for water at 298 K, or 25°C. Thus, the equilibrium voltage, the stability range of water, is 1.23 V at that temperature. If there is a bit of salt present to provide ionic conductivity, the application of sufficiently higher voltages causes current to flow, and the two gases bubble off the electrodes.

The reactions at the electrodes when water is being electrolyzed can be written as

$$H_2O + e^- = \frac{1}{2}H_2 + OH^- \tag{8.7}$$

on the negative side, and

$$H_2O = e^- + \frac{1}{2}O_2 + 2H^+ \tag{8.8}$$

on the positive side of the cell. Inside the water, the result is the formation of OH^- ions at the negative electrode, and H^+ ions at the positive electrode.

But it must be realized that in order for these reactions to occur there must be ionic transport through the water. This would normally be either by the transport of H^+ ions in acid solutions, or the transport of OH^- ions in alkaline solutions.

But where does the water actually disappear? It cannot go away at both electrodes.

If the water is acidic, it disappears from the solution at the positive electrode. The protons formed on that side, along with the O_2 gas, are transported to the negative side. When they reach the negative electrode, they react with the OH^- ions that are generated there to form H_2O. Thus there is a net loss of H_2O at the negative side in this case, not at the positive side.

The total cell voltage E_t when current is flowing is

$$E_t = E_{eq} + E_{neg} + E_{pos} + iR_{electrolyte} \tag{8.9}$$

where E_{neg} is the voltage loss at the negative electrode, E_{pos} is the voltage loss at the positive electrode, and $iR_{electrolyte}$ is the voltage loss across the water electrolyte.

This can also be written as

$$E_t = E_{eq} + iZ_{neg} + iZ_{pos} + iR_{electrolyte} \tag{8.10}$$

where i is the current, and the Z values are the impedances at the two electrodes. These can be thought of as resistances whose values may depend upon the value of the current. On the other hand, the resistance of the electrolyte, $R_{electrolyte}$, is essentially current-independent.

The energy consumed during the electrolysis of the water is

$$Energy = iE \tag{8.11}$$

and the energy efficiency is

$$Efficiency = iE_{eq}/iE_t \tag{8.12}$$

As mentioned already, although the electrolytic production of hydrogen is significantly more expensive than obtaining it from natural gas, it has the advantage that the resulting gas can be of significantly greater purity. This can be especially important when the hydrogen is used in low temperature fuel cells with polymeric solid electrolyte membranes. Even minor amounts of impurity species, such as CO, can cause problems by absorbing on the surfaces of the platinum catalysts that are typically used to assist the conversion of H_2 molecules to H^+ ions and electrons at the negative electrode.

This requires additional treatment, and results in higher costs. The degree of CO adsorption on the catalyst surface decreases as the temperature is raised, so that this type of poisoning is not present in high temperature fuel cells. This is one of the reasons for interest in the further development of that type of fuel cell.

Current low temperature fuel cells actually operate at temperatures somewhat above ambient temperature in order to increase the overall kinetics.

As with any fuel, in addition to its acquisition, there must also be methods to transport hydrogen to the locations at which it will be used. Because such matters are typically not fully coordinated, there must also be methods for its storage.

Large commercial electrolyzers now produce hydrogen at about 30 bar pressure and a temperature of 80°C, and have energy efficiencies of 80–90%. The major source of loss is connected with the processes that take place at the positive electrode, where oxygen is evolved.

The decomposition of water requires that the oxygen species change from oxide ions in the water, which carry a charge of -2, to oxygen molecules, O_2, in which the charge on the oxide ions is effectively zero. An intermediate state, with peroxide O^{-1} ions, must be present on or near the catalyst surface. High surface area nickel or nickel alloy electrodes are typically used on the oxygen side of water electrolysis cells. They have a surface layer of nickel oxy-hydroxide, NiOOH. The properties of this material will be discussed in a later chapter.

Electrolysis cells generally operate at a voltage of about 2 V, which is substantially greater than the open circuit thermodynamic value of 1.23 V, and there is a continual effort to reduce the magnitude of this excess voltage.

8.2.4   Thermal Decomposition of Water to Produce Hydrogen

An additional method that can be used to produce hydrogen from water is to thermally decompose it by heating to a very high temperature. The stability of water relative to its two components, hydrogen and oxygen, is expressed in terms of its standard Gibbs free energy of the formation, ΔG_f^0, as mentioned above. This value is temperature-dependent, and this can be expressed as

$$\Delta G_f^0 = \Delta H_f^0 - T\Delta S_f^0 \tag{8.13}$$

where ΔH_f^0 and ΔS_f^0 are the enthalpy and entropy values for the formation of water, respectively. The values of both of these terms are relatively independent of the temperature. Using values from [1], the temperature dependence of the standard Gibbs free energy of formation of water is shown in Fig. 8.5.

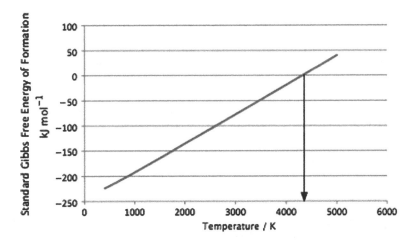

Fig. 8.5 Temperature dependence of the standard Gibbs free energy of formation of water

It is seen that water does not become thermodynamically unstable until the temperature reaches about 4,300 K. Thus, this method is hardly a practical alternative for splitting water into hydrogen and oxygen, for the required temperature is very high.

8.2.5 Chemical Extraction of Hydrogen from Water

It is also possible to produce hydrogen by chemically decomposing water. Many species form oxides when in contact with water. In general, however, the oxide that forms produces a protective surface layer that prevents further reaction with the water. There are a few exceptions in which a nonprotective product is formed. One of these is lithium, which forms soluble LiOH. Another is aluminum, that can form soluble aluminum hydroxide, or hydrated oxide, instead of the simple oxide Al_2O_3, in basic solutions. This has led to the development of the so-called aluminum/air battery, which utilizes air as the reactant in the positive electrode. This will be discussed briefly in Chap. 21.

The chemical reaction of aluminum with water can also be used to produce hydrogen, as was first shown by Cuomo and Woodall at the IBM Laboratory in 1968 [2]. This is accomplished by the use of a solution of aluminum in a low melting point liquid alloy based on gallium. The aluminum–gallium phase diagram is shown in Fig. 8.6. It can be seen that the melting point of gallium is 30°C, and slightly below that, there is an eutectic reaction at 26°C. There is a gradual increase in the solubility of aluminum in the liquid phase as the temperature increases, reaching complete miscibility at the melting point of aluminum at 660°C. There is also a modest amount of solubility of gallium in solid aluminum all the way up to the melting point.

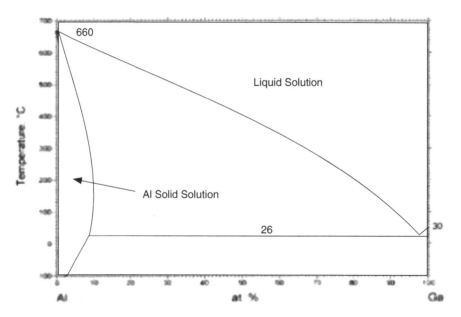

Fig. 8.6 The aluminum–gallium phase diagram

When this alloy is in contact with water at slightly above room temperature, the dissolved aluminum will not form a protective skin, but reacts vigorously to form hydrogen and finely dispersed aluminum oxide, which floats on the top of the liquid gallium alloy, according to the equation

$$Al + 3H_2O = \frac{3}{2}H_2 + Al(OH)_3 + heat \qquad (8.14)$$

Aluminum hydroxide is generally described as a hydrated version of Al_2O_3, called Gibbsite, $Al_2O_3 \cdot 3\,H_2O$. There is also a version with a slightly different crystal structure, called Bayerite.

The behavior of this ternary system can be understood by the use of the ternary phase stability diagram for the H–Al–O system shown in Fig. 8.7. The principles and methods that can be used to draw and quantitatively interpret such figures will be described in Chap. 12.

The driving force for this reaction, assuming that all species are at unit activity can be calculated from the values of their Gibbs free energy of formation. Using the data in Table 8.5, this is found to be -443.8 kJ per mol of aluminum at 300 K. This value will vary somewhat with temperature, of course.

Table 8.5 Thermodynamic data related to the formation of Gibbsite

Species	ΔG_f^0 at 300 K (kJ mol^{-1})	ΔH_f^0 at 300 K (kJ mol^{-1})
H_2O	-236.8	-285.8
Al_2O_3	-1581.7	-1675.7
$Al_2O_3 \cdot 3\,H_2O$	-2308.4	-2586.6

Similarly, the magnitude of the heat produced by this reaction at 300°C can be calculated from the respective values of their standard enthalpies. This is found to be 435.9 kJ per mol of aluminum.

In addition to the heat released in this reaction, the hydrogen product can also be subsequently oxidized to produce further heat. This can be done either at the same location or elsewhere, of course.

This second reaction, the combustion of the hydrogen formed, is simply

$$\frac{3}{2}H_2 + \frac{3}{4}O_2 = \frac{3}{2}H_2O + \text{heat} \tag{8.15}$$

From the enthalpy data in Table 8.5, it can be found that the amount of heat released in this second reaction is 428.7 kJ for the 1.5 mol of hydrogen produced at 300°C. Thus, the overall amount of heat produced from the aluminum is 864.6 kJ per mol of aluminum.

It is useful to consider why it is possible to produce hydrogen and Gibbsite by the reaction of aluminum with water, rather than forming Al_2O_3, which would block further reaction. This can be seen by looking at the ternary phase stability diagram in Fig. 8.7. As will be discussed in Chap. 12, the overall composition during the reaction of aluminum with water moves along the dashed line in Fig. 8.7, starting at the composition of water.

As aluminum is added, the overall composition first moves into the composition triangle that has H, H_2O, and $Al(OH)_3$ at its corners. In this composition regime, the activity of the aluminum is low. When the overall composition crosses the line between $Al(OH)_3$ and hydrogen and enters the triangle whose corners are Al_2O_3, $Al(OH)_3$, and H, the aluminum activity is much greater. Upon crossing into the triangle Al, Al_2O_3, and H, the aluminum activity becomes unity. The phase

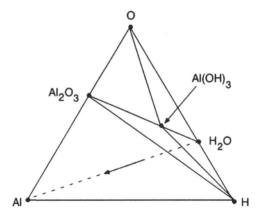

Fig. 8.7 Ternary phase stability diagram for the H–Al–O system

Al_2O_3 is present in either of the latter triangles, and will form, blocking further reaction.

Thus, the condition for the absence of a protective layer of Al_2O_3 is that the aluminum activity must be smaller than a critical value. This is accomplished by dissolving it into the liquid alloy.

The use of this very powerful approach to understanding reactions in ternary phases will be discussed much more completely in Chap. 12, and used in both understanding and predicting the potentials and compositional ranges of electrodes in battery systems in subsequent chapters.

Since aluminum from the alloy is consumed in the reaction to form hydrogen and Gibbsite, it must be replenished in order for the reaction to continue. This can easily be done merely by putting solid aluminum in contact with the alloy. The protective Al_2O_3 skin is decomposed and aluminum dissolved into the liquid alloy.

Subsequent to the discovery that dilute low temperature liquid aluminum alloys can react with water to produce hydrogen gas, it has been found that solid aluminum alloys containing some 5% of a ternary Ga–In–Sn alloy will also react with water to form hydrogen and finely divided Gibbsite [3–6].

Although gallium dissolves somewhat in solid aluminum, this is essentially not true in the cases of indium and tin. On the other hand, aluminum dissolves slightly in a liquid solution containing all three of these elements. By the use of compositions in this ternary system, it is possible to obtain aluminum-containing liquids at ambient temperature.

When used as components in alloys that are mostly aluminum, this low-melting phase resides at the aluminum grain boundaries. This grain boundary region becomes liquid at temperatures not far above ambient. When such alloys are in contact with water, it is believed that aluminum in the liquid grain boundary phase reacts with water, forming hydrogen and Gibbsite. As the grain boundary aluminum is reacted, it is replenished by the solution of more of the adjacent solid aluminum phase of the alloy.

The use of such solid aluminum alloys, instead of aluminum-containing low melting metals, has significant practical advantages. Such alloys can be readily stored and transported, so that hydrogen and heat can be generated at any location where water is available. The weight of the water is about twice that of the alloy per unit of hydrogen produced.

There are a number of aspects of this method that are quite favorable for the production of hydrogen. It is not necessary that either the aluminum or the alloy constituents be of high purity. No additional materials or electrical power are required. And it can be used as a reserve system to generate hydrogen (and heat) only when needed.

Aluminum is relatively abundant and inexpensive. It is used in many products. This results in the accumulation of a large amount of scrap aluminum. This scrap is generally not reprocessed, for it is cheaper to produce more aluminum from the alumina (aluminum oxide) that is extracted from bauxite ore. The result is that there is an immense, and growing, amount of inexpensive scrap aluminum in the world. One estimate was that this currently amounts to about 400 billion kilograms [6].

The Al_2O_3 that is formed by this process can be recycled by electrolysis in the same way that aluminum is produced from its natural oxide ore, bauxite. The Ga–In–Sn alloy is not consumed in this reaction, so that it is completely recoverable.

The specific energy of this system is attractive, 1,170 Wh/kg, counting the alloy and water weights, but not considering the weight of the container and any other system components.

However, because of the weights of the aluminum and water, the weight efficiency of the production of hydrogen is not especially attractive – only 3.6% hydrogen by weight, so such a system is not interesting for on-board vehicle use.

8.2.6 Additional Approaches

Another option that has been pursued somewhat involves passing hydrocarbons through an electric arc, whereby they decompose to form carbon and hydrogen at temperatures over 1,600°C.

There has also been a considerable amount of research on the possibility of either the use of *photoelectrolysis*, in which solar energy is used to decompose water directly, or *photoelectrochemical cells*, in which solar energy is employed to reduce the necessary applied voltage in electrically-driven cells. Although this is a very active area of investigation [7–11] it will not be discussed here.

8.3 Governmental Promotion of the Use of Hydrogen

The President of the United States announced a major Hydrogen Fuel Initiative in 2003 to accelerate the research and development of technologies needed to support hydrogen-powered fuel cells for use in transportation and electricity generation.

The underlying objective of this program was to decrease the degree of air pollution resulting from the use of petroleum. This program resulted in a significant increase in the amount of research and development on both hydrogen-based fuel cells and on-vehicle hydrogen storage. The proposed budget was $1.2 billion over a span of 5 years, to be used to develop hydrogen production, delivery, storage, and fuel cell technologies to enable the automobile and energy industries to commercialize fuel cell vehicles and the hydrogen fuel infrastructure. The general assumption was that hydrogen-powered vehicles should have performance that is equal or superior to current gasoline-consuming vehicles.

One aspect of this is a practical driving range of at least 300 miles for light-duty vehicles. This has resulted in the development of targets for the performance of on-board hydrogen storage systems in terms of weight, volume, and cost, as well as operating parameters. The more important targets for 2010 are shown in Table 8.6. It was assumed that meeting these goals would make it possible for some smaller and lighter vehicles to achieve the desired performance. Even more challenging targets were proposed for the year 2015 that would be appropriate for the full range of light-duty vehicles in North America.

Table 8.6 US Department of energy hydrogen storage system performance targets for year 2010

System parameters	Year 2010 targets
Specific energy from H_2	2 kWh/kg or 6 wt% H_2
Energy density from H_2	1.5 kW/L
Operating ambient temperature	$-30°C$ to $50°C$
H_2 delivery temperature	$-40°C$ to $85°C$
Cycle life (25–100%)	1,000 cycles
Minimum delivery pressure	4 atm (fuel cell) or 35 atm (ICE)
Recharging time (for 5 kg H_2)	3 min
Minimum flow rate	0.02 g/s per kW

In arriving at these targets, it was assumed that fuel cell power plants would have a factor of 2.5–3 times greater efficiencies than current gasoline-powered vehicles. It was also assumed that 5–13 kg of hydrogen would be necessary for fuel cell-driven vehicles, and that 1 kg of hydrogen can contribute about the same amount of energy as 1 gallon of gasoline.

Consideration of the magnitudes of some of these parameters is instructive. For example, recharging some of the hydrogen storage systems being considered can involve a large amount of heat generation. The reaction of hydrogen with some metal hydride materials, for example, is highly exothermic. A material that has a reaction enthalpy of 30 kJ/mol of hydrogen will generate 400 kW of heat if it reacts with 8 kg of hydrogen in 5 min. This heat will obviously have to be removed somehow.

Likewise, a fuel cell operating at 80 kW requires a hydrogen flow rate of 1.6 g/s. Therefore, the generation of this hydrogen from metal hydrides will require an equivalent amount of heat input. As a result, system thermal management can become critical.

These Department of Energy targets are very ambitious, and will require materials with properties significantly better than those presently available as well as innovative system designs. A rough picture of the current status is given in Fig. 8.8.

As will be discussed in Chap. 15, it is likely that serious attention will have to be given to compromises that are similar to some of those presently being considered for battery-propelled vehicles, such as plug-in hybrids.

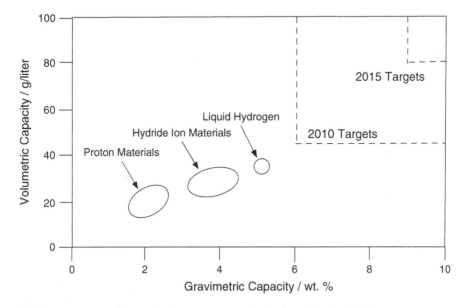

Fig. 8.8 Parameters of current hydrogen storage technologies, related to DOE targets

As mentioned above, one of the problems that could limit the rapid expansion of the use of hydrogen-propelled vehicles is the current lack of a large-scale hydrogen support infrastructure. Another is the amount of hydrogen that would be needed if a significant number of vehicles were to be converted to hydrogen propulsion. It was estimated [12] that one million fuel cell-propelled vehicles would consume about 0.4 million tons of hydrogen per year. The magnitude of this becomes evident when it is compared with the fact that the total hydrogen production in the United States is currently about 10 million tons. It should also be remembered that the production of hydrogen from natural gas causes about as much pollution as burning it directly.

8.4 Current On-Board Hydrogen Storage Alternatives

It is currently recognized that the problem of carrying the hydrogen fuel in the vehicle is the critical issue. Three different approaches have been pursued by the auto industry to date. Each of these has potential advantages and disadvantages as will be discussed below.

8.4.1 Storage of Gaseous Hydrogen in High Pressure Tanks

One of the traditional methods for storing hydrogen involves containment at high pressure in tanks. Although such tanks have traditionally been made of steel, fiber-reinforced composite materials have recently been developed that can withstand internal pressures up to 5,000–10,000 psi (340–680 atm). Hydrogen gas, even at these high pressures, has a relatively large volume, about twice that of liquid hydrogen.

This topic was discussed briefly in Chap. 6, where it was pointed out that according to the ideal gas law, the amount of gas that is stored in a fixed volume tank is proportional to the pressure. In addition, attention must be given to the thermal effects. Rapid charging of a tank with hydrogen is almost adiabatic (without heat exchange with the surroundings), so that a large amount of heat can be generated, giving rise to high temperatures. If the discharge of the gas from the tank into a fuel cell, for example, is relatively slow, so that heat transfer to the environment can occur, this reverse process is not so adiabatic, and the resulting cooling may not be important.

8.4.2 Storage of Liquid Hydrogen in Insulated Tanks

Hydrogen can be liquefied by cooling. Its boiling point at one atmosphere is 20.3 K. It can be contained and transported in liquid form in thermally-insulated containers.

Hydrogen gas can be readily obtained by applying heat so as to raise the temperature above the liquid/gas transition. Demonstration vehicles were constructed by the auto firm BMW in which this method was used for the storage of hydrogen. The hydrogen was combusted in a slightly modified internal combustion engine. A significant disadvantage in this approach is that the process of liquefaction requires some 30–40% of the final energy content of the hydrogen.

8.4.3 Storage of Hydrogen as Protons in Solids; Metal Hydrides

The third approach is to reversibly absorb hydrogen in solid metal-hydrogen compounds, called *metal hydrides*.

Hydrogen reversibly dissolves in a number of solids, and in some cases, to surprisingly high concentrations. The reason why this is possible is that the hydrogen is present in them as protons (H^+) ions, not as hydrogen atoms or hydride (H^{-1}) ions. The electrical charge in solids must always be balanced, so the charge due to the presence of protons is balanced by the presence of an equal number of extra electrons. Thus, these materials are always good electronic conductors.

The capacities of such materials depend upon the amount of hydrogen that they can absorb in their crystal structures. This is directly analogous to the absorption of lithium into the electrode materials commonly used in lithium batteries, a topic that is discussed in Chap. 11.

Protons are very small, and easily fit into the intersticies (spaces) between the other atoms present. An example of this is the metal alloy $LaNi_5$. It can absorb up to six hydrogen atoms per molecule, to form $LaNi_5H_6$. Another example is the alloy FeTi, which can absorb two hydrogen atoms per molecule, forming $FeTiH_2$. The hydrogen densities of these materials, which can absorb up to one hydrogen atom per metal atom, are 5.5×10^{22} and 5.8×10^{22} atoms of hydrogen per cm^3, respectively. These are greater than the density of liquid hydrogen, which is 4.2×10^{22} atoms cm^{-3}. Thus, the storage of hydrogen in such materials is very attractive from a volumetric standpoint.

Despite the fact that the hydrogen is present as positively-charged ions (protons) such hydrogen-containing metallic alloys are generally called metal hydrides. A number of them are widely used as negative electrode reactants in aqueous electrolyte batteries. The most common examples are the hydride/"nickel" cells that are used in many small applications, as well as the battery component of hybrid automobiles. One of their especially attractive properties is the high rate at which they can be charged and discharged. This topic will be discussed in some detail in Chap. 10.

Such metal alloy hydrides are certainly candidates for use for hydrogen storage in vehicles. Their major disadvantage, however, is that the presently known materials are too heavy and do not meet the weight requirements. As an example, the solid hydride materials that are now employed in the ubiquitous small hydride/"nickel" batteries used in many electronic devices store only about 2–3% hydrogen by weight. This is far from the 6% target of the Department of Energy. There is also some concern about the cost and large-scale availability of some such materials.

Vehicles with hydrogen-powered internal combustion engines, in which the hydrogen was stored in metal hydrides, were demonstrated some time ago by Mercedes Benz. In addition to the matter of cost, the primary problem with hydrogen storage in metal hydrides is the relatively small amount of hydrogen that they can store per unit weight. Brief mention was also made in Chap. 1 of the use of such metal hydrides as hydrogen storage media in fuel cell-propelled submarines. In that case, the volume of the hydrogen-absorbing material, a titanium–manganese alloy, is important, but the weight is not.

8.5 Other Approaches to Hydrogen Storage

In addition to the approaches that have been followed to- date in the automobile industry, there are some other approaches to hydrogen storage that might deserve some consideration, for either the vehicle, or other, applications. Two of these are described briefly below.

8.5.1 Hydrogen from the Decomposition of Materials Containing Hydride Anions

Another alternative strategy that is being employed for hydrogen storage involves materials in which the hydrogen is present in the form of hydride (H^-) ions, instead of protons. Whereas protons are very small, and can readily dissolve in a number of metal alloys, hydride ions are large, with an ionic radius of 146 pm, which is close to the size of oxide (O^{-2}) ions. Materials containing hydride ions are quite ionic, rather than metallic, in character. The large hydride ions typically have very low mobilities within their crystal structures.

There are two families of materials containing hydride ions that have received a lot of attention as possible hydrogen storage media. One of these is the *borohydride* family, which can be represented by the general formula $M^+BH_4^-$, where the species M^+ can be Li^+, Na^+ or K^+ or NH_4^+. $LiBH_4$ can theoretically store 13.9 wt% hydrogen, whereas $NaBH_4$ contains 7.9 wt% hydrogen.

There is an analogous family of materials, in which the boron is replaced by aluminum. These materials are called *alanates*. A further group of materials are the boranes, which have the general formula NH_nBH_n, where n can range from 1 to 4. One member of this group is amine borohydride, NH_4BH_4, which has a hydrogen mass ratio of 24% has received attention [12–14]. This material decomposes in several stages as it is heated, giving off about 6% of its mass in each step. This is shown in Table 8.7.

Table 8.7 Multistep decomposition of amine borohydride

Reaction	Wt% H_2 change	Temperature ($^\circ$C)
$NH_4BH_4 = NH_3BH_3 + H_2$	−6.1	<25
$NH_3BH_3 = NH_2BH_2 + H_2$	−6.5	<120
$NH_2BH_2 = NHBH + H_2$	−6.9	<155
$NHBH = BN + H_2$	−7.3	>500

Although their hydrogen capacities are quite high, there is a serious practical problem with materials of this general type, for they are not reversible. Hydrogen cannot be simply reacted with them so as to return them to their initial state, as can be done with the common proton-containing metal hydride materials. There is some hope that a method might be found for this, however, based on some recent work on titanium-based catalysts [15].

If that cannot be done, they must undergo a chemical reconstitution. This will require their removal from the vehicle and replacement by a new chemical charge. The chemical process to regenerate such hydride-ion materials will probably most effectively be done in large central chemical plants. But then there must also be a system for the transportation of the spent and renewed materials. This requirement for external chemical regeneration constitutes a major disadvantage of this approach.

A similar chemical regeneration requirement led to the demise of the highly touted proposal to use zinc/air batteries for vehicle propulsion some years ago. Although the reaction between zinc and oxygen to form ZnO provides a lot of energy per unit weight, a method for this cell to be electrically recharged was not known. The result was that the discharged battery electrodes had to be shipped to a chemical plant for the conversion of the ZnO product back to elemental zinc. The shipping and chemical processes to reconstitute the zinc electrodes combined to make this approach unfeasible.

8.5.2   Ammonia and Related Materials as Hydrogen Storage Media

Another possible approach to consider is the use of ammonia, NH_3, which is 17% hydrogen by weight. It is possible to thermally decompose ammonia at modest temperatures. This is seen from the data shown in Fig. 8.9. It can be seen that at one atmosphere pressure, ammonia decomposes into its elements at about 460 K (187°C).

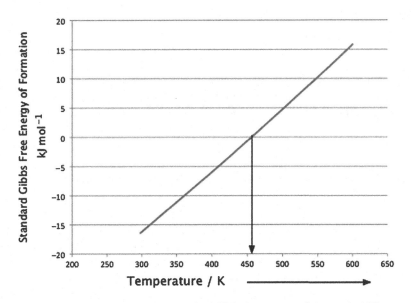

Fig. 8.9 Temperature dependence of the standard Gibbs free energy of Ammonia, NH_3

Consideration is also being given to more complex materials that can decompose to form NH_3, which can then be further decomposed to provide hydrogen. Metal ammines are one type of such materials, in which ammonia is coordinated to a metal ion. An example is $Mg(NH_3)_6Cl_2$ [16].

Another concept is the use of a material containing hydrogen that might react with another material and give off hydrogen as one of the reaction products. One example that has been proposed is lithium amide, $LiNH_2$, with the expectation that it would react with LiH to form lithium imide, Li_2NH and hydrogen.

The feasibility of this concept can be evaluated from thermodynamic data related to the phases concerned. Values of the standard Gibbs free energy of formation for the pertinent phases in the lithium–hydrogen–nitrogen system [17] are included in Table 8.8.

Table 8.8 Standard Gibbs free energy of formation data for phases in the Li–H–N system

Phase	ΔG_f^0 at 298 K (kJ mol^{-1})
Li_3N	-122.2
LiH	-70.0
NH_3	-16.4
Li_2NH	-169.9
$LiNH_2$	-140.6
H_2	0

The proposed reaction can be written as

$$LiNH_2 + LiH = Li_2NH + H_2 \tag{8.16}$$

The sum of the ΔG_f^0 values of the phases on the left is -210.6 kJ mol^{-1}, and of the phases on the right is -169.9 kJ mol^{-1}. Therefore, this reaction will tend to go to the left, not the right. Thus this concept makes no sense.

8.5.3   Storage of Hydrogen in Reversible Organic Liquids

In addition to solid materials that absorb hydrogen as protons to become metal hydrides, or already contain hydrogen in the form of hydride anions, it is reasonable to also consider the use of liquids. Many organic liquids have large, and often variable, amounts of hydrogen present in their structures.

If everything else were the same, one potential advantage of the use of liquids as hydrogen-containers is the assumption that such materials could be inexpensively stored in tanks, and pumping them in and out should be no great problem.

In some countries, such as Switzerland, that have surplus hydropower during part of the year, a large scale method for the storage of some of this energy for use at other times of the year could be valuable. One possible method for such large scale and long-term energy storage by the use of organic hydrides was proposed some time ago [18]. This concept involves the electrolytic production of hydrogen when electricity is readily available and inexpensive, and reacting it with a simple organic molecule to produce a product containing more hydrogen. The species with the higher hydrogen content could later be treated such that the hydrogen is released.

The simplest example would be the conversion of benzene (C_6H_6) to form cyclohexane (C_6H_{12}). Benzene is a simple aromatic molecule, in which each of the carbon atoms in the 6-member ring is bonded to one hydrogen atom. In cyclohexane, each carbon has two hydrogen neighbors. This possibility cannot be seriously considered for actual use, however, because benzine is considered to be a carcinogen.

Thus, it is preferable to use toluene (C_7H_8), and to produce methylcyclohexane (C_7H_{14}) by the reversible addition of six hydrogen atoms per molecule. Both of these liquids can readily be stored in large tanks. They are inexpensive and convenient liquids that are easy to transport and store, with freezing temperatures that are convenient, 178 K and 146.4 K, respectively.

It can readily be seen that the weight of the hydrogen stored in this case is 6.5% of the weight of its carrier, the toluene. This is substantially better than the comparable values for the solid metal hydrides.

The volumetric density of hydrogen is also important. Methylcyclohexane contains 47.4 g of H_2 per liter. The hydrogen density in gaseous hydrogen varies with the pressure and temperature according to the ideal gas equation, of course. At 200 bar, it is only 18 g H_2 per liter.

Both the hydrogen-addition and the hydrogen-deletion processes require catalysts. The conversion of toluene to methylcyclohexane, a well-known large-scale industrial process, is exothermic, and the reverse reaction, which takes place at 400°C, is endothermic, so heat has to be supplied at a relatively high temperature when extracting the hydrogen. This hydrogenation cycle has a round trip efficiency of about 80%.

An example of the use of this concept [19] involved installation in a 17 ton truck in Switzerland. The hydrogen obtained from the onboard catalytic splitting of methylcyclohexane was used as the fuel for its 150 kW internal combustion engine. Hydrogen was injected at 10 bar pressure, and resulted in an engine efficiency of 32%. In an analysis of costs [20], this scheme for the storage and use of energy was found to be economically competitive with other carbon-free large-scale energy storage methods, as well as with the cost of the construction of additional hydropower facilities.

Other covalent organic materials with larger molecular structures can also be considered for this purpose, including perhydrofluorene and several species in the carbazole family [21].

An additional interesting feature of the reversible liquid hydrogen system has to do with its potential use for the transmission of energy over long distances using simple, low-cost pipelines. This has been pioneered as a method to get energy to geographically remote areas in Russia and Brazil, and is illustrated schematically in Fig. 8.10.

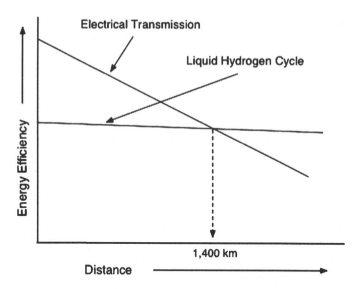

Fig. 8.10 Energy transmission efficiency versus distance

Electrical power transmission involves both resistive and capacitive losses that increase linearly with distance, reducing the overall efficiency of the electrical transmission system. These can be appreciable if energy is sent over long distances. On the other hand, the losses that are involved in the transport of liquids through pipelines are relatively small.

As a result, there is a critical distance over which energy can be transmitted more economically by fluids in pipelines than by electrical lines. In one study, the breakeven point was at 1,400 km. At longer distances, the use of the reversible liquid cycle is more effective.

8.6   The Question of Safety

Almost everyone knows that hydrogen readily burns in air, and with a flame that is almost colorless. Those working with hydrogen are always cautioned to keep it away from open flames. Attention to this potential problem rapidly became widespread after the widely publicized burning of the large Hindenburg zeppelin in New Jersey in 1937.

It is important to give this potential safety problem some attention. The ignition temperature of hydrogen in air is 585°C at one atmosphere pressure. Thus, there is no problem with having a mixture of them at ambient temperature. At elevated temperatures, the composition range within which flammability is possible is between 4 and 75% hydrogen by volume.

The density of hydrogen gas is low, only 0.08, so that it rises rapidly in air. This reduces the possibility of reaching the flammability compositional range in open spaces. In enclosed spaces, it tends to accumulate at the highest locations, for example, near the ceilings of rooms.

This question was raised some years ago, when BMW was demonstrating a hydrogen-powered demonstration auto in which the hydrogen was carried as a liquid in an insulated tank. It was contended that any hydrogen that got free would rise so rapidly that its possible combustion would not be a threat.

References

1. I. Barin, Thermochemical Data of Pure Substances, 3rd Edition, VCH, New York (1995)
2. J.J. Cuomo and J.M. Woodall, US Patent 4,358,291, November 9, 1982
3. J.M. Woodall, J. Ziebarth and C.R. Allen, Proc. 2nd Energy Nanotechnology International Conference, Santa Clara, CA, Sept. 5, 2007
4. J.M. Woodall, J.T. Ziebarth, C.R. Allen, D.M. Sherman, J. Jeon and G. Choi, Proc. Hydrogen 2008, Feb. 2008
5. J.M. Woodall, J. Ziebarth, C.R. Allen, J. Jeon, G. Choi and R. Kramer, Clean Technology, June 1, 2008
6. J.M. Woodall, Presentation at the Electrochemical Society Meeting, San Francisco, May 26, 2009. To be published
7. M. Graetzl, Nature *414*, 15 (2001)
8. G.W. Crabtree, M.S. Dresselhaus and M.V. Buchanan, Phys Today *57*, 39 (2004)
9. M. Graetzl, Inorg Chem *44*, 6841 (2005)
10. G.W. Crabtree and M.S. Dresselhaus, MRS Bull *33*, 421 (2008)
11. National Hydrogen Energy Roadmap, US Department of Energy, November, 2002, http://www.hydrogen.energy.gov/pdfs/national_h2_roadmap.pdf
12. A. Gutowska, L. Li, Y. Shin, C.M. Wang, X.S. Li, J.C. Linehan, R.S. Smith, B.D. Kay, B. Schmid, W. Shaw, M. Gutowski and T. Autrey, Angew Chem Int Ed *44*, 3578 (2005)
13. M.H. Matus, K.D. Anderson, D.M. Camaioni, S.T. Autrey and D.A. Dixon, J Phys Chem *111*, 4411 (2007)
14. C.W. Yoon and L.G. Sneddon, J Am Chem Soc *128*, 13992 (2006)
15. E. Muller, E. Sutter, P. Zahl, C.V. Ciobanu, P. Suttera, Appl Phys Lett *90*, 151917 (2007)
16. C.H. Christensen, T. Johannessen, R.Z. Soerensen and J.K. Norskov, Catal Today *111*, 140 (2006)
17. B.A. Boukamp and R.A. Huggins, Phys Lett *72A*, 464 (1979)
18. M. Taube and P. Taube, in Proc. of 3rd World Hydrogen Energy Conference, Tokyo (1980)
19. M. Taube, D. Rippin, W. Knecht, B. Milisavijevic and D. Hakimifard, Hydrogen Energy Progress V, ed. By T.N. Veziroglu and J.B. Taylor, Pergamon Press, New York (1984), p. 1341
20. G.W.H. Scherer, E. Newson and A. Wokaun, J Hydrogen Energy *24*, 1157 (1999)
21. A. Cooper, A. Scott, D. Fowler, F. Wilhelm, V. Monk, H. Cheng and G. Pez, Presentation at 2008 DOE Hydrogen Program Meeting, June, 2008; http://www.hydrogen.energy.gov/pdfs/review08/stp_25_Cooper.pdf

Chapter 9
Introduction to Electrochemical Energy Storage

9.1 Introduction

Among the various methods that can be used for the storage of energy that are discussed in this text, electrochemical methods, involving what are generally called *batteries*, deserve the most attention. They can be used for a very wide range of applications, from assisting the very large-scale electrical grid down to tiny portable devices used for many purposes. Battery-powered computers, phones, music players, etc. are everywhere, and one of the currently hot topics involves the use of batteries in the propulsion of vehicles, hybrid autos, plug-in hybrids, and fully electric types.

Many students are put off from discussions of electrochemical systems because of unfamiliarity with electrochemistry. It will be shown in this text that one can understand the major phenomena and issues in electrochemical systems without considering their truly electrochemical features in detail. As an example, it will be shown that the driving forces of electrochemical cells are related to the driving forces between the electrically neutral components in the electrodes. Electrochemical considerations only come into play in certain features of their mechanisms.

Electrochemical energy storage involves the conversion, or *transduction*, of chemical energy into electrical energy, and vice versa. In order to understand how this works, it is first necessary to consider the *driving forces* that cause electrochemical transduction in electrochemical cells as well as the major types of *reaction mechanisms* that can occur.

This is followed by a brief description of the important practical parameters that are used to describe the behavior of electrochemical cells, and how the basic properties of such electrochemical systems can be modeled by the use of simple equivalent electrical circuits.

Also included in this chapter is a brief discussion of the principles that determine the major properties of electrochemical cells, their voltages, and capacities.

R.A. Huggins, *Energy Storage*,
DOI 10.1007/978-1-4419-1024-0_9, © Springer Science+Business Media, LLC 2010

9.2 Simple Chemical and Electrochemical Reactions

Consider a simple *chemical reaction* between two metallic materials A and B, which react to form an electronically conducting product AB. As discussed in Chap. 4, this can be represented simply by the relation:

$$A + B = AB \qquad (9.1)$$

The driving force for this reaction is the difference in the values of the *standard Gibbs free energy* of the products, only AB in this case, and the standard Gibbs free energies of the reactants, A and B.

If A and B are simple elements, this is called a *formation reaction*, and since the standard Gibbs free energy of formation of elements is zero, the value of the Gibbs free energy change that results per mole of the reaction is simply the *Gibbs free energy of formation* per mole of AB, that is:

$$\Delta G_r{}^0 = \Delta G_f{}^0(AB) \qquad (9.2)$$

Values of this parameter for many materials can be found in a number of sources, for example, [1].

While the morphology of such a reaction can take a number of forms, consider a simple one-dimension case in which the reactants are placed in direct contact and the product phase AB forms between them. The time sequence of the evolution of the microstructure during such a reaction is shown schematically in Fig. 9.1. Later times are at the bottom.

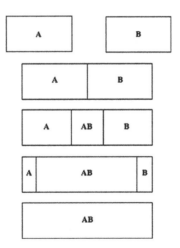

Fig. 9.1 Simple schematic model of the chemical reaction of A and B to form AB, indicating how the microstructure of the system varies with time

It is obvious that in order for the reaction product phase AB to grow, atoms of either A or B must move (*diffuse*) through it to reach the other side to come into contact with the other reactant. If, for example, A moves through the AB phase to the B side, additional AB will form at the AB/B interface. Since some B is consumed, the AB/B interface will move to the right. Also, since the amount of A on the A side has decreased, the A/AB interface will likewise move to the left. The AB will grow in width in the middle. One should note that the same thing will happen in the case that the species B, rather than the species A, moves through the AB phase in this process. There are experimental ways in which one can determine the identity of the moving species, but it is not necessary to be concerned with them here.

Now suppose that this process occurs by an *electrochemical mechanism*. The time dependence of the microstructure in this case is shown schematically in Fig. 9.2. As in the chemical reaction case, the product AB must form as the result of a reaction between the reactants A and B. But there is an additional phase present in the system, an electrolyte.

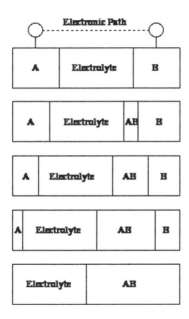

Fig. 9.2 Simple schematic model of the time evolution of the microstructure during the electro-chemical reaction of A and B to form AB, a mixed conductor. In this case, it is assumed that A^+ ions are the predominant ionic species in the electrolyte. To simplify the figure, the external electronic path is shown only at the start of the reaction

The *function of the electrolyte* is to *act as a filter* that allows the passage of ionic, but not electronic species. This means that the electrolyte contains ions of either A or B, or both, and is an *electronic insulator*.

But the reaction between A and B involves *electrically neutral atoms*, not just ions. This means that in order for the reaction to proceed there must be another path whereby electrons can also move through the system. This is typically an external electrical circuit that connects A and B. In the case that it is A that is transported in the system, and that the electrolyte contains A^+ ions, negatively charged electrons, e^-, must pass through the external circuit in equal numbers, or at an equal rate, to match the charge flux due to the passage of A^+ ions through the electrolyte to the other side.

During an electrochemical discharge reaction of the type illustrated in Fig. 9.2 the reaction at the interface between the phase A and the electrolyte can be written as:

$$A = A^+ + e^- \tag{9.3}$$

with the A^+ ions moving into the electrolyte phase and the electrons entering the external circuit through a *current collector*. At the same time, there will be a corresponding reaction on the other side of the electrolyte

$$A^+ + e^- = A \tag{9.4}$$

with ions arriving at the interface from the electrolyte and electrons coming to the interface from the external circuit through the electronic current collector. The result is to deposit A atoms onto the adjacent solid phase AB. The result is that the A/electrolyte interface and the electrolyte/AB interface both move incrementally to the left in Fig. 9.2. There must be *interdiffusion* of A and B atoms within the phase AB so that its surface does not have only A atoms. In addition, this phase must be an electronic conductor.

The fact that the overall reaction is between neutral species, and that this requires the concurrent motion of either A or B ions through the electrolyte, and electrons through external circuit, has several important consequences. One is that if flow in either the electronic path or the ionic path does not occur, the whole reaction must stop. For example, if the external electrical circuit is opened so that no electrons can flow through it, no ions can flow through the electrolyte, and the reaction stops. Likewise, if the flow of ions in the electrolyte is impeded – for example, by the presence of some material with a very high resistance for the moving ionic species, or a loss of contact between the electrolyte and the two materials on its sides – there will be no electronic current in the external circuit.

When the electronic circuit is open, and there is no current flowing, there must be a force balance operating upon the electrically charged ions in the electrolyte. A *chemical driving force* upon the mobile ionic species within the electrolyte in one direction is simply balanced by *an electrostatic driving force* in the opposite direction.

The *chemical driving force* across the cell is due to the difference in the chemical potentials of its two electrodes. It can be expressed as the *standard Gibbs free energy change per mole of reaction*, ΔG_r^0. This is determined by the difference in the *standard Gibbs free energies of formation* of the products and the reactants in the *virtual chemical reaction* that would occur if the *electrically neutral* materials in the two electrodes were to react chemically. It makes no difference that the reaction actually happens by the transport of ions and electrons across the electrochemical system from one electrode to the other.

The electrostatic energy per mole of an electrically charged species is $-zFE$, where E is the voltage between the electrodes, and z is the *charge number* of the mobile ionic species involved in the virtual reaction. The charge number is the number of elementary charges that they transport. F is the *Faraday constant* (96,500 C per equiv). An *equivalent* is *Avogadro's number* (1 mol) of electronic charges.

The balance between the chemical and electrical forces upon the ions under open circuit conditions can thus be simply expressed as a chemical energy – electrostatic energy balance

$$\Delta G_r^0 = -zFE \qquad (9.5)$$

Here the value of ΔG_r^0 is in Joules per mole of reaction, as 1 J is the product of 1 C and 1 V.

Thus, this is an interesting situation in which a chemical reaction between neutral species in the electrodes determines the forces upon charged particles in the electrolyte in the interior of an electrochemical system.

If it is assumed that the electrodes on the two sides of the electrolyte are good electronic conductors, there is an externally measurable voltage E between the points where the external electronic circuit contacts the two electrodes. As the result of this voltage, electrical work can be done by the passage of electrons in an external electric circuit if ionic current travels through the electrolyte inside the cell.

Thus, this simple electrochemical cell can act as a *transducer* between chemical and electrical quantities; forces, fluxes, and energy. In the ideal case, the chemical energy reduction due to the chemical reaction that takes place between A and B to form mixed-conducting AB is just compensated by the electrical energy transferred to the external electronic circuit.

The flow of both internal ionic species and external electrons can be reversed if a voltage is imposed in the electronic path in the opposite direction that is larger than the voltage that is the result of the driving force of the chemical reaction. Since this causes current to flow in the reverse direction, electrical energy will be consumed and the chemical energy inside the system will increase. This is what occurs when an electrochemical system is recharged.

From these considerations, it is obvious that it is not important whether the ionic species are related to element A or to element B. However, the answer to this question will influence the configuration of the cell. The example illustrated schematically in Fig. 9.2 deals with the case in which there are predominantly A^+ ions in the electrolyte. The chemical reaction proceeds by the transport of A^+ ions in the electrolyte and electrons in the external circuit from the left (A) side of the cell to the right hand side. This involves two electrochemical reactions. On the left side, A atoms are converted to A^+ ions and electrons at the A/electrolyte interface. The electrons travel back through the metallic A and go out into the external electronic circuit. The reverse electrochemical reaction takes place on the other side of the cell. A^+ ions from the electrolyte combine with electrons that have come through the external circuit to form neutral A at the electrolyte/AB interface.

As before, the physical locations of the interfaces, the A/electrolyte interface, the electrolyte/AB interface and the AB/B interface, will move with time as the amounts of the various species vary with the extent of the reaction.

It must be recognized that the reaction product AB will not form unless there is a mechanism that allows the newly-arrived A to react with B atoms to form AB. Thus the transport of either A or B atoms within the AB product phase is necessary in this case, as it was in the chemical reaction case illustrated in Fig. 9.1 above. If this did not happen, pure A would be deposited at the right hand electrolyte interface. The chemical composition on both sides of the electrolyte would then be the same, and there would be no driving force to cause further transport of ionic species through the electrolyte, and thus no external voltage.

If B^+ ions, rather than A^+ ions, are present in the electrolyte, so that B species can flow from right to left, the direction of electron flow, and thus the voltage polarity, in the external circuit will be opposite from that discussed above, and the reaction product will form on the left side, rather than on the right side.

It is also possible, of course, that the ions in the electrolyte are negatively charged. In that case, the direction of electron flow in the external circuit will be in the opposite direction.

In any case, it is important to realize that the basic driving force in an electrochemical cell is a *chemical reaction of neutral species* to form an *electrically neutral product*. This is why one can use standard chemical thermodynamic data to understand the equilibrium (no current, or open circuit) potentials, and voltages in electrochemical cells.

For any given chemical reaction, the open circuit voltage is independent of the identity of the species in the electrolyte and the details of the reactions that take place at the electrode/electrolyte interfaces.

The situation becomes different if one considers the kinetic behavior of electrochemical cells, for then one has to be concerned with phenomena at all of the interfaces, as well as in the electrodes, the electrolyte, and the external circuit. Such matters will be discussed in some detail later.

9.3 Major Types of Reaction Mechanisms in Electrochemical Cells

The operation of electrochemical cells involves the transport of neutral chemical species into and out of the electrodes, their ionic parts move through the electrolyte, and the charge-balancing electrons move through the external electrical circuit. In many, but not all, cases, this results in changes in the chemical constitution of electrodes, that is, the amounts and chemical compositions of the phases present.

The result is that the microstructure of one or more of the electrode materials gets significantly changed, or *reconstituted*. There are a number of important chemical, and therefore possible electrochemical, reactions in which some phases grow and others disappear.

Reactions in which there is a change in the identity or amounts of the phases present are designated as *reconstitution reactions*.

Phase diagrams are useful *thinking tools* to help understand these types of phenomena. As discussed in Chap. 4, they are graphical representations that indicate the phases and their compositions that are present in a materials system under equilibrium conditions, and were often called *constitution diagrams* in the past. In Chap. 4, the discussion was focused upon phenomena that occur as the result of changes in the temperature. Electrochemical systems, on the other hand, generally operate at a constant temperature, that is, isothermally. This involves the consideration of what occurs as the composition moves horizontally, rather than vertically, across phase diagrams.

Two major types of reconstitution reactions that are relevant to electrochemical systems, *formation reactions* and *displacement reactions*, will be briefly mentioned here. This will be followed by an introduction to *insertion reactions*, which play a major role in the operation of electrodes in a number of important modern battery systems.

9.3.1 Formation Reactions

The simple example that was discussed earlier in this chapter, represented by the equation:

$$A + B = AB \tag{9.6}$$

is a *formation reaction*, in which a new phase *AB* is formed in one of the electrodes from its atomic constituents. This can result from the transport of one of the elements, for example, *A*, passing across an electrochemical cell through the electrolyte from one electrode to react with the other component in the other electrode. Since this modifies the microstructure, it is an example of one type of *reconstitution* reaction.

There are many examples of this type of formation reaction. There can also be subsequent additional formation reactions whereby other phases can be formed by further reaction of an original product.

As an example, consider the reaction of lithium with aluminum. Lithium–aluminum alloys were explored for use as electrodes in high temperature lithium batteries some time ago [2, 3], and their critical thermodynamic and kinetic properties were studied by the use of molten salt electrolyte electrochemical cells [4–6].

The reactions in this alloy system can be understood by the use of the *lithium–aluminum system* phase diagram, as shown in Fig. 9.3.

Assume that the negative electrode is lithium and the positive electrode is initially pure aluminum. Upon the imposition of current by making lithium positive, and aluminum negatively charged, lithium ions pass across the cell and react with the aluminum in the positive electrode, changing its chemical composition. If this

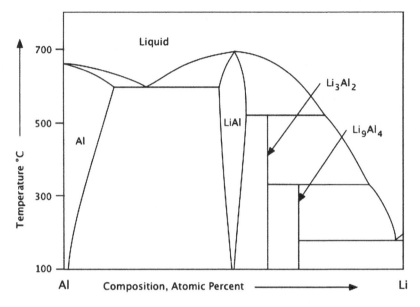

Fig. 9.3 Equilibrium phase diagram of the aluminum–lithium system

were to happen at 100°C, the lowest temperature in Fig. 9.3, and it could be assumed that equilibrium conditions can readily be attained, it can be seen that a solid solution is initially formed, in which a small amount of lithium dissolves into the aluminum.

As more lithium is passed across the cell, the solubility limit of lithium in aluminum is reached, and the composition moves into a region of the phase diagram in which both the lithium–saturated aluminum phase Al_{sat} and a new phase, "LiAl", are present in the positive electrode. The quotation marks are used here, for the composition of the phase is not exactly 1:1 Li/Al. Thus, in this part of the phase diagram, the formation reaction:

$$Li + Al_{sat} = \text{"LiAl"} \tag{9.7}$$

takes place. As more lithium passes across the electrochemical cell, the overall composition traverses the two-phase Al_{sat} + "LiAl" region, more "LiAl" is formed, and the amount of Al_{sat} decreases. By the time the overall composition reaches the low-Li boundary of the "LiAl" region, there is no more of the Al_{sat} present.

The addition of more lithium causes the overall composition to go across the range of the "LiAl" phase. Thereafter, there is another two-phase formation reaction

$$Li + 2\text{"LiAl"} = Li_3Al_2 \tag{9.8}$$

that is later followed by the reaction

$$3Li + 2Li_3Al_2 = Li_9Al_4 \tag{9.9}$$

as more lithium reacts with the structure that results from reaction (9.8).

The electrical potential varies with the chemical composition of electrodes in the Li–Al alloy system, as well as others that exhibit either ranges of solid solution or multiphase reactions. This will be discussed in substantial detail in later chapters.

It is also not necessary that both reactants in formation reactions are solids or liquids, of course. For example, the phase LiCl can result from the reaction of lithium with chlorine gas, and ZnO can form as the result of the reaction of zinc with oxygen in the air. Zn/O_2 cells, in which ZnO is formed, are commonly used as the power source in hearing aids.

9.3.2 Displacement Reactions

As discussed in Chap. 4, another type of *reconstitution reaction* involves a *displacement* process, which can be simply represented as

$$A + BX = AX + B \tag{9.10}$$

in which species A displaces species B in the simple binary phase BX, to form AX instead. A new phase consisting of elemental B will be formed in addition. There will be a driving force causing this reaction to tend to occur if phase AX has a greater stability, that is, it has a more negative value of ΔG_f^0, than the phase BX. An example of this type that was discussed in Chap. 4 was

$$Li + Cu_2O = Li_2O + Cu \tag{9.11}$$

in which the reaction of lithium with Cu_2O results in the formation of two new phases, Li_2O and elemental copper.

A change in the chemical state in the electrode results in a change in its electrical potential, of course. The relation between the chemical driving forces for such reactions, and the related electrical potentials, will be discussed for this case in later chapters.

9.3.3 Insertion Reactions

Again, as mentioned in Chap. 4, a quite different type of reaction mechanism can also occur in materials in chemical and electrochemical systems. This involves the *insertion* of guest species into normally unoccupied interstitial sites in the crystal structure of an existing stable host material. Although the chemical composition of the host phase initially present can be changed substantially, this type of reaction does not result in a change in the identity, the basic crystal structure, or amounts of the phases in the microstructure. However, in most cases, the addition of interstitial species into previously unoccupied locations in the structure causes a change in volume. This involves mechanical stresses and mechanical energy. The mechanical energy related to the insertion and extraction of interstitial species plays a significant role in the hysteresis, and thus energy loss, observed in a number of reversible battery electrode reactions.

In the particular case of the insertion of species into materials with layer-type crystal structures, insertion reactions are sometimes called *intercalation reactions*. Such reactions, in which the composition of an existing phase is changed by the incorporation of guest species, can also be thought of as a solution of the guest into the host material. Therefore, such processes are also sometimes called *solid solution reactions*.

Generally, the incorporation of such guest species occurs *topotactically*. This means that the guest species tend to be present at specific (low energy) locations inside the crystal structure of the host species, rather than being randomly distributed.

A simple reaction of this type might be the reaction of an amount x of species A with a phase BX to produce the product A_xBX. This can be written as

$$xA + BX = A_xBX \qquad (9.12)$$

for such a case. The solid solution phase can have a range of composition, that is, a range of values of x. As an example, the incorporation of lithium into TiS_2 produces a product in which the value of x can extend from 0 to 1. This was an important early example of this type of insertion reaction [7], and it can be simply represented as

$$x\text{Li} + TiS_2 = \text{Li}_x TiS_2 \qquad (9.13)$$

It is also possible to have a *displacement reaction* occur by the replacement of one interstitial species by another inside a stable host material. In this case, only one additional phase is formed, the material that is displaced. The term *extrusion* is sometimes used to describe this process.

In some cases, the new element or phase that is formed by such an *interstitial displacement process* is *crystalline*, whereas in other cases, it can be *amorphous*.

9.4 Important Practical Parameters

When considering the use of electrochemical energy storage systems in various applications, it is important to be aware of the properties that might be relevant, for they are not always the same in every case.

The energy and power available per unit weight, called the *specific energy* and *specific power*, are of great importance in some applications such as vehicle propulsion.

On the other hand, the amount of energy that can be stored per unit volume, called the *energy density*, can be more important in some other areas of application. This is often the case when such devices are being considered as power sources in portable electronic devices, such as cellular telephones, portable computers, and video camcorders.

The power per unit volume, called the *power density*, can also be especially important for some uses, such as cordless power tools, whereas in others the *cycle life* – the number of times that a device can be effectively recharged before its performance, for example, its capacity, is degraded too far – is critical. In addition, cost is always of concern, and sometimes can be of overriding importance, even at the expense of reduced performance.

Methods will be described later that allow the determination of the maximum theoretical values of some of these parameters, based upon the properties of the materials in the electrodes alone. However, practical systems never achieve these maximum theoretical values, but instead, often provide much lower performance. One obvious reason is that a practical battery has a number of passive components that are not involved in the basic chemical reaction that acts as the energy storage mechanism. These include the electrolyte, a separator that mechanically prevents the electrodes from coming into contact, the current connectors that transport electrical current to and from the interior of the cell, and the container. In addition, the effective utilization of the active components in the chemical reaction is often less than optimal. Electrode reactant materials can become electronically disconnected, or shielded from the electrolyte. When that happens, they cannot participate in the electrochemical reaction and have to be considered passive. They add to the weight and volume, but do not contribute to the transduction between electrical and chemical energy.

A rule of thumb that was used for a number of the conventional aqueous electrolyte battery systems was that a practical cell could only produce about 1/5–1/4 of the maximum theoretical specific energy (MTSE). Optimization of a number of factors has made it now possible to exceed such values in a number of cases. In addition, the maximum theoretical values of some of the newer electrochemical systems are considerably higher than what was available earlier.

Some rough values of the practical *energy density* (Wh/l) and *specific energy* (Wh/kg) of several of the common rechargeable battery systems are listed in Table 9.1. These particular values should not be taken as definitive, for they depend upon a number of operating factors and vary with the designs of different manufacturers. Nevertheless, they indicate the wide range of these parameters available commercially from different technologies.

In addition to their energy capacity, another important parameter relating to the practical use of batteries is the amount of power that they can supply. This is often expressed as specific power, the amount of power per unit weight, and it is very dependent upon the details of the design of the cell, as well as the characteristics of the reactive components. Therefore, values vary over a wide range.

Table 9.1 Approximate values of the practical specific energy
and energy density of some common battery systems

System	Specific energy (Wh/kg)	Energy density (Wh/l)
Pb/PbO$_2$	40	90
Cd/Ni	60	130
Hydride/Ni	80	215
Li-Ion	135	320

The characteristics of batteries are often graphically illustrated by the use of
Ragone plots, in which the specific power is plotted versus the specific energy. This
type of presentation was named after D.V. Ragone, who was the chairman of a
governmental committee that wrote a report on the relative properties of different
battery systems many years ago. Such a plot, including very approximate data on
three current battery systems, is shown in Fig. 9.4.

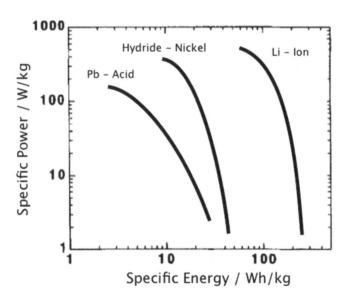

Fig. 9.4 Ragone plot showing approximate practical values of specific power and specific energy
of three common battery systems

9.4.1 The Operating Voltage and the Concept of Energy Quality

In addition to the amount of energy stored, another important parameter of a battery
system is the voltage at which it operates, both during discharge, when it supplies
electrical energy and power, and also when it is being recharged.

As discussed earlier in this chapter, the open circuit, or equilibrium, cell voltage is primarily determined by the thermodynamics of the chemical reaction between the components in the electrodes, for they supply the driving force for the transport of ions through the electrolyte, and electrons through the external circuit. During actual use, however, the operating voltage will vary from these theoretical values, depending upon various kinetic factors. These will be discussed extensively later in this text.

In discussing electrochemical energy storage, it is useful to consider another parameter, its *quality*, and how it matches the expected applications. The concept of *energy quality* is analogous to the concept of *heat quality* that is well known in engineering thermodynamics.

As discussed earlier, it is widely recognized that high temperature heat is often more useful (for example, has higher quality) than low temperature heat in many applications. Similarly, the usefulness of electrical energy is often related to the voltage at which it is available. High voltage energy is often more useful (has higher quality) than low voltage energy. For example, in simple resistive applications, the electrical power P is related to the practical (not just theoretical) voltage E and the resistance R by

$$P = E^2/R \qquad (9.14)$$

Thus, the utility of an electrochemical cell in powering a light source or driving an electric motor is particularly voltage-sensitive. Because of the square relation, high voltage stored energy has a much higher quality for such applications than low voltage stored energy.

Rough energy quality rankings can be tentatively assigned to electrochemical cells on the basis of their output voltages as follows:

3.0–5.5 V	High quality energy
1.5–3.0 V	Medium quality energy
0–1.5 V	Low quality energy

There are a number of applications in which a high voltage is required. One example is the electrical system used to propel either hybrid- or all-electric vehicles. Auto manufacturers typically wish to operate such systems at over 200 V. For this type of high voltage application, it is desirable that individual cells produce the highest possible voltage, for the greater the voltage of each individual cell, the fewer cells are necessary. There is also the movement toward the use of 36–42 V systems for the starter, lighting, and ignition systems in normal internal combustion engine automobiles, as mentioned earlier.

Despite the implications of this matter of energy quality, it is important that the voltage characteristics of electrochemical energy storage systems *match the requirements of the intended application*. It is not always best to have the highest possible cell voltage, for it can be wasteful if it is too high in some applications.

A further matter that can becomes especially important in some applications is safety, and this can be a potential problem with some high potential electrode materials. As a result, development efforts aimed at large batteries for vehicle traction applications have been investigating materials that sacrifice cell voltage to obtain greater safety.

9.4.2 The Charge Capacity

The *energy contained* in an electrochemical system is the integral of the voltage multiplied by the *charge capacity*, that is, the amount of charge available. That is,

$$\text{Energy} = \int E dq \tag{9.15}$$

where E is the output voltage, which can vary with the state of charge as well as kinetic parameters, and q is the amount of electronic charge that can be supplied to the external circuit.

Thus, it is important to know the maximum capacity, the amount of charge that can theoretically be stored in a battery. As in the case of the voltage, the maximum amount of charge available under ideal conditions is also a thermodynamic quantity, but it is of a different type. Whereas voltage is an *intensive quantity*, independent of the amount of material present, charge capacity is an *extensive quantity*. The amount of charge that can be stored in an electrode depends upon the amount of material in it. Therefore, capacity is always stated in terms of a measure such as the number of Coulombs per mole of material, per gram of electrode weight, or milliliter of electrode volume.

The *state of charge* is the current value of the fraction of the maximum capacity that is still available to be used.

9.4.3 The Maximum Theoretical Specific Energy

Consider a simple insertion or formation reaction that can be represented as

$$xA + R = A_xR \tag{9.16}$$

where x is the number of mols of A that reacts per mol of R. It is also the number of elementary charges per mol of R. If E is the average voltage of this reaction, the theoretical energy involved in this reaction follows directly from (9.15). If the energy is expressed in Joules, it is the product of the voltage in volts and the charge capacity, in Coulombs, involved in the reaction.

If W_t is the sum of the molecular weights of the reactants engaged in the reaction, the maximum theoretical specific energy (MTSE), the energy per unit weight, is simply

$$\text{MTSE} = (xE/W_t)F \tag{9.17}$$

MTSE is in J/g, or kJ/kg, when x is in equivalents per mol, E is in volts, and W_t is in g/mol. F is the Faraday constant, 96,500 C per equiv.

Since 1 W is 1 J per second, 1 Wh is 3.6 kJ, and the value of the *MTSE* can be expressed in Wh/kg as:

$$\text{MTSE} = 26,805(xE/W_t) \tag{9.18}$$

9.4.4 Variation of the Voltage as Batteries are Discharged and Recharged

Looking into the literature, it is seen that the voltage of most – but not all – electrochemical cells varies as their chemical energy is deleted. That is, as they are discharged. Likewise, it changes in the reverse direction when they are recharged. That may not be a surprise. However, not only the voltage ranges, but also the characteristics of these state of charge-dependent changes vary widely between different electrochemical systems. It is important to understand what causes these variations.

A characteristic way to present this information is in terms of *discharge curves* and *charge curves*, in which the cell voltage is plotted as a function of the state of charge. These relationships can vary greatly, depending upon the rate at which the energy is extracted from, or added to, the cell.

It is useful to consider the relation between the cell voltage and the state of charge under equilibrium or near-equilibrium conditions. In this case, a very useful experimental technique, known as *Coulometric titration*, can provide a lot of information. This will be described in a later section.

Some examples of discharge curves under low current, or near-equilibrium, conditions are shown in Fig. 9.5. These are presented here to show the cell voltage as a function of the state of charge parameter. However, different battery systems have different capacities. Thus, one has to be careful to not compare the energies stored in different systems in this manner.

The reason for presenting the near-equilibrium properties of these different cells in this way is to show that there are significant differences in the *types* of their behavior, as indicated by the shapes of their curves. It is clear that some of these discharge curves are essentially flat. Some have more than one flat region, and others have a slanted and stretched S-shape, sometimes with an appreciable slope, and sometimes not. These variants can be simplified into three basic types of discharge curve shapes, as depicted in Fig. 9.6. The reasons behind their general characteristics will be discussed later.

9.4.5 Cycling Behavior

In many applications, a battery is expected to maintain its major properties over many discharge–charge cycles. This can be a serious practical challenge, and is often given

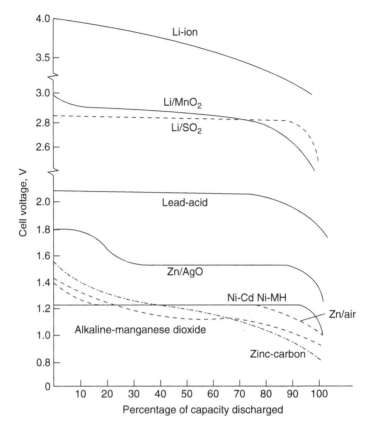

Fig. 9.5 Examples of battery discharge curves, showing the variation of the voltage shown as a function of the fraction of their available capacity

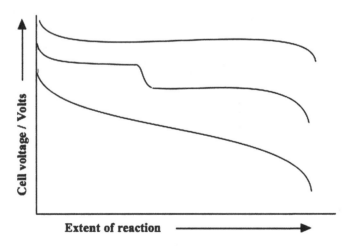

Fig. 9.6 Schematic representation of different types of discharge curves

a lot of attention during the development and optimization of batteries. Figure 9.7 shows how the initial capacity is reduced during cycling, assuming three different values of the *Coulombic efficiency* – the fraction of the prior charge capacity that is available during the following discharge. This depends upon a number of factors, especially the current density and the depth of discharge in each cycle.

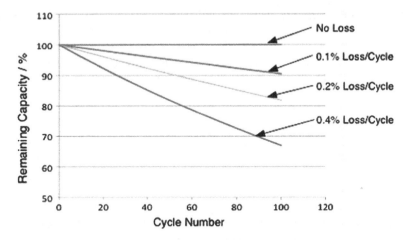

Fig. 9.7 Influence of Coulombic efficiency upon available capacity during cycling

It is seen that even a minor amount of inefficiency per cycle can have important consequences. For example, 0.1% loss per cycle causes the available capacity to drop to only 90% of the original value after 100 cycles. The situation is worse if the Coulombic efficiency is lower.

Applications that involve many cycles of operation require that cells are designed and constructed such that the capacity loss per cycle is extremely low. This typically means that compromises must be made in other properties. *Super-capacitors* are expected to be used over a very large number of cycles, and they typically have much lower values of specific energy than electrochemical cells, that are used for applications in which the amount of energy stored is paramount.

9.4.6   Self-Discharge

Another property that can be of importance in practical cells is called *self-discharge*. Evidence for this is a decrease in the available capacity with time, even without energy being taken from the cell by the passage of current through the external circuit. This is a serious practical problem in some systems, and is negligible in others.

The main point to understand at this juncture is that the capacity is a property of the electrodes. Its value at any time is determined by the remaining extent of the chemical reaction between the neutral species in the electrodes. Thus, any self-discharge mechanism that reduces the remaining capacity must involve either the transport of neutral species, or the concurrent transport of neutral combinations of charged species, through the cell. If such a process involves the transport of charged species, it is *electrochemical self-discharge*.

There are also several ways in which individual neutral species can move across a cell. These include transport through an adjacent vapor phase, through cracks in a solid electrolyte, or as a dissolved gas in a liquid electrolyte. Since the transport of charged species is not involved, these processes produce *chemical self-discharge*.

It is also possible that impurities can react with constituents in the electrodes or the electrolyte so as to reduce the available capacity with time.

9.5   General Equivalent Circuit of an Electrochemical Cell

It is often useful to devise electrical circuits whose electrical behavior is analogous to important phenomena in physical systems. By the examination of the influence of changes in the parameters in such *equivalent circuits*, they can be used as *thinking tools* to obtain useful insight into the significance of particular phenomena to the observable properties of complex physical systems. By utilizing this approach, the techniques of circuit analysis, which have been developed for use in various branches of electrical engineering, can be very helpful in the analysis of inter-dependent physical phenomena.

This procedure has proven to be very useful in some areas of electrochemistry, and will be utilized later in this text for a number of purposes. At this point, however, it will only be considered for the case of an ideal electrochemical cell. It will show what happens if the electrolyte is not a perfect filter, and also allows the flow of some electronic current in addition to the expected ionic current. An electrochemical cell can be simply modeled as shown in Fig. 9.8, and the basic equivalent circuit is shown in Fig. 9.9.

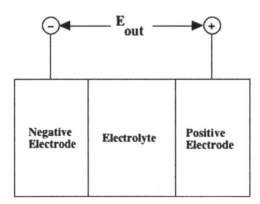

Fig. 9.8 Simplified physical model of electrochemical cell

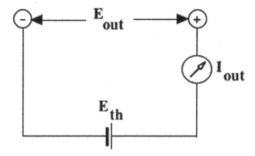

Fig. 9.9 Simple equivalent circuit model of an ideal electrochemical cell

The value of the electrical equivalent of the theoretical chemical driving force is E_{th}, which is given by

$$E_{th} = -\Delta G_r^{\ 0}/zF \tag{9.19}$$

as the result of the balance between the chemical and electrical forces acting upon the ionic species in the electrolyte, as mentioned earlier. If there are no impedances or other loss mechanisms, the externally measurable cell voltage E_{out} is simply equal to E_{th}.

9.5.1 *Influence of Impedances to the Transport of Ionic and Atomic Species Within the Cell*

In practical electrochemical cells, E_{out} is not always equal to E_{th}. There can be several possible reasons for this disparity. There will always be some impedance to the transport of the electroactive ions and the related atomic species across the cell. One source is the resistance of the electrolyte to ionic transport. There may also be significant impeding effects at one or both of the two electrolyte/electrode interfaces. In addition, there can be a further impedance to the progress of the cell reaction in some cases related to the time-dependent solid state diffusion of the atomic species into, or out of, the electrode microstructure.

Note that *impedances* are used in this discussion instead of resistances, because they can be time-dependent, if time-dependent changes in structure or composition are occurring in the system. The impedance is the instantaneous ratio of the applied force (e.g., voltage) E_{appl} and the response (e.g., current) across any circuit element. As an example, if a voltage E_{appl} is imposed across a material that conducts electronic current I_e, the electronic impedance Z_e is given by:

$$Z_e = E_{appl}/I_e \tag{9.20}$$

The inverse of the impedance is the *admittance*, which is the ratio current/voltage. Under steady state (time-independent) DC conditions, the impedance and resistance of a circuit element are equivalent.

If current is flowing through the cell, there will be a voltage drop related to each of the impedances to the flow of ionic current within the cell. Thus, if the sum of these internal impedances is Z_i, the output voltage can be written as

$$E_{out} = E_{th} - I_{out}Z_i \tag{9.21}$$

This relationship can be modeled by the simple circuit in Fig. 9.10.

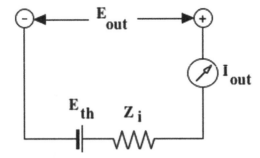

Fig. 9.10 Simple equivalent circuit for a battery or fuel cell indicating the effect of the internal ionic impedance Z_i upon the output voltage

9.5.2 Influence of Electronic Leakage Within the Electrolyte

The output voltage E_{out} can also be different from the theoretical electrical equivalent of the thermodynamic driving force of the reaction between the neutral species in the electrodes E_{th} even if there is no external current I_{out} flowing. This can be the result of electronic leakage through the electrolyte that acts to short-circuit the cell. This effect can be added to the previous equivalent circuit to give the circuit shown in Fig. 9.11.

It is evident that, even with no external current, there is an internal current related to the transport of the electronic species through the electrolyte I_e. Since the current must be the same everywhere in the lower loop, there must be a current through the electrolyte I_i with the same magnitude as the electronic current. There must be *charge flux balance* so that there is no net charge buildup at the electrodes.

The current through the internal ionic impedance Z_i generates a voltage drop, reducing the output voltage E_{out} by the product $I_i Z_i$, which is equal to $I_e Z_e$.

$$E_{out} = E_{th} - I_i Z_i \tag{9.22}$$

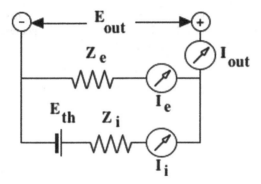

Fig. 9.11 Modified circuit including electronic leakage through the electrolyte

In addition, the fact that both ionic and electronic species flow through the cell means that this is a mechanism of *self-discharge*. This results in a decrease of the available charge capacity of the cell.

9.5.3 Transference Numbers of Individual Species in an Electrochemical Cell

If more than one species can carry charge in an electrolyte it is often of interest to know the relative conductivities or impedances of different species. The parameter that is used to describe the contributions of individual species to the transport of charge when an electrical potential difference (voltage) is applied across an electrolyte is the transference number. This is defined as the fraction of the total current that passes through the system that is carried by a particular species.

In the simple case that electrons and one type of ion can move through the electrochemical cell, the transference number of ions is t_i, and electrons is t_e, where

$$t_i = I_i/(I_i + I_e) \qquad (9.23)$$

and

$$t_e = i_e/(I_i + I_e) \qquad (9.24)$$

and I_i and I_e are their respective partial currents upon the application of an external voltage E_{appl} across the system. It can readily be seen that the sum of the transference numbers of all mobile charge-carrying species is unity. In this case:

$$t_i + t_e = 1 \qquad (9.25)$$

Instead of expressing transference numbers in terms of currents, they can also be written in terms of impedances. For the case of these two species, the transport of charge by the motion of the ions under the influence of an applied voltage E_{appl}, is

$$t_i = (E_{appl}/Z_i)/[E_{appl}/Z_i) + (E_{appl}/Z_e)] = Z_e/(Z_i + Z_e) \qquad (9.26)$$

and likewise for electrons:

$$t_e = Z_i/(Z_i + Z_e) \qquad (9.27)$$

Whereas these parameters are often thought of as properties of the electrolyte, in actual experiments, they can also be influenced by what happens at the interfaces between the electrolyte and the electrodes and thus are properties of the whole electrode–electrolyte system. They are only properties of the electrolyte alone if there is no impedance to the transfer of either ions or electrons across the electrolyte/electrode interface or atomic and electronic species within the electrodes.

9.5.4 Relation Between the Output Voltage and the Values of the Ionic and Electronic Transference Numbers

Making the simplifying assumption that the internal impedance is primarily due to the behavior of the ions, the general equivalent circuit of Fig. 9.10 can be rearranged to look like that in Fig. 9.12.

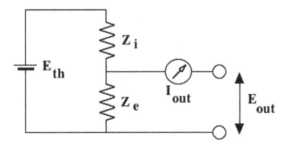

Fig. 9.12 Different representation of general equivalent circuit

When drawn this way, it can be readily seen that the series combination of Z_i and Z_e acts as a simple voltage divider.

If no current passes out of the system, that is, under open circuit conditions, the output voltage is equal to the product of E_{th} and the ratio $Z_e/(Z_i + Z_e)$.

$$E_{out} = E_{th}Z_e/(Z_i + Z_e) \qquad (9.28)$$

Introducing (9.26), the output voltage can then be expressed as

$$E_{out} = E_{th} t_i \tag{9.29}$$

or

$$E_{out} = E_{th}(1 - t_e) \tag{9.30}$$

These are well-known relations that can be derived in other ways, as will be shown later. It is clear that the output voltage is optimized when t_i is as close to unity as possible.

9.5.5 Joule Heating Due to Self-Discharge in Electrochemical Cells

Electrochemical self-discharge causes heat generation, often called *Joule heating*, due to the transport of charged species through the cell. The *thermal power* P_{th} caused by the passage of a current through a simple resistance R is given by

$$P_{th} = I^2 R \tag{9.31}$$

But as shown earlier, if self-discharge results from the leakage of electrons through the electrolyte, there must be both electronic and ionic current, and they must have equal values. Thus, the thermal power due to this type of self-discharge is:

$$P_{th} = I_i^2 Z_i + I_e^2 Z_e = I_e^2(Z_i + Z_e) \tag{9.32}$$

Measurements of the rate of heat generation by Joule heating under open circuit conditions can be used to evaluate the rate of self-discharge in practical cells.

9.5.6 What if Current is Drawn from the Cell?

If current is drawn from the cell into an external circuit, the normal mode of operation when chemical energy is converted into electrical energy, it flows through the ionic impedance, Z_i. This results in an additional voltage drop of $I_{out} Z_i$, further reducing the output voltage. If there were no electrochemical self-discharge, this can be written as

$$E_{out} = E_{th} t_i - I_{out} Z_i \tag{9.33}$$

The value of the ionic impedance of the system, Z_i, may increase with the value of the output current as the result of current-dependent impedances at the electrolyte/electrode interfaces. The difference between E_{th} and E_{out} is often called *polarization* in the electrochemical literature.

The result of the presence of current-dependent interfacial impedances to the passage of ionic species that increase Z_i is that the *effective transference number* of the ions t_i is reduced, since $t_i = Z_e/(Z_i + Z_e)$. This causes an additional reduction in the output voltage.

But, in addition to a *reduced output voltage*, there will also be further *heat generation*. The total amount of Joule heating is the sum of that due to the passage of current into the external circuit I_{ext} and that due to electrochemical self-discharge.

$$P_{th} = I_{ext}^2 Z_i + I_e^2 (Z_i + Z_e) \qquad (9.34)$$

In most cases, the first term is considerably larger than the second term.

Measured discharge curves vary with the current density as conditions deviate farther and farther from equilibrium. This is shown schematically in Fig. 9.13.

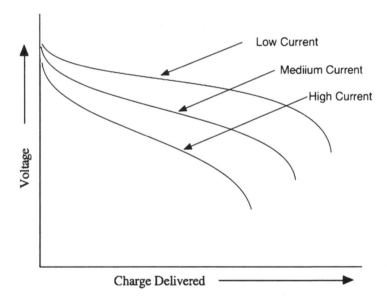

Fig. 9.13 Schematic drawing showing the influence of the current density upon the discharge curve

A parameter that is often used to indicate the rate at which a battery is discharged is the so-called *C-Rate*. The discharge rate of a battery is expressed as C/R, where R is the number of hours required to completely discharge its nominal capacity.

As an example, if a cell has a nominal capacity of 5 Ah, discharge at the rate of $C/10$ means that it would be fully discharged in 10 h. Thus, the current is 0.5 A. And if the discharge rate is $C/5$ the discharge current is 1 A.

Although the *C-Rate* is often specified when either complete cells or individual electrodes are evaluated experimentally, and the current can be specified, this parameter is often not time-independent during real applications. If the electrical load is primarily resistive, for example, the current will decrease as the output voltage falls. This means that the *C-Rate* drops as the battery is discharged. Nevertheless, it is often important to consider the *C-Rate* when comparing the behavior of different materials, electrodes, and complete cells.

It is obvious that not only the average voltage, but also the charge delivered can vary appreciably with changes in the *C-Rate*. But in addition, the amount of energy that can be supplied, which will be seen in later chapters to be related to the area under the discharge curve, is strongly *C-Rate* dependent.

A further point that should be kept in mind is that not all of the stored energy may be useful. If the load is resistive, the output power is proportional to the square of the voltage according to (9.14), so that the energy that is available at lower voltages may not be of much benefit.

This behavior can be understood in terms of the equivalent circuit of the battery. The internal ionic impedance Z_i, the sum of the impedances in the electrolyte and at the two electrode/electrolyte interfaces, is a function of the local current density in the cell. This impedance typically also varies with the state of charge. The mechanisms responsible for this behavior will be discussed later in the text.

References

1. I. Barin, *Thermochemical Data of Pure Substances*, (2 volumes), VCH Verlagsgesellschaft mbH (1989)
2. N.P. Yao, L.A. Heredy and R.C. Saunders, J. Electrochem. Soc. *118*, 1039 (1971)
3. E.C. Gay, et al. J. Electrochem. Soc. *123*, 1591 (1976)
4. C.J. Wen, B.A. Boukamp, W. Weppner and R.A. Huggins, J. Electrochemical Soc. *126*, 2258 (1979)
5. C.J. Wen, W. Weppner, B.A. Boukamp and R.A. Huggins, Metall. Trans. *11B*, 131 (1980)
6. C.J. Wen, C. Ho, B.A. Boukamp, I.D. Raistrick, W. Weppner and R.A. Huggins, Int. Met. Rev. *5*, 253, (1981)
7. M.S. Whittingham, Science *192*, 1126 (1976)

Chapter 10
Principles Determining the Voltages and Capacities of Electrochemical Cells

10.1 Introduction

In the prior chapter it was shown that the fundamental driving force across an electrochemical cell is the virtual chemical reaction that would occur if the materials in the two electrodes were to react with each other. If the electrolyte is a perfect filter that allows the passage of ionic species, but not electrons, the cell voltage when no current is passing through the system is determined by the difference in the electrically neutral chemical compositions of the electrodes. The identity and properties of the electrolyte and the phenomena that occur at the electrode/electrolyte interfaces play no role. Likewise, it is the properties of the electrodes that determine the capacity of an electrochemical cell.

These general principles will be extended further in this chapter. Emphasis will be placed upon the equilibrium, or near-equilibrium state. This will address the ideal properties of such systems, which provide the upper limits of various important parameters.

Real systems under load deviate from this behavior. As will be shown later, this is primarily because of kinetic factors. Such factors vary from one system to the next, and are highly dependent upon both the details of the materials present and the cell construction, and the experimental conditions. As a result, it is difficult to obtain reproducible and quantitative experimental results. Such matters will appear later in this text. First, the factors that determine the equilibrium, or near-equilibrium, behavior will be discussed.

10.2 Thermodynamic Properties of Individual Species

It was shown in Chap. 9 that the overall driving force across a simple electrochemical cell is determined by the change in the standard Gibbs free energy, $\Delta G_r{}^0$ of the virtual chemical reaction that would occur if the materials in the electrodes were to react with one another. If there is no current flowing, this chemical driving force is just balanced by an electrical driving force in the opposite direction.

Individual species within the electrolyte in the cell will now be considered. Under open circuit conditions (and no electronic leakage), there is no net current flow. Thus, there must be a *force balance* acting on all mobile species.

The thermodynamic properties of a material can be related to those of its constituents by using the concept of the *chemical potential* of an individual species. The chemical potential of species i in a phase j is defined as

$$\mu_i = \partial G_j / \partial n_i, \tag{10.1}$$

where G_j is the molar Gibbs free energy of phase j, and n_i is the mol fraction of the i species in phase j. In integral form, this is

$$\Delta \mu_i = \Delta G_j. \tag{10.2}$$

Since the free energy of the phase changes with the amount of species i, it is easy to see that the chemical potential has the same dimension as the free energy. Thus, gradients in the chemical potential of species i produce chemical forces causing i to tend to move in the direction of lower μ_i. It was shown in Chap. 1 that when there is no net flux in the electrolyte, this chemical force must be balanced by an electrostatic force due to the voltage between the electrodes. The energy balance in the electrolyte, and thus in the cell, can be written in terms of the single species i.

$$\Delta \mu_i = -z_i F E, \tag{10.3}$$

where z_i is the number of elementary charges carried by particles (ions) of species i.

The chemical potential of a given species is related to another thermodynamic quantity, its *activity*, a_i. The defining relation is

$$\mu_i = \mu_i^0 + RT \ln a_i, \tag{10.4}$$

where μ_i^0 is a constant, the value of the chemical potential of species i in its standard state. R is the gas constant (8.315 J/mol degree), and T is the absolute temperature.

The activity of a species can be thought of as its *effective concentration*. If the activity of species i, a_i, is equal to unity, it behaves chemically as though it is pure i. If a_i is 0.5, it behaves chemically as though it is composed of half species i, and half something else that is chemically inert. In the case of a property such as vapor pressure, a material i with an activity of 0.5 will have a vapor pressure half of that of pure i.

Consider an electrochemical cell in which the activity of species i is different in the two electrodes, $a_i(-)$ in the negative electrode, and $a_i(+)$ in the positive electrode. The difference between the chemical potential on the positive side and that on the negative side can be written as

$$\mu_i(+) - \mu_i(-) = RT[\ln a_i(+) - \ln a_i(-)] = RT \ln[a_i(+)/a_i(-)]. \qquad (10.5)$$

If this chemical potential difference is balanced by the electrostatic energy from (10.2):

$$E = -(RT/z_iF) \ln[a_i(+)/a_i(-)]. \qquad (10.6)$$

This relation, which is often called the *Nernst equation*, is very useful, for it relates the measurable cell voltage to the chemical difference across an electrochemical cell. That is, it transduces between the chemical and electrical driving forces. If the activity of species i in one of the electrodes is a standard reference value, the Nernst equation provides the relative electrical potential of the other electrode.

10.3   A Simple Example: The Lithium/Iodine Cell

As an initial example, the thermodynamic basis for the voltage of a lithium/iodine cell will be considered. Primary (non-rechargeable) cells based upon this chemical system were invented by Schneider and Moser in 1972 [1, 2], and they are currently widely used to supply the energy in cardiac pacemakers.

The typical configuration of this electrochemical cell employs metallic lithium as the negative electrode and a composite of iodine with about 10 wt% of poly-2-vinylpyridine (P2VP) on the positive side. The composite of iodine and P2VP is a charge transfer complex, with the P2VP acting as an electron donor, and the iodine as an acceptor. The result is that the combination has a high electronic conductivity and the chemical properties are essentially the same as pure iodine. Reaction between the Li and the (iodine, P2VP) composite produces a layer of solid LiI. This material acts as a solid electrolyte in which Li^+ ions move from the interface with the negative electrode to the interface with the positive electrode, where they react with iodine to form more LiI. The transport mechanism involves a flux of lithium ion vacancies in the opposite direction. Although LiI has relatively low ionic conductivity, it has negligible electronic transport, meeting the requirements of an electrolyte.

This system can be represented simply as:

$$(-)Li/electrolyte/I_2(+). \qquad (10.7)$$

The *virtual reaction* that determines the voltage is thus

$$Li + 1/2I_2 = LiI. \qquad (10.8)$$

More and more LiI forms between the lithium electrode and the iodine electrode as the reaction progresses. The time evolution of the microstructure during discharge is shown schematically in Fig. 10.1.

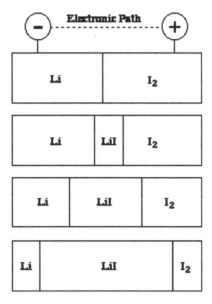

Fig. 10.1 Schematic representation of the microstructure of a Li/I$_2$ cell at several stages of discharge

The voltage across this cell under open circuit conditions can be readily calculated from the balance between the chemical and electrical driving forces, as shown in Chap. 9.

$$E = -\Delta G_r / z_i F, \tag{10.9}$$

where

$$\Delta G_r = \Delta G_j(\text{LiI}), \tag{10.10}$$

and z_i is $+1$, for the electroactive species are the Li$^+$ ions.

According to the data in Barin [3], the standard Gibbs free energy of formation of LiI is -269.67 kJ/mol at $25°C$. Since the value of the Faraday constant is 96,500 Coulombs per equivalent (mol of electronic charge), the open circuit voltage can be calculated to be 2.795 V at $25°C$ and 1 atmosphere pressure.

Data on the properties of commercial Li/I$_2$ cells are shown in Fig. 10.2 [4]. It is seen that during most of the life of this battery the voltage corresponds closely to that which was calculated above. It is also seen in this figure that the resistance across the cell increases with the extent of reaction, due to the increasing thickness

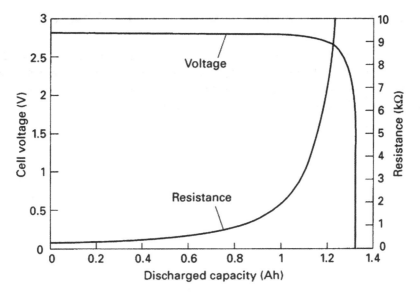

Fig. 10.2 Output voltage and internal resistance of a typical Li/I$_2$ battery of the type used in cardiac pacemakers. After [4]

of the solid electrolyte product that grows as the cell is discharged. Such cells are typically designed to be *positive-electrode-limited*. This means that the positive electrode capacity is somewhat less than the negative electrode capacity, and therefore is the part of the cell that determines the overall capacity.

10.3.1 Calculation of the Maximum Theoretical Specific Energy

The value of the maximum theoretical specific energy of a Li/I$_2$ cell can now be calculated from this information and the weights of the reactants. It was shown in Chap. 9 that the MTSE, in Wh/kg, is given by:

$$\text{MTSE} = 26,805(xE/W_i). \tag{10.11}$$

The reactant weight W_t is the weight of a mol of Li (6.94 g) plus half a mol of I$_2$ (126.9), or 133.84 g. The value of x is 1, and E was calculated to be 2.795 V. Thus, the value of the MTSE is 559.77 Wh/kg.

This is a large number, about 15 times the value that is typical of the Pb-acid cells that are so widely used as SLI batteries in automobiles, as well as for a number of other purposes. The lack of rechargeability as well as the cost of the ingredients and the low discharge rate unfortunately limits the range of application of Li/I$_2$ cells, however.

10.3.2   The Temperature Dependence of the Cell Voltage

As it has been seen, the quantity that determines the voltage is the Gibbs free energy change associated with the virtual cell reaction between the chemical species in the electrodes. That quantity is, however, temperature dependent. This can be seen by dividing the Gibbs free energy into its enthalpy and entropy components.

$$\Delta G_r = \Delta H_r - T\Delta S_r \tag{10.12}$$

so that

$$d(\Delta G_r)/dT = -\Delta S_r \tag{10.13}$$

and

$$dE/dT = \Delta S_r/z_i F. \tag{10.14}$$

The value of ΔS for the formation of LiI is given by:

$$\Delta S_r(\text{LiI}) = S(\text{LiI}) - S(\text{Li}) - 1/2S(\text{I}_2). \tag{10.15}$$

Entropy data for these materials, as well as a number of others, are given in Table 10.1. Note that these entropy values are in J/mol deg, whereas Gibbs free energy values are typically in kJ/mol. From these data, the value of ΔS_r for the formation of LiI is -1.38 J/K mol. Thus, from (10.13), the cell voltage varies only -1.43×10^{-5} V/K. This is very small. As will be seen later, the temperature dependence of the voltage related to many other electrochemical reactions, and thus of other batteries, is often significantly greater. An example is the small Zn/O_2 battery that is commonly used in hearing aids, where it is -5.2×10^{-4} V/K.

Table 10.1 Entropy data for some species at 25 and 225°C

Species	S at 25°C (J/K mol)
Li	29.08
Zn	41.63
H_2	130.68
O_2	205.15
Cl_2	304.32
I_2	116.14
LiF	35.66
LiCl	59.30
LiBr	74.06
LiI	85.77
H_2O (liquid)	69.95
ZnO	43.64
H_2	145.74
O_2	220.69
H_2O (gas)	206.66

The data in Table 10.1 show that the entropy values of simple solids are considerably lower than those of liquids, which, in turn, are lower than gases. This is reflected, of course, in the temperature dependence of electrochemical cells.

An interesting example is the H_2/O_2 fuel cell. In that case, the voltage varies -1.7×10^{-3} V/K near room temperature where water, the product of the reaction, is a liquid. But at 225°C, where the product of the cell reaction is a gas, steam, the variation is only -0.5×10^{-3} V/K. The resultant variation of the cell voltage with temperature from about room temperature to the operating temperature of high temperature oxide-electrolyte fuel cells is shown in Fig. 10.3. Operation at a high temperature results in a significantly lower voltage. The theoretical open circuit voltage is 1.23 V at 25°C, but only 0.91 V at 1,025°C.

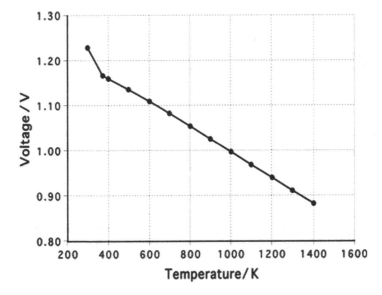

Fig. 10.3 Theoretical open circuit voltage of a H_2/O_2 fuel cell as a function of the absolute temperature

10.4 The Shape of Discharge Curves and the Gibbs Phase Rule

It was shown earlier that the voltage of batteries often varies with the state of charge, and it was pointed out that their discharge curves typically have one or another of three general shapes. Some are relatively flat, others have more than one relatively flat portion, and others have a slanted or stretched-S shape, sometimes with a relatively large slope. The data in Fig. 10.2 show that the Li/I_2 cell falls in the first category.

To understand how the voltage across an electrochemical cell varies with the state of charge, and why it is essentially flat in the case of the Li/I_2 cell, it is useful to consider the application of the *Gibbs Phase Rule* to such systems.

The *Gibbs Phase Rule* is often written as

$$F = C - P + 2 \tag{10.16}$$

in which C is the *number of components* (e.g., elements) present, and P is the *number of phases present* in this materials system in a given experiment. The quantity F may be more difficult to understand. It is the *number of degrees of freedom*; that means the number of *intensive thermodynamic parameters* that must be specified in order to *define the system* and *all of its associated properties*. One of these properties is, of course, the electric potential.

To understand the application of the Phase Rule to this situation, it must be determined what thermodynamic parameters should be considered. They must be intensive variables, which means that their values are independent of the amount of material present. For this purpose, the most useful thermodynamic parameters are the temperature, overall pressure, and either chemical potential or chemical composition of each of the phases present.

How does this apply to the Li/I_2 cell? Starting with the negative electrode; there is only one phase present, Li, so P is 1. It is a single element, with only one type of atom. Thus, the number of components C is also equal to 1. Thus, F must be equal to 2.

What is the meaning of $F = 2$? It means that if the values of two intensive thermodynamic parameters, such as the temperature and the overall pressure, are specified, there are no degrees of freedom left over. Thus, the *residual value of F* is zero. This means that all of the intensive properties of the negative electrode system are fully defined. That is, they have fixed values.

Thus, in the case of the lithium negative electrode the chemical potentials of all species (i.e., the pure lithium), as well as the electrical potential, have fixed values, *regardless of the amount* of lithium present. The amount of lithium in the negative electrode decreases as the cell becomes discharged and the product LiI is formed. That is, the amount of lithium varies with the state of charge of the Li/I_2 battery. But since $F = 2$, and thus the residual value of F, if the temperature and total pressure are held constant, is zero, none of the intensive properties change. This means that the electrical potential of the lithium electrode is independent of the state of charge of the cell. This is shown schematically in Fig. 10.4.

On the other hand, if some iodine *could* dissolve into the lithium, forming a solid solution, *which it does not*, the number of components in the negative electrode would be two. In a solid solution, there is only one phase present. Thus $C = 2$, $P = 1$ and $F = 3$.

In this hypothetical case, the system would not be fully defined after fixing the temperature and the overall pressure. There would be a residual value of F, that is, one. Thus, the electrical potential of the lithium–iodine alloy would not be fixed, but would vary, depending upon some other parameter, such as the amount of iodine in the Li–I solid solution. This is shown schematically in Fig. 10.5.

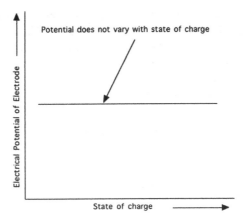

Fig. 10.4 The potential of a pure lithium electrode does not vary with the state of charge of the Li/I$_2$ cell

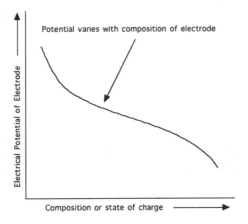

Fig. 10.5 Schematic representation of the variation of the electrical potential of an electrode as a function of its composition for the case in which the residual value of F is not zero

Although it is *not true* in the lithium/iodine cell, it is quite common in other electrochemical cells for the electrical potential of electrodes to vary with the composition, and thus with the state of charge. A number of examples will be discussed in subsequent chapters.

Now consider the positive electrode. There is only one active component (element) present, iodine. There is also only one electrochemically active phase, iodine. Thus both C and P have values of 1. The number of degrees of freedom is thus again 2. Therefore, the values of all intensive variables and associated properties, such as the electrical potential, of the iodine electrode will be determined if the values of the two independent thermodynamic parameters, the temperature and the total pressure are fixed.

This means that the potential of the I_2 electrode does not vary with its state of charge. Since both the negative and positive electrode potentials are independent of the state of charge, the voltage across the cell must also be independent of the state of charge of the Li/I_2 battery. This was illustrated in Fig. 10.2.

The earlier discussion showed that the chemical potential of an element depends upon its activity, and for the case of the iodine electrode

$$\mu(I_2) = \mu^0(I_2) + RT \ln a(I_2), \qquad (10.17)$$

where $\mu^0(I_2)$ is the chemical potential of iodine in its standard state, that is, pure iodine at a pressure of one atmosphere at the temperature in question. When the activity is unity, that is, for pure I_2,

$$\mu(I_2) = \mu^0(I_2). \qquad (10.18)$$

Now consider the voltage of the Li/I_2 cell. This is determined by the Gibbs free energy of formation of the LiI phase, as given in (10.8) and (10.9). But it is also related to the difference in the chemical potential of iodine at the two electrode/electrolyte interfaces according to the relation

$$E = -\Delta\mu(I_2)/z_i F \qquad (10.19)$$

where the value of z_i is -2. Therefore, the activity of iodine at the positive side of the electrolyte is unity, but it is very small at the interface on the negative electrode side. Likewise, the cell voltage is related to the difference in the chemical potential of lithium at the two electrode/electrolyte interfaces:

$$E = -\Delta\mu(Li)/z_i F \qquad (10.20)$$

in which the value of z_i is $+1$. In this case, the activity of lithium is unity at the negative interface, and very small at the positive interface, where the electrolyte is in contact with I_2.

Whereas this discussion has focused on the potential of a single electrode, the shape of the equilibrium discharge curve (voltage versus state of charge) of an electrochemical cell is the result of the change of the potentials of both electrodes as the overall reaction takes place. If the potential of one of the electrodes does not vary, the variation of the cell voltage is obviously the result of the change of the potential of the other electrode as its overall composition changes.

There are a number of materials that are used as electrodes in electrochemical cells in which more than one reaction can occur in sequence as the overall discharge process takes place. In some cases, these reactions are of the same type, whereas in others they are not.

As one example, a *series of multi-phase reactions* in which the number of residual degrees of freedom is zero can result in a discharge curve with a set of constant voltage plateaus. This is illustrated schematically in Fig. 10.6.

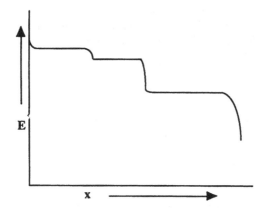

Fig. 10.6 Schematic equilibrium discharge curve of an electrode that undergoes a series of multi-phase reactions in which the residual value of F is zero

It is also possible for an electrode to undergo sequential reactions that are not of the same type. An example of this is the reaction of lithium with a spinel phase in the Li-Ti-O system. Experimental data are shown in Fig. 10.7 [5].

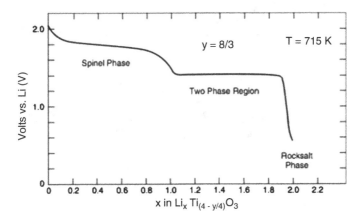

Fig. 10.7 Equilibrium discharge curve of a material in the Li–Ti–O system that initially had a composition with a spinel type of crystal structure

In this case, approximately one Li per mol could be inserted into the host spinel phase as a *solid solution reaction*. Thus, the potential varied continuously as a function of composition.

The introduction of additional lithium caused the nucleation, and subsequent growth, of a second phase that has the rocksalt structure and a composition of approximately two Li per mol of the original host. Thus, a *reconstitution reaction* took place when more than one lithium was added. During a reconstitution reaction, there are two regions within the material with different Li contents. As the reaction proceeds, the compositions of the two phases do not change, but the relative amount of the phase with the higher Li content increases, and that of the initial solid solution phase is reduced. This occurs by the *movement of the interface* between them. This type of a *moving interface reconstitution reaction* can be schematically represented as shown in Fig. 10.8.

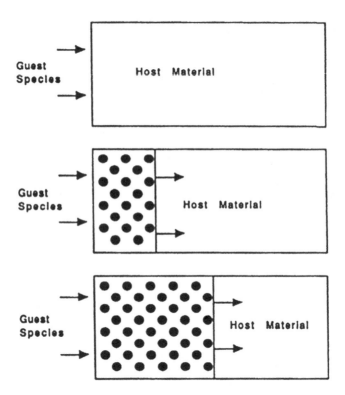

Fig. 10.8 Schematic representation of a one-dimensional moving interface reaction

An example in which a series of reactions occur as the overall composition is changed is the Li–Mn–O system. In this case, there is a series of three different reactions. This is seen from the shape of the equilibrium discharge curve in Fig. 10.9 [6]. There is a two-phase plateau, followed by a single-phase solid solution region, and then by another two-phase plateau.

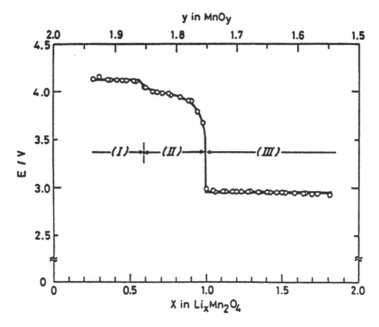

Fig. 10.9 Equilibrium discharge curve for $Li_xMn_2O_4$. After [6]

This interpretation was reinforced by the results of x-ray diffraction experiments, which are shown in Fig. 10.10 [6]. It is seen that the lattice parameters remain constant within two-phase regions, and vary with the composition within the single-phase solid solution region.

10.5 The Coulometric Titration Technique

The simple examples that have been discussed so far in this chapter assume that the requisite thermodynamic data are already known. One can calculate the open circuit voltage of an electrochemical cell from the value of the Gibbs free energy of the appropriate virtual reaction, and the ideal capacity can be determined from the *reaction's stoichiometry*.

It is also possible to do the opposite, using electrochemical measurements to obtain thermodynamic information. A useful tool for this purpose is the *coulometric titration technique*, which was first introduced by Wagner [7] to study the phase $Ag_{2+x}S$, which exists over a relatively narrow range of composition x. Its composition, or stoichiometry (the relative amounts of silver and sulfur) depends upon the value of the activity of silver within it. One can use a simple electrochemical cell to both change the stoichiometry and evaluate the activity of one of the species, for example, silver in this case.

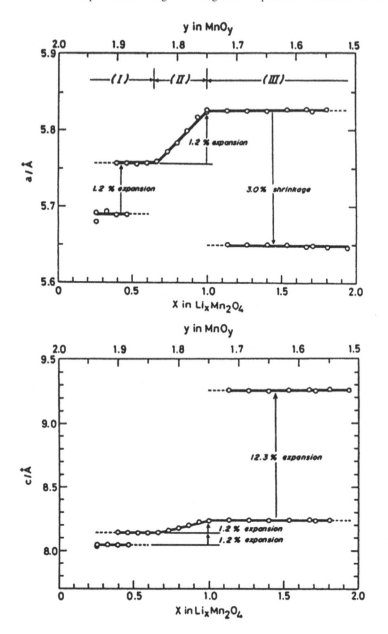

Fig. 10.10 Changes in unit cell dimensions as a function of composition in $Li_xMn_2O_4$ [6]

This method was further developed and applied by Weppner and Huggins [8] to the investigation of poly-phase alloy systems. It was demonstrated that the phase diagram can be determined, as well as the thermodynamic properties of the individual phases within it, by the use of this technique.

Consider the use of the following simple electrochemical cell to investigate the properties of the *vario-stoichiometric* (the stoichiometry can have a range of values) *phase* A_yB. This can be represented schematically as

$$A/\text{Electrolyte that transports } A^+ \text{ ions}/A_yB$$

In this case, the element A acts as both a source and sink for the electroactive species A and a thermodynamic reference for component A. For simplicity, it can be assumed that this electrode is pure A, and thus has an activity of unity. It can also be assumed that both A and A_yB are good electronic conductors, that the ionic transference number in the electrolyte is unity, and that the system is under isobaric and isothermal conditions.

Under these conditions, the open circuit voltage E is a direct measure of the chemical potential and the activity of A in the phase A_yB according to (10.5) that appeared earlier. As the electrode of pure A has an activity of unity, this relation can be written as

$$E = -\Delta\mu_A/(z_{A+}F) = -(RT/Z_{A+}F)\ln a(A) \tag{10.21}$$

where z_{A+} is the charge number of the A^+ ions in the electrolyte, which is 1.

If a positive current is passed through the cell by the use of an electronic source, A^+ ions will be transported through the electrolyte from the left electrode to the right electrode. An equal current of electrons will go through the outer circuit because of the *requirement for charge flux balance*. The result is that the value of y in the A_yB phase will be increased.

For the case that a steady value of current I is applied for a fixed time t, the amount of charge Q that is passed across the cell is simply

$$Q = It \tag{10.22}$$

The number of mols of species A that are transported during this current pulse is

$$\Delta m(A) = Q/z_{A+}F \tag{10.23}$$

so that the change in the value of y, the mol fraction of species A, is

$$\Delta y = \Delta m(A)/m(B) = Q/(z_{A+}Fm(B)) \tag{10.24}$$

where m(B) is the number of moles of B present in the electrode.

This method can be used to make *very minute* changes in the composition of the electrode material. One can see how sensitive this procedure is by putting some numbers into this relation.

Assume that the electrode has a weight of 5 grams, and the component B has a molecular weight of 100 g/mol. The value of m(B) is thus 0.05 mols. Now suppose that a current of 0.1 mA is run through the cell for 10 seconds. The value of Q is then 0.001 Coulombs. With $z_{A+} = 1$ equivalent per mol and $F = 96,500$ Coulombs/equivalent, then Δy is only about 2×10^{-7}.

This is very small. Thus, it is possible to investigate the compositional dependence of the properties of phases with very narrow compositional ranges. It is very difficult to get such a high degree of compositional resolution by other techniques.

By waiting for a sufficiently long time to allow the composition to become homogeneous throughout the electrode material, as evidenced by reaching a steady-state value of open circuit voltage, information can be obtained about the equilibrium chemical potential and activity of the mobile electroactive species as a function of composition. This technique has been used to investigate a wide variety of materials of potential interest in battery systems, and numerous examples will be discussed in later chapters.

The success of this method depends upon a number of assumptions. One is that the electrolyte is essentially only an ionic conductor, that is, the ionic transference number is very close to unity. Another is that there can be no appreciable loss of either component from the electrode material $A_y B$ by evaporation, dissolution, or interaction with the electrical lead materials, the so-called *current collectors*. Furthermore, the rate of *compositional equilibration via chemical diffusion* in the electrode material must be sufficiently fast. This means that it may be necessary to use thin samples as electrodes in order to reduce the time necessary for concentration homogenization.

It should also be recognized that this Coulometric titration technique gives information about the influence of compositional changes, but not the absolute composition. That will have to be determined by some other method.

References

1. J.R. Moser, US Patent 3,660,163 (1972)
2. A.A. Schneider and J.R. Moser, US Patent 3,674,562 (1972)
3. I. Barin, *Thermochemical Data of Pure Substances* (2 volumes), VCH Verlag (1989)
4. Courtesy of Catalyst Research Corp.
5. B.E. Liebert, W. Weppner and R.A. Huggins (1977) in J.D.E. McIntyre, S. Srinivasan and F.G. Will (eds.), *Proceedings of the Symposium on Electrode Materials and Processes for Energy Conversion and Storage*, Electrochemical Society, Princeton, p. 821.
6. T. Ohzuku, M. Kitagawa and T. Hirai (1990) J. Electrochem. Soc. 137, 769
7. C. Wagner (1953) J. Chem. Phys. 21, 1819
8. W. Weppner and R.A. Huggins (1978) J. Electrochem. Soc. 125, 7

Chapter 11
Binary Electrodes Under Equilibrium or Near-Equilibrium Conditions

11.1 Introduction

The theoretical basis for understanding and predicting the composition-dependence of the potentials, as well as the capacities, of both binary (two element) and ternary (three element) alloys has now been established. The relevant principles will be discussed for the case of binary systems in this chapter. Ternary systems will be treated in the next chapter.

Under equilibrium and near-equilibrium conditions, these important practical parameters are directly related to the thermodynamic properties and compositional ranges of the pertinent phases in the respective *phase diagrams*. Phase diagrams, which were touched upon briefly in both Chaps. 4 and 9, are graphical representations that indicate the phases and their compositions that are present in a materials system under equilibrium conditions. They can be useful *thinking tools* to help understand the fundamental properties of electrodes in electrochemical systems.

One can often understand the behavior under dynamic conditions in terms of simple deviations from the equilibrium conditions assumed in phase diagrams. In other cases, however, *metastable* phases may be present in the microstructure of an electrode whose properties are considerably different from those of the *absolutely-stable* phases. The influence of *metastable* microstructures will be discussed in a later chapter. In addition, it is possible that the compositional changes occurring in an electrode during the operation of an electrochemical cell can cause *amorphization* of its structure. This will also be discussed later.

11.2 Relationship Between Phase Diagrams and Electrical Potentials in Binary Systems

In order to demonstrate the relationship between phase diagrams and some of the important features of electrodes in electrochemical systems, a schematic phase diagram for a hypothetical binary alloy system A–B is shown in Fig. 11.1. In this

R.A. Huggins, *Energy Storage*,
DOI 10.1007/978-1-4419-1024-0_11, © Springer Science+Business Media, LLC 2010

case, there are four 1-phase regions. The solid phases are designated as phases α, β, and γ. In addition, there is a liquid phase at higher temperatures. It can be seen in the figure that the single phases are all separated by 2-phase regions as the composition moves horizontally (isothermally) across the diagram.

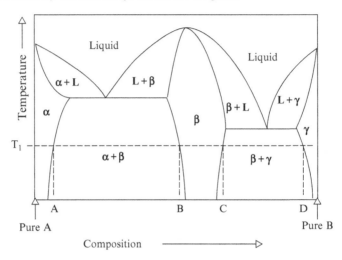

Fig. 11.1 Schematic binary phase diagram with an intermediate phase β, and solid solubility in terminal phases α and γ

It was shown earlier that according to the Gibbs Phase Rule all intensive properties, including the electrical potential, vary continually with the composition within single-phase regions in a binary system. Correspondingly, the intensive properties are composition-independent when two phases are present in a binary system. Since the equilibrium electrical potential of such an electrode, E, in an electrochemical cell is determined by the chemical potential or activity of the electroactive species, it also varies with composition within single-phase regions, and is composition-independent when there are two phases present under the equilibrium conditions that are assumed here.

The variation of the electrical potential with the overall composition in this hypothetical system at temperature T_1 is shown in Fig. 11.2. It is seen that it alternately goes through composition regions in which it is constant (*potential plateaus*) and those in which it varies. If B atoms are added to pure element A, the overall composition is initially in the solid solution phase α and the electrical potential varies with composition. When the α solubility limit is reached, indicated as composition A, the addition of more B causes the nucleation and growth of the β phase. Two phases are then present, and the potential maintains a fixed value. When the overall composition reaches composition B, all of the α phase will have been consumed and there will only be phase β present. Upon further compositional change the electrical potential again becomes composition-dependent. At composition C, the upper compositional limit of the β phase at that temperature, the overall composition again enters a 2-phase (β and γ) range and the potential is again composition-independent. Upon reaching composition D the potential again varies with composition.

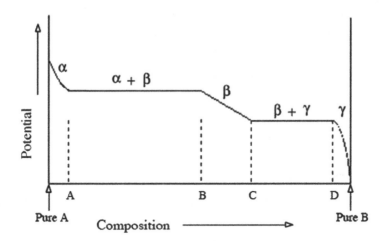

Fig. 11.2 Schematic variation of electrical potential with composition across the binary phase diagram shown in Fig. 11.1

It is also possible for the composition ranges of phases to be quite narrow, and then they are sometimes called *line phases*. As an example, a variation upon the phase diagram presented in Fig. 11.1 is shown in Fig. 11.3.

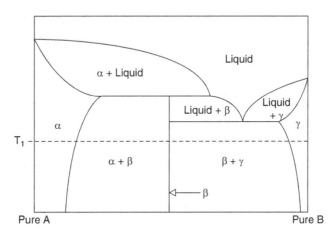

Fig. 11.3 A hypothetical binary phase diagram in which the intermediate β phase has a small range of composition

The corresponding variation of the electrical potential with composition is shown in schematically in Fig. 11.4. The potential drops abruptly, rather than gradually, across the narrow β phase in this case.

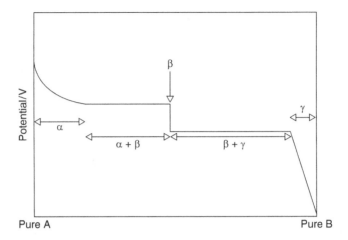

Fig. 11.4 Schematic variation of electrical potential with composition across the binary phase diagram shown in Fig. 11.3

11.3 A Real Example, The Lithium: Antimony System Again

As a concrete example to demonstrate these principles consider the Li–Sb system. It has been studied both experimentally and theoretically in some detail [1–3]. The phase diagram is shown in Fig. 11.5.

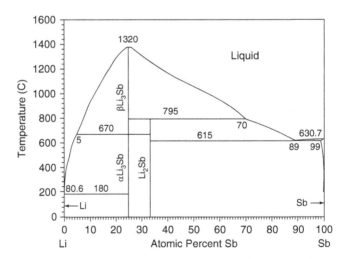

Fig. 11.5 Lithium–antimony phase diagram

Below 615°C, there are two intermediate phases between Sb and Li, Li_2Sb and Li_3Sb. Both have rather narrow ranges of composition and are represented simply as vertical lines in the phase diagram. Thus, if an electrode starts as pure Sb and lithium is added, it successively goes through two different reactions. The first involves the formation of the phase Li_2Sb, and can be written as

$$2Li + Sb = Li_2Sb. \tag{11.1}$$

Upon the addition of more lithium, a second reaction will occur that results in the formation of the second intermediate phase from the first. This can be written as

$$Li + Li_2Sb = Li_3Sb. \tag{11.2}$$

This process can be studied experimentally by the use of an electrochemical cell whose initial configuration is similar to that shown schematically in Fig. 11.6.

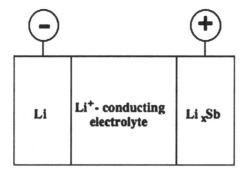

Fig. 11.6 Schematic drawing of electrochemical cell to use to study the Li–Sb system

By driving current through this cell from an external source, which causes the voltage between the two electrodes to be reduced from the open circuit value that it has when the positive electrode is pure antimony, lithium will leave the negative electrode, pass through the electrolyte, and arrive at the positive electrode. If the chemical diffusion rate within the Li_xSb electrode is sufficiently high, relative to the rate at which lithium ions arrive at the positive electrode surface, this lithium will be incorporated into the bulk of the electrode crystal structure, changing its composition. That is, the value of x in the positive electrode material Li_xSb will increase.

If the lithium is either added very slowly, or stepwise, allowing equilibrium to be attained within the positive electrode material after each step, the influence of the lithium concentration in the positive electrode upon its potential under equilibrium or near-equilibrium conditions can be investigated. This is the *Coulometric titration technique* discussed earlier. Data from such an experiment at 360°C are shown in Fig. 11.7.

These results can be understood by consideration of the Gibbs phase rule, which was discussed in Chap. 10.

After an initial, invisibly narrow, range of solid solution, the first plateau in Fig. 11.7 corresponds to compositions in the phase diagram in which both (almost pure) Sb and the phase Li_2Sb are present. Thus, it is related to the reaction in (11.1).

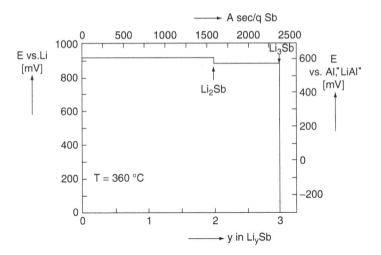

Fig. 11.7 Results from a coulometric titration experiment on the Li–Sb system at 360°C. After [3]

There is also a very narrow composition range in which only one phase, Li_2Sb, is present and the potential varies. Upon the addition of further Li, the overall composition moves into the region of the phase diagram in which two phases are again present, in this case Li_2Sb and Li_3Sb, and the potential follows along a second plateau, related to (11.2).

The potentials of these two plateaus can be calculated from thermodynamic data on the standard Gibbs free energies of formation of the two phases, Li_2Sb and Li_3Sb. According to [3], these values are -176.0 and -260.1 kJ/mol, respectively, at that temperature.

The standard Gibbs free energy change, ΔG_r^0, related to virtual reaction (11.1), is simply the standard Gibbs free energy of formation of the phase Li_2Sb, ΔG_f^0 (Li_2Sb). From this the potential of the first plateau can calculated from

$$E - E^0 = -\Delta G_r^0/2F, \tag{11.3}$$

where E^0 is the potential of pure Li. This was found to be 912 mV in the experiment.

The potential of the second plateau is related to virtual reaction (11.2), where

$$\Delta G_r^0 = \Delta G_f^0(Li_3Sb) - \Delta G_f^0(Li_2Sb). \tag{11.4}$$

and in this case

$$E - E^0 = -\Delta G_r^0/F, \tag{11.5}$$

The result is that the potential of this plateau was experimentally found to be 871 mV versus pure Li.

The maximum theoretical energy that can be obtained from this alloy system is the sum of the energies involved in the two reactions. These relationships are shown schematically in Fig. 11.8.

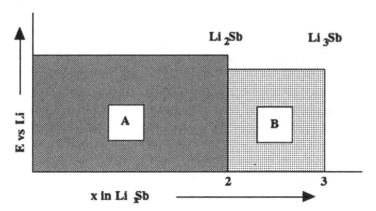

Fig. 11.8 Relation between energy stored and the titration curve in the Li–Sb system

The total energy that can be stored is proportional to the total area under the titration curve. The energy released in the first reaction is the product of the voltage of the first plateau times its capacity, that is , the charge passed through the cell in connection with that reaction. That corresponds to the area inside rectangle A. The energy released in the second discharge reaction step is the product of its voltage and its capacity, and corresponds to the area inside rectangle B. The total energy is the sum of the two areas.

These energy values can be converted into *specific energy*, energy per unit weight. In the case of the first plateau, the *maximum theoretical specific energy*, MTSE, is simply the standard Gibbs free energy of the reaction divided by the sum of the atomic weights in the product. This was found to be 1,298 kJ/kg. This can also be expressed as 360 Wh/kg, since 3.6 kJ is equal to 1 Wh.

The MTSE can also be calculated if the composition were to only vary between the compositions Li_2Sb and Li_3Sb. In that case, the voltage was 871 mV and the capacity was only 1 mol of lithium per mole of original Li_2Sb. When calculating the MTSE, the weight of the product is then that of Li_3Sb, 142.57 g/mol. The result is 589 kJ/kg and 164 Wh/kg for a cell operated in that composition range.

However, if the experiment is performed starting with pure Li and pure Sb, and the energy relating to both plateaus is used, the relevant weight for both steps is the final weight, that of Li_3Sb. Thus the MTSE of the first reaction in the two-reaction scheme is less than it would be if it were used alone. Instead of 1,298 kJ/kg, it is only 1,234 kJ/kg. The total MTSE is then 1,234 + 589 = 1,823 kJ/kg.

This is the same result that would be found if it were assumed that the intermediate phase, Li_2Sb, did not form, and that there is only a single voltage plateau between Li and Li_3Sb.

If the electrochemical titration curve were calculated from the experimental value of the total energy, it would have only a single plateau, and at a voltage that is the weighted average of the voltages of the two reactions that actually take place. This is a false result, due to the lack of recognition of the existence of the intermediate phase. Thus one has to be careful to be aware of all of the stable phases when making voltage predictions from thermodynamic data.

11.4 Stability Ranges of Phases

Whereas emphasis has been placed upon the potentials at which reactions take place in this discussion thus far, there is another important type of information that is available from equilibrium electrochemical titration curves. The potential ranges over which the various intermediate phases are stable can be readily obtained. Since they are present at compositions between two plateaus, they are stable at all potentials between the two plateau potentials. This can be important information if they are to be used as *mixed-conductors*, as will be described later.

11.5 Another Example, The Lithium: Bismuth System

The lithium–bismuth binary system has also been extensively explored by the use of the coulometric titration technique. The phase diagram is shown in Fig. 11.9. Note that this diagram is drawn with lithium on the right hand side, that is, in the opposite direction from the Li–Sb diagram. This difference is, in principle, not important, however.

The titration curve that resulted from measurements made at 360°C is shown in Fig. 11.10.

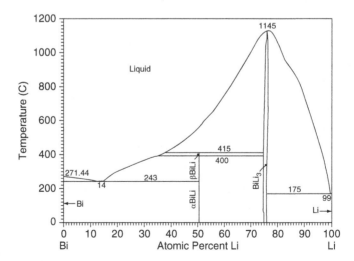

Fig. 11.9 The lithium–bismuth binary phase diagram

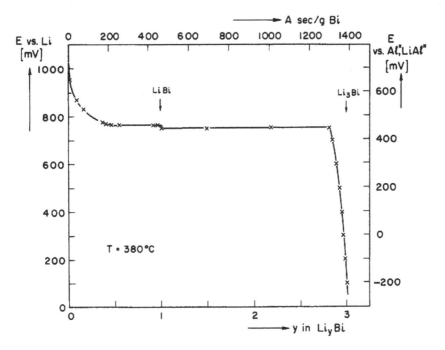

Fig. 11.10 Results from a coulometric titration experiment on the Li–Bi system at 360°C. After [3]

It can be seen that there are three differences from the Li–Sb system. The phase diagram shows that there is a considerable amount of solubility of bismuth in liquid lithium at that temperature. This results in the appearance of a single-phase region in the titration curve. Also, there is a phase LiBi in the Li–Bi case, and Li_2Sb in the Li–Sb case.

In addition, the phase diagram in Fig. 11.9 indicates that the solid phase "Li_3Bi" has an appreciable range of composition. This can also be seen in the titration curve. Because of the very high sensitivity of the coulometric titration technique, the electrochemical properties of this phase could be explored in much more detail. This is shown in Fig. 11.11.

11.6   Coulometric Titration Measurements on other Binary Systems

Coulometric titration experiments have been made on a number of other binary metallic systems at several temperatures. In order to obtain reliable data, it is important that the experiments be undertaken under conditions such that equilibrium can be reached within a reasonable time. This requirement is fulfilled much more easily at elevated temperatures, but in some cases, equilibrium data can also be obtained at ambient temperatures, albeit with a bit more patience. A number of further examples will be discussed in later chapters.

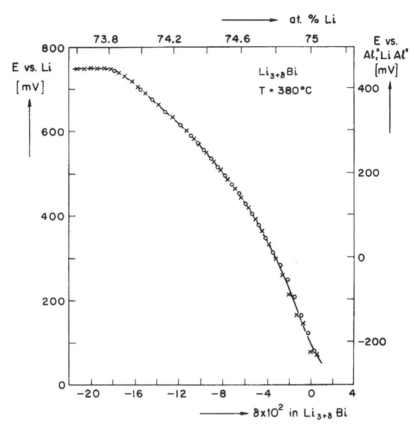

Fig. 11.11 Coulometric titration measurements within the composition range of the phase "Li$_3$Bi". After [3].

11.7   Temperature Dependence of the Potential

The early measurements of the equilibrium electrochemical properties of binary lithium alloys and their relationship to the relevant phase diagrams were made at elevated temperatures using a LiCl–KCl molten salt electrolyte. These included experiments on the Li–Al, Li–Bi, Li–Cd, Li–Ga, Li–In, Li–Pb, Li–Sb, Li–Si, and Li–Sn systems [4–10]. This molten salt electrolyte was being used in research efforts aimed at the development of large-scale batteries for electric vehicle propulsion and load leveling applications. Subsequently, measurements were made with lower temperature molten salts, LiNO$_3$–KNO$_3$ [11], and at ambient temperatures with organic solvent electrolytes. The latter will be discussed later.

It has been found, as expected, that the temperature dependence of the potentials and capacities can be explained in terms of the relevant phase diagrams and thermodynamic data in all of these cases.

To demonstrate the principles involved, experimental results on materials in the Li–Sb and Li–Bi systems over a wide range of temperature will be described. The results are shown in Fig. 11.12.

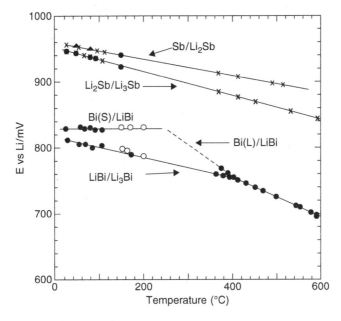

Fig. 11.12 Temperature dependence of the potentials of the two-phase plateaus in the Li–Sb and Li–Bi systems. After [12]

Each of these systems has two intermediate phases at low temperatures. The temperature dependence of the potentials of the plateaus due to the presence of two-phase equilibria in the Li–Sb system fall upon two straight lines, corresponding to the reactions

$$2Li + Sb = Li_2Sb, \tag{11.6}$$

and

$$Li + Li_2Sb = Li_3Sb. \tag{11.7}$$

In the Li–Bi case, however, where the comparable reactions are

$$Li + Bi = LiBi, \tag{11.8}$$

and

$$2Li + LiBi = Li_3Bi, \tag{11.9}$$

the temperature dependence of the plateau potentials is different from the Li–Sb case. There is a change in slope at the *eutectic melting point* (243°C), and the data for the two plateaus converge at about 420°C, which corresponds to the fact that the LiBi phase is no longer stable above that temperature. These can be seen in the phase diagram for that system shown in Fig. 11.9. At higher temperatures, there is only a single reaction,

$$3Li + Bi = Li_3Bi. \tag{11.10}$$

In addition, the potentials of the second reaction fall along two straight line segments, depending upon the temperature range. There is a significant change in slope at about 210°C, resulting in a negligible temperature dependence of the potential at low temperatures, due to the melting of bismuth.

The potentials are related to the standard Gibbs free energy change ΔG_r^0 relating to the relevant reaction, and the temperature dependence of the value of ΔG_r^0 can be seen from the relation between the Gibbs free energy, the enthalpy, and the entropy

$$\Delta G_r^0 = \Delta H_r^0 - T\Delta S_r^0, \tag{11.11}$$

where ΔH_r^0 is the change in the standard enthalpy, and ΔS_r^0 is the change in the standard entropy resulting from the reaction. Thus, it can be seen that

$$d\Delta G_r^0/dT = \Delta S_r^0. \tag{11.12}$$

From these data, one can obtain values of the standard molar entropy changes involved in these several reactions. They are shown in Table 11.1. Thus, the potentials at any temperature within this range can be predicted.

Table 11.1 Reaction entropies in the lithium–antimony and lithium–bismuth systems

Reaction	Molar entropy of reaction (J/K mol)	Temperature range (°C)
$2Li + Sb = Li_2Sb$	−31.9	25–500
$Li + Li_2Sb = Li_3Sb$	−46.5	25–600
$Li + Bi = LiBi$	0	25–200
$2Li + LiBi = Li_3Bi$	−36.4	25–400

11.8   Application to Oxides and Similar Materials

This discussion thus far has been concerned with binary metallic alloys. However, the same principles can be applied to binary metal–oxygen systems. The Nb–O system will be used as an example. The niobium–oxygen phase diagram is shown in Fig. 11.13.

There are three intermediate phases in this system, and thermodynamic data can be used to calculate the potentials of the various two-phase plateaus at any

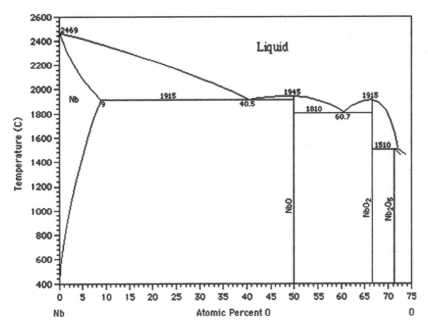

Fig. 11.13 Niobium–oxygen phase diagram

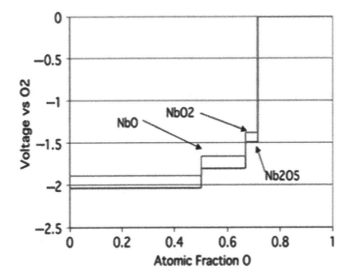

Fig. 11.14 Equilibrium potentials of the three two-phase plateaus in the Nb–O system at two temperatures

temperature, as before. In addition, these potentials can be converted into values of the oxygen activity at the respective temperatures. Data on the plateau potentials, which define the limiting values for the stability of each of the phases, are shown for two temperatures in Fig. 11.14. In this case, the potentials are shown as voltages relative to the potential of pure oxygen at one atmosphere at the respective temperatures.

Similar procedures can be employed to the analysis of other metal–gas systems, such as iodides and chlorides.

11.9 Ellingham Diagrams and Difference Diagrams

Another thinking tool that is sometimes used to help understand the behavior of metal–oxygen systems is the so-called *Ellingham diagram*. These are plots of the Gibbs free energy of formation of their oxides as a function of temperature.

There are two different ways in which this information can be presented. The Ellingham diagrams that are most often seen in textbooks are of the *integral*

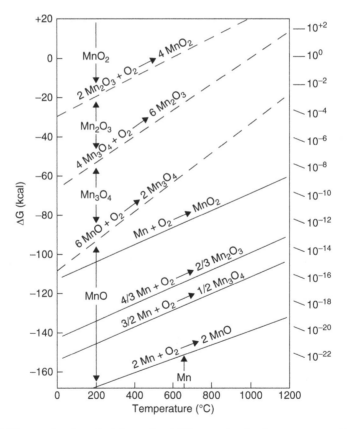

Fig. 11.15 Diagram that shows both integral and difference data for the manganese oxide system. After [14]

type, for they indicate the temperature dependence of the Gibbs free energy necessary for the formation of a particular oxide from its component elements. The formation reactions are generally written on a "per mol oxygen" basis, so that the lines relating to different oxides are generally parallel, as the entropy of gaseous oxygen makes the major contribution to the entropy of the formation reaction.

An equilibrium oxygen pressure scale is generally added on the right side to provide a simple graphical means to determine the oxygen partial pressure of the oxides as a function of temperature [13].

However, this information is only valid for the direct formation of an oxide from its elements, and in many cases more than one oxide can be formed from a given metal, depending upon the oxygen partial pressure. As an example, there are several manganese oxides: MnO, Mn_3O_4, Mn_2O_3, and MnO_2. Except for the lowest oxide, MnO they do not form directly by the oxidation of manganese. The higher oxides form by the reaction of oxygen with lower oxides, and the relevant oxygen pressure for the formation of a given oxide is related to the reaction of oxygen with its next lower oxide.

Therefore, it is much more useful to have a *difference diagram* that provides information about the oxygen pressure for the formation of given phases from their neighbors. An example that shows both types of information is given in Fig. 11.15.

11.10 Liquid Binary Electrodes

Although the discussion here has involved solid binary alloys and oxides, similar principles apply to liquids in binary systems.

An example that was of great interest some years ago involved the so-called *sodium–sulfur battery* that operates at about 300°C. In this case, both electrodes are liquids, and the electrolyte is a solid sodium ion conductor called *sodium beta alumina*. This can be described as an *L/S/L system*. It is the inverse of conventional systems, which have solid electrodes and a liquid electrolyte, that is, *S/L/S systems*. The negative electrode is molten sodium, and the positive electrode is the product of the reaction of sodium with liquid sulfur. Thus, the basic reaction can be written as

$$x\text{Na} + \text{S} = \text{Na}_x\text{S}. \tag{11.13}$$

The potential of the elemental sodium is constant, independent of the amount of sodium present. The potential of the positive electrode changes as the sodium concentration varies by its transport across the cell.

The relevant portion of the Na–S phase diagram is shown in Fig. 11.16. It is seen that at about 300°C a relatively small amount of sodium can be dissolved in liquid sulfur. When this concentration is exceeded, a second liquid phase, with a composition of about 78 atomic percent Na, is nucleated. This has a composition that is roughly $Na_{0.4}S$. As more sodium is added, the overall composition traverses the two-phase region, and the amount of this liquid phase increases relative to the

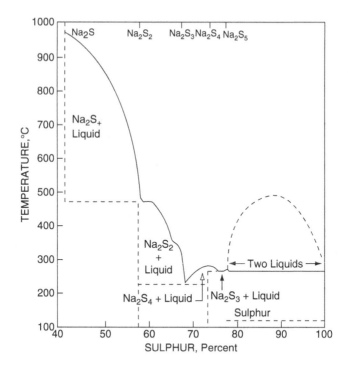

Fig. 11.16 Part of the sodium–sulfur phase diagram

amount of the sulfur-rich liquid phase. Thus, a potential plateau is expected over this composition range. When the sodium concentration exceeds that corresponding to about $Na_{0.4}S$, the overall composition moves into a single-phase liquid range, and thus the potential varies. The maximum amount of sodium that can be used in this electrode corresponds roughly to $Na_{0.67}S$.

At higher sodium concentrations a solid second phase begins to form from the liquid solution. This tends to form at the interface between the solid electrolyte and the liquid electrode, and prevents the ingress of more sodium, and thus blocks further reaction. The variation of the potential with the composition of the electrode is shown in Fig. 11.17.

11.11 Comments on Mechanisms and Terminology

It has been shown that the incorporation of species into solid electrodes can involve either a change in the composition of a single phase that is already present in the microstructure, or the nucleation and growth of an additional phase. When that species is deleted, the same two types of phenomena can occur, but in the opposite sense. Consider how this can happen. If, for example, lithium is added to an existing phase, it forms a solid solution and the composition will change, becoming more

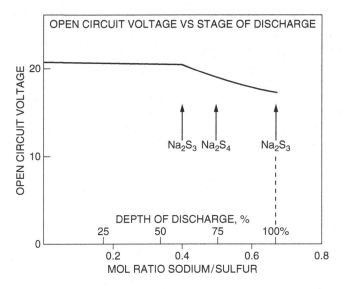

Fig. 11.17 Voltage versus pure sodium as a function of composition

lithium-rich. If this involves more than merely the surface layer, it must involve the diffusive motion of lithium atoms or ions into the crystal structure.

In many metallic alloys and ceramic materials, the inserted species occupies the same type of lattice positions as the host material. This is called a *substitutional solid solution*. In order for the composition to change, there must be a mechanism that allows atomic, or ionic, motion through the crystal structure. In substitutional solid solutions, this typically involves the presence and motion of empty lattice sites, *vacancies*. Atoms can jump into these vacant lattice sites from adjacent positions in the structure. The result of a single jump is the effective motion of the vacancy in one direction, and the atom in the opposite direction. Compositional changes occur by this vacancy mechanism if one type of atom has a greater probability of jumping into adjacent vacancies than the other type and a gradient in the composition is present.

Another type of diffusion mechanism is often present in crystal structures in which one type of atom or ion is appreciably smaller than the other types present. This is often the case in lithium and hydrogen systems, as these species are quite small. The smaller atoms can occupy interstitial sites between the other atoms in the crystal lattice. They can move about by jumping from one interstitial site to the next. Diffusion by this *interstitial mechanism* does not require the motion of either the other atoms or vacancies, and it is typically very much faster than the vacancy diffusion mechanism. That is, interstitial *diffusion coefficients*, or *diffusivities*, are typically greater than vacancy mechanism diffusion coefficients, or diffusivities.

Because of the large concentration of interstitial positions in most crystal structures, it is structurally possible for a large number of guest species to be inserted into the host crystal structure, if other factors, such as the electronic energy spectrum, are favorable.

If the basic structure of the host material is not appreciably altered by the *insertion* of additional guest species, so that there is a definite relation between the initial crystal structure and the structure that results, the reaction is called *topotactic*. This is often the case in materials of interest as electrode reactants in lithium battery systems.

Topotaxy implies a three-dimensional (3D) relation between the structures of the *parent* and *product* structures, whereas the term *epitaxy* is used to describe a two-dimensional (2D) correspondence between two structures. Likewise, an *insertion reaction* that has a 2D character is often called *intercalation*.

From this structural viewpoint, it can be readily understood that there can be a limit to the concentration of interstitial guest species that can be inserted into a host crystal structure. This limit can be due to either crystallographic or electronic factors, and will not be discussed further here.

If there is a thermodynamic driving force for the incorporation of *additional* guest species than can be accommodated interstitially, this must occur by a different mechanism. A *second phase*, with a different crystal structure as well as a higher solute concentration, must be nucleated. As more and more of the guest species atoms arrive, the extent of this second phase increases, gradually replacing the interstitial phase that was initially present. This change in the microstructure, in which one phase is gradually replaced by another phase, is an example of a *reconstitution* reaction.

When a reconstitution reaction is taking place, the initial and product phases are both present in the microstructure. This is sometimes called a *heterophase structure*, in contrast to a *homophase* structure, in which only a single phase is present. Thus, this range of compositions must be in a two-phase region of the corresponding phase diagram.

Phase diagrams are expected to provide information about the *absolutely stable* phases that tend to be present in a chemical system as a function of intensive thermodynamic variables such as temperature and composition. The term *absolutely stable* has been used to describe the *most stable equilibrium structure* possible for a given composition. On the other hand, a phase that is *stable relative to small perturbations*, and thus meets the general requirement for equilibrium, yet is not the most stable variation, is termed *metastable*.

11.12 Summary

Many batteries use binary systems as either negative or positive, or both, electrode reactants today. The theoretical limits of the potentials and capacities of such electrodes can be determined from a combination of thermodynamic data and phase diagrams. This has been demonstrated here for several examples of binary systems.

There are two general types of reactions that can take place, *homophase* reactions, in which guest atoms are inserted into an existing phase, often *topotactically*, and *reconstitution* reactions in which phases nucleate and grow in *heterophase* microstructures. The potential varies with the overall composition of an electrode in the insertion reaction homophase case, but is composition-independent when reconstitution reactions take place in heterophase microstructures.

The electrochemical titration method can be used to investigate the relevant parameters experimentally. When the composition is within a single-phase region of the relevant phase diagram, the potential varies as guest species are inserted or extracted during an electrochemical reaction. One the other hand, when the composition is within two-phase regions of the relevant phase diagram, reconstitution reactions take place, and the potential is independent of composition. Experimental results are now available for a number of systems of each type, both at elevated temperatures and at ambient temperatures.

Under equilibrium, or near-equilibrium, conditions the potentials are directly related to the values of the standard Gibbs free energies of formation of the phases involved. Thus thermodynamic data can predict experimental results. Likewise, experiments can provide thermodynamic data. As an example, the temperature dependence of potential plateaus can be used to determine the standard entropy changes in the relevant reaction. These experimental data also correlate with the stability of phases in the phase diagram. Furthermore, the MTSE of an electrochemical system can also be determined from the equilibrium electrochemical titration curve and the related thermodynamic data.

These principles are also applicable to metal oxides, as well as liquid binary materials, as illustrated by the Nb–O and the Na–S systems. The latter is the basis for a high temperature L/S/L battery system that uses sodium beta alumina as a solid electrolyte.

References

1. W. Weppner, and R.A. Huggins. In Proceedings of the Symposium on Electrode Materials and Processes for Energy Conversion and Storage, ed. by J.D.E. McIntyre, S. Srinivasan and F.G. Will. Electrochemical Society, Princeton, NJ (1977), p. 833
2. W. Weppner, and R.A. Huggins. Z. Phys. Chem. N.F. *108*, 105 (1977)
3. W. Weppner, and R.A. Huggins. J. Electrochem. Soc. *125*, 7 (1978)
4. C.J. Wen, et al. J. Electrochem. Soc. *126*, 2258 (1979)
5. C.J. Wen. Ph.D. Dissertation. Chemical Diffusion in Lithium Alloys, Stanford University (1980)
6. C.J. Wen and R.A. Huggins. J. Electrochem. Soc. *128*, 1636 (1981)
7. C.J. Wen and R.A. Huggins. Mater. Res. Bull. *15*, 1225 (1980)
8. M.L. Saboungi, et al. J. Electrochem. Soc. *126*, 322 (1979)
9. C.J. Wen and R.A. Huggins. J. Solid State Chem. *37*, 271 (1981)
10. C.J. Wen and R.A. Huggins. J. Electrochem. Soc. *128*, 1181 (1981)
11. J.P. Doench and R.A. Huggins. J. Electrochem. Soc. *129*, 341C (1982)
12. J. Wang, I.D. Raistrick, and R.A. Huggins. J. Electrochem. Soc. *133*, 457 (1986)
13. F.D. Richardson and J.H.E. Jeffes. J. Iron Steel Inst. *160*, 261 (1948)
14. N.A. Godshall, I.D. Raistrick and R.A. Huggins. J. Electrochem. Soc. *131*, 543 (1984)

Chapter 12
Ternary Electrodes Under Equilibrium or Near-Equilibrium Conditions

12.1 Introduction

The previous chapter described binary electrodes, in which the microstructure is composed of phases made up of two elements. It was pointed out that there are also cases in which three elements are present, but only partial equilibrium can be obtained in experiments, so the electrode behaves as though it were composed of two, rather than three, components.

This chapter will discuss active materials that contain three elements, but have kinetic behavior such that they behave as true ternary systems. As before, it will be seen that phase diagrams and equilibrium electrochemical titration curves are very useful *thinking* tools in understanding the potentials and capacities of electrodes containing such materials.

It is generally more difficult to obtain complete equilibrium in ternary systems than in binary systems, so that much of the available equilibrium or near-equilibrium information stems from experiments at elevated temperatures. Selective or partial equilibrium is much more common at ambient temperatures. This will be discussed in another chapter.

12.2 Ternary Phase Diagrams and Phase Stability Diagrams

In order to represent compositions in a three-component system, one must have a figure that represents the concentrations of three components. This can be done by using a two-dimensional figure, as will be seen shortly. However, if information about the influence of temperature is also desired, a three-dimensional figure is required. This is often done in metallurgical and ceramic systems in which experiments commonly involve changes in the temperature. Most electrochemical systems operate at or near constant temperatures, so three-dimensional figures are not generally considered necessary.

R.A. Huggins, *Energy Storage*,
DOI 10.1007/978-1-4419-1024-0_12, © Springer Science+Business Media, LLC 2010

Compositions in isothermal ternary systems can be represented on paper by using a triangular coordinate system. The method that is commonly employed in materials systems involves the use of *isothermal Gibbs triangles*. This scheme is illustrated in Fig. 12.1.

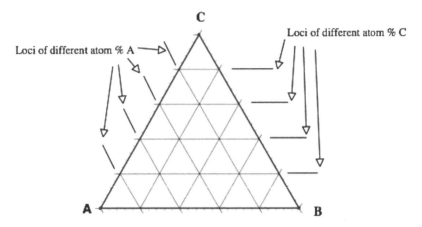

Fig. 12.1 General coordinate scheme used to depict compositions and phase equilibria in ternary systems on isothermal Gibbs triangles

Compositions are expressed in terms of the atomic percent of each of the three components, indicated as A, B, and C in this case. For the purposes of this discussion, it is desirable to have elements as components, so that three elements are placed at the corners, and the atomic percent of an element varies from zero along the opposite side to 100% at its corner. Thus the position of each point within the triangle represents the atomic fraction of each of the elements present in the system.

Although phases in ternary systems generally have ranges of composition, as they do in binary systems, it is often useful to simplify the phase equilibrium situation by assuming that they act as *point phases*. That is, that they have very narrow composition ranges. The term *phase stability diagram* will be used in this discussion to describe this approximation to the actual ternary phase diagram. It will be seen that it is possible to get a large amount of useful information by the use of such an approximate isothermal Gibbs triangle.

If there are phases inside the Gibbs triangle, the influence of the Gibbs phase rule must be considered. It was shown earlier that the Gibbs phase rule can be written as

$$F = C - P + 2 \qquad (12.1)$$

If the temperature and total pressure are kept constant, the number of residual degrees of freedom F will be zero when there are three phases present in a ternary system. Three phases are in equilibrium with each other within triangles inside the overall Gibbs triangle. Two phases are in equilibrium if their compositions are connected by a line, called a *two-phase tie line*. As shown in Fig. 12.2, if intermediate ternary phases are present, the total area within the Gibbs triangle is divided into *sub-triangles* whose sides are *two-phase tie lines*.

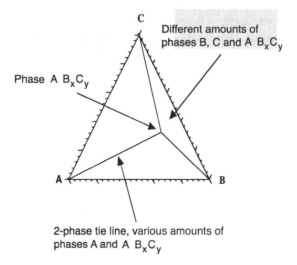

Fig. 12.2 Isothermal phase stability diagram ABC for the case in which there is a single intermediate phase whose composition is A_xB_yC

All of the compositions that lie within a given triangle have microstructures that are composed of mixtures of the three phases that are at the corners of that triangle. The overall composition determines the amounts of these different phases present, but not their compositions, for the latter are specified by the locations of the points at the corners of the triangle. Any materials having compositions that fall along one of the sides of a triangle will have microstructures composed of the two phases at the ends of that tie line. The amounts are determined by the position along the tie line. Points closer to a given end have greater amounts of the phase whose composition is at the end.

Because the compositions of the phases present within triangles are constant, determined by the locations of the corner points, all of the intensive (amount-independent) thermodynamic parameters and properties are the same for all compositions inside the triangle. Important intensive properties include the chemical potentials and activities of all of the components and the electrical potential.

12.3 Comments on the Influence of Sub-triangle Configurations in Ternary Systems

Binary systems can be changed to ternary systems by the addition of another element. As an example, consider a lithium-based binary system Li–M, in which the lithium composition can be varied. The addition of an additional element X converts this to a ternary Li–M–X system. The presence of X can result in a significant change in the potentials in the Li–M system, even if X does not react with lithium itself.

Consider a simple case. Assume that the thermodynamic properties of this system lead to a ternary phase stability diagram of the type shown in Fig. 12.3, in which it is assumed that there are two stable binary phases, LiM, and MX.

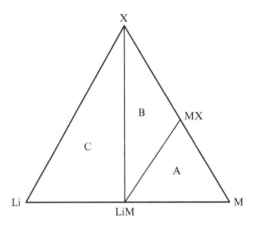

Fig. 12.3 Schematic ternary phase diagram for the Li–M–X system in which there are intermediate phases in the centers of both the Li–M and M–X binary systems

If there is no X present, the composition moves along the Li–M edge of the ternary diagram, which is simply the binary Li–M system, and there will be a constant potential plateau for all compositions between pure M and LiM. The voltage vs pure lithium in this compositional range, and therefore in triangle A of the ternary system, will be given by

$$E_A = \Delta G_f^0 (LiM)/F \tag{12.2}$$

What happens if lithium reacts with a material that has an original composition containing some X? The overall composition will follow a trajectory that starts at that position along the X–M side of the triangle and goes in the direction of the lithium corner of the ternary diagram. The addition of X to the M will not change the plateau potential for all compositions in triangle A. Therefore, there will be a plateau at that potential. Its length, however, will vary, depending upon the starting composition.

In addition, an additional plateau will appear at higher lithium concentrations as the overall composition enters and traverses triangle B. The potential of all compositions in that triangle will be given by

$$E_B = (\Delta G_f^0 (MX) - \Delta G_f^0 (LiM))/F \tag{12.3}$$

As in the case of the binary Li–M system, when the overall composition gets into triangle C the potential will be the same as that of pure lithium. These effects are illustrated in Figs. 12.4 and 12.5.

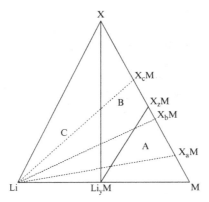

Fig. 12.4 The ternary Li–M–X system shown in Fig. 12.3, showing the loci of the overall composition as lithium reacts with three different initial compositions

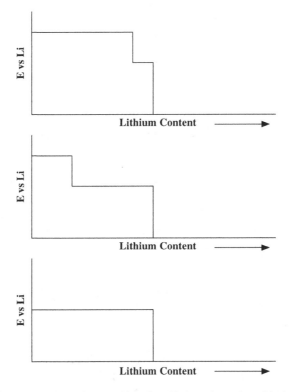

Fig. 12.5 Variation of the potential as lithium is added to electrodes with the three different starting compositions shown in Fig. 12.4. Top: X_aM, middle: X_bM, and bottom: X_cM

In Fig. 12.5, the variation of the electrode potential with overall composition is shown schematically for three different starting electrode compositions in Fig. 12.4.

In all three cases, the number of mols of lithium stored per mol of M does not change, but the weight of the electrode will change, depending upon the relative weights of M and X. In addition, the average electrode potential becomes closer to that of pure lithium. This can be either advantageous or disadvantageous, depending upon whether the material is used as a negative electrode or as a positive electrode in a lithium-based cell.

Another ternary phase configuration is shown in Fig. 12.6. In this case, it is assumed that there is also an intermediate phase in the Li–X binary system. The weight of the electrode per mol of Li will be reduced if the weight of Li_xX per mol of Li is less than that of Li per mol of Li_yM.

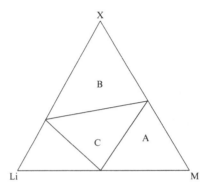

Fig. 12.6 Hypothetic ternary phase diagram in which there is one intermediate phase in each of the binary systems

In practice, an alloy system containing several intermediate phases may not be useful over its entire range of lithium composition, due to the change of the potential with composition. Poor diffusion kinetics in one of the intermediate phases or the terminal phase can also be deleterious.

There are many other possible configurations in ternary systems, including those containing ternary phases in the interior of the diagram. In screening possible systems for study, however, a logical starting point is to examine systems with known binary and ternary phases.

12.4   An Example: The Sodium/Nickel Chloride "ZEBRA" System

Some years ago an interesting battery system suddenly appeared that had been initially developed secretly in South Africa and England. It is based upon the use of the solid electrolyte Na β-alumina, as is the Na/Na_xS system. It soon became known as the *ZEBRA cell*.

It operates at 250–300°C and uses liquid sodium as the negative electrode, which is enclosed in a solid β-aluminum tube. At this temperature, sodium is liquid, and the ionic conductivity of the β-alumina is quite high. When the cell is fully charged, the positive electrode reactant is finely powdered $NiCl_2$, which is present adjacent to the β-alumina inside a solid container. Because the contact between the solid β-alumina tube and the particles of $NiCl_2$ is only where they touch, a second (liquid) electrolyte, $NaAlCl_4$, is also present in the outer, positive electrode compartment, part of the cell. Thus, the full surface area of the $NiCl_2$ particles acts as the electrochemical interface, which greatly increases the kinetics.

This electrochemical system, when charged, has the configuration:

$$Na/Na\ \beta\text{-alumina}/NaAlCl_4/NiCl_2$$

The physical arrangement of this cell is shown schematically in Fig. 12.7.

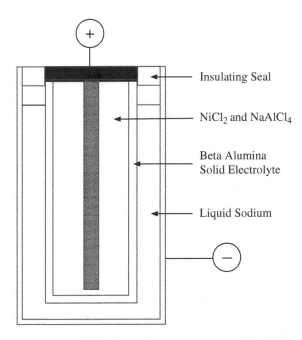

Insulating Seal

$NiCl_2$ and $NaAlCl_4$

Beta Alumina
Solid Electrolyte

Liquid Sodium

Fig. 12.7 Schematic view of the "ZEBRA" cell, which operates at 250–300°C

The electrochemical behavior of a ZEBRA cell can be understood by considering the Na–Ni–Cl ternary phase diagram. Thermodynamic data indicate that there are only two binary phases in this ternary system, $NiCl_2$, and NaCl. They lie on two different sides of the ternary Na–Ni–Cl phase diagram. Since the total area must be divided into triangles, it is evident that there are two possibilities. There is either a tie line from $NiCl_2$ to the Na corner, or there is one from NaCl to the Ni corner. The decision as to which of these is stable can be determined by the direction of the virtual reaction

$$2Na + NiCl_2 = 2NaCl + Ni \qquad (12.4)$$

The Gibbs free energy change in this virtual reaction is given by

$$\Delta G_r^0 = 2\Delta G_f^0 (NaCl) - \Delta G_f^0 (NiCl_2) \qquad (12.5)$$

Values of the standard Gibbs free energies of formation of NaCl and $NiCl_2$ at 275°C are -360.25 and -221.12 kJ/mol, respectively. Therefore, the reaction in (12.4) will tend to go to the right, and the tie line between NaCl and Ni is more stable than the one between $NiCl_2$ and Na.

As a result, the phase stability diagram must be as shown by the solid lines in Fig. 12.8. As Na reacts with $NiCl_2$, the overall composition of the positive electrode follows the dotted line in that figure. When it reaches the composition indicated by the small circle, all of the $NiCl_2$ will have been consumed, and only NaCl and Ni are present.

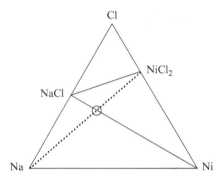

Fig. 12.8 The Na–Ni–Cl ternary phase diagram, showing the locus of the overall composition as Na reacts with $NiCl_2$

So long as the overall composition remains in the NaCl–$NiCl_2$–Ni triangle, the potential is constant. Its value can be calculated from the Gibbs free energy of reaction value corresponding to (12.5). The voltage of the positive electrode with respect to the pure Na negative electrode is given by

$$\Delta E = -\Delta G_r^0 / zF \qquad (12.6)$$

where $z = 2$, according to reaction (12.4). The result is that the potential of all compositions within that triangle in the ternary diagram, and also across the ZEBRA cell, is constant, and equal to 2.59 V. This is also what is observed experimentally.

12.5 A Second Example: The Lithium–Copper–Chlorine Ternary System

The Li–Cu–Cl system will be used as a further example to illustrate these principles, and show how useful information can be derived from a combination of a ternary phase diagram and thermodynamic data concerning the stable phases within it.

Thermodynamic information shows that there are three stable phases within this system at 298 K: LiCl, CuCl, and $CuCl_2$. Values of their standard Gibbs free energies of formation are given in Table 12.1.

Table 12.1 Gibbs free energies of formation of phases in the Li–Cu–Cl system

Phase	ΔG_f^0 at 298 K (kJ/mol)
LiCl	−384.0
CuCl	−138.7
$CuCl_2$	−173.8

All of these phases fall on the edges of the isothermal Gibbs triangle. If they are assumed to be point phases, the phase stability diagram can be constructed by following a few simple rules and procedures.

1. The total area must be divided into sub-triangles. Their edges are tie lines between pairs of phases.
2. No more than three phases can be present within a triangle. Their compositions must be at the corners.
3. Tie lines cannot cross.

The first task is to determine the stable tie lines in this system. This can be done by drawing all the possible tie lines between the stable phases on a trial basis, and then determining which of them are stable. The end result must be that the overall triangle is divided into sub-triangles.

The line between LiCl and $CuCl_2$ must be stable, as there are no other possible lines that could cross it. There are four additional possibilities, lines between Li and $CuCl_2$, Li and CuCl, LiCl and CuCl, and Li and Cu. A method that can be used to determine which of these is actually stable is to write the virtual reactions between the phases at the ends of conflicting (crossing) tie lines. Which of the two pairs of phases are more stable in each case can be determined from the available thermodynamic data.

As an example, consider whether there is a tie line between LiCl and Cu or one between CuCl and Li. Both cannot be stable, for they would cross.

The virtual reaction between the pairs of possible end phases can be written as

$$LiCl + Cu = CuCl + Li. \qquad (12.7)$$

As before, the direction in which this virtual reaction would tend to go can be determined from the value of the standard Gibbs free energy of reaction. In this case, it is given by

$$\Delta G_r^0 = \Delta G_f^0 \,(\text{CuCl}) - \Delta G_f^0 \,(\text{LiCl}) \qquad (12.8)$$

The result is that ΔG_r^0 is $(-138.7)-(-384.0) = +245.3$ kJ/mol. Thus, this reaction would tend to go to the left. This means that the combination of the phases LiCl and Cu is more stable than the combination of CuCl and Li. Thus, the tie line between LiCl and Cu is stable in the phase diagram.

This implies that the tie line between LiCl and Cu is also more stable than one between CuCl_2 and Li, and also that a line between LiCl and CuCl exists. These conclusions can be verified by consideration of the virtual reaction between LiCl and Cu, and CuCl_2 and Li. This reaction would be written as

$$2\text{LiCl} + \text{Cu} = \text{CuCl}_2 + 2\text{Li} \qquad (12.9)$$

for which the standard Gibbs free energy of reaction is $(-173.8)-2(-384.0) = +594.2$ kJ/mol. Thus these conclusions were correct.

The resulting isothermal phase stability diagram for this system is shown in Fig. 12.9.

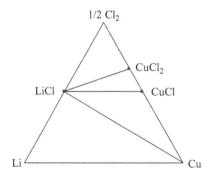

Fig. 12.9 Isothermal phase stability diagram for the Li–Cu–Cl ternary system at 25°C

12.5.1 Calculation of the Voltages in This System

From this diagram and the thermodynamic data, the voltages and capacities of electrodes in this system can also be calculated. As the first example, consider the reaction of lithium with CuCl. This reaction can be understood in terms of the ternary phase diagram as shown in Fig. 12.10.

By the addition of lithium, the overall composition moves from the initial composition at the CuCl point along the dotted line toward the Li corner, as shown in Fig. 12.10. In doing so, it moves into and across the LiCl–CuCl–Cu triangle. So long as it is inside this triangle its voltage remains constant.

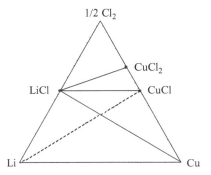

Fig. 12.10 Use of ternary phase diagram to understand the reaction of lithium with CuCl

This voltage can be calculated from the virtual reaction that takes place by the addition of lithium as the overall composition moves into, and through, the LiCl–CuCl–Cu triangle.

$$Li + CuCl = LiCl + Cu \tag{12.10}$$

The standard Gibbs free energy change as the result of this reaction is $(-384.0 - (-138.7) = -245.3$ kJ/mol. The voltage can be calculated from

$$E = -(-245.3)/[(1)(96.5)] \tag{12.11}$$

The result is 2.54 V vs. pure Li. This voltage remains constant as long as the overall composition stays in the LiCl–CuCl–Cu triangle. It is obvious from (12.10) and the phase diagram in Fig. 12.10 that up to 1 mol of Li can participate in this reaction. Thus, the equilibrium titration curve, the variation of the voltage of a cell of this type as a function of composition can be drawn as in Fig. 12.11.

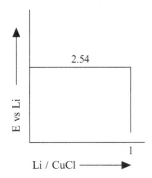

Fig. 12.11 Variation of the equilibrium voltage of Li/CuCl cell as a function of the extent of reaction

If, on the other hand, the positive electrode were to initially consist of $CuCl_2$ instead of CuCl, the overall composition would move along the dotted line shown in Fig. 12.12.

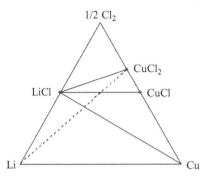

Fig. 12.12 Use of ternary phase stability diagram to understand the reaction of Li with $CuCl_2$

The overall composition first enters the $LiCl$–$CuCl_2$–$CuCl$ triangle. The relevant virtual reaction for this triangle is

$$Li + CuCl_2 = LiCl + CuCl \qquad (12.12)$$

The standard Gibbs free energy change as the result of this reaction is $(-384.0) + (-138.7) - (-173.8) = -348.9$ kJ/mol. The voltage with respect to pure Li can be calculated from

$$E = -\Delta G_r^0 / zF = 348.9/96.5 \qquad (12.13)$$

or 3.615 V vs. Li

There will be a plateau at this voltage in the equilibrium titration curve. The LiCl cannot react further with Li. But the CuCl that is formed in this reaction can undergo a further reaction with additional lithium. When this happens, the overall composition moves into and across the second triangle, whose corners are at LiCl, CuCl, and Cu. Although the reaction path is different, this is the same triangle whose voltage was calculated above for the reaction of lithium with CuCl. Thus the same voltage will be observed, 2.54 V vs. Li, in this second reaction, written in (12.5). The equilibrium titration curve will therefore have two plateaus, related to the two triangles that the overall composition traverses as lithium reacts with $CuCl_2$. This is shown in Fig. 12.13.

12.5.2 Experimental Arrangement for Copper Chloride Cells

Cells based upon the reaction of lithium with either of the copper chloride phases can be constructed at ambient temperature using an electrolyte with a non-aqueous

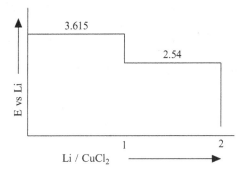

Fig. 12.13 Equilibrium titration curve for the reaction of lithium with $CuCl_2$ to form LiCl and CuCl, and then more LiCl and Cu

solvent, such as propylene carbonate, containing a lithium salt such as $LiClO_4$. There are a number of alternative solvents, as well as alternative salts, and this topic will be discussed in a later chapter. The important thing at the present time is that water and oxygen must be avoided, and the salt should have a relatively high solubility in the non-aqueous solvent.

12.6 Calculation of the Maximum Theoretical Specific Energies of Li/CuCl and Li/CuCl₂ Cells

The maximum values of specific energies that might be obtained from electrochemical cells containing either CuCl or $CuCl_2$ as positive electrode reactants can be calculated from this information.

As shown in an earlier chapter, the general relation for the maximum theoretical specific energy (MTSE) is

$$MTSE = (xV/W)(F) \, kJ/kg \qquad (12.14)$$

where x is the number of mols of Li involved in the reaction, V the average voltage, and W the sum of the atomic weights of the reactants. F is the Faraday constant, 96,500 Coulombs per mol.

In the case of a positive electrode that starts as CuCl and undergoes reaction (5), the sum of the atomic weights of the reactants is $(7 + 63.55 + 35.45) = 106.0$ g. The value of x is unity, and the average cell voltage is 2.54 V. Thus, the MTSE is 2,312.4 kJ/kg.

This can be converted to Wh/kg by dividing by 3.6, the number of kJ per Wh. The result is that the MTSE can also be written as 642.3 Wh/kg for this reaction.

If the positive electrode starts as $CuCl_2$ and undergoes reaction (7) to form LiCl and CuCl, the weight of the reactants is $(7 + 63.55 + (2 \times 35.45)) = 141.45$ g. The value of x is again unity, and the cell voltage was calculated to be 3.615 V.

This then gives a value of MTSE of 2,466.2 kJ/kg. Alternatively, it could be expressed as 685.1 Wh/kg.

If further lithium reacts with the products of this reaction, the voltage will proceed along the lower plateau, as was the case for an electrode whose composition started as CuCl. Thus additional energy is available. However, the total specific energy is not simply the sum of the specific energies that have just been calculated for the two plateau reactions independently. The reason for this is that the weight that must be considered in the calculation for the second reaction is the starting weight before the first reaction in this case.

Then, for the second plateau reaction:

$$MTSE = (1)(2.54)(96,500)/141.45 = 1,732.8 \text{ kJ/kg} \tag{12.15}$$

This is less than for the second plateau alone, starting with CuCl, which was shown above to be 2,312 kJ/kg. Alternatively, the specific energy content of the second plateau for an electrode that starts as $CuCl_2$ is 481.3 Wh/kg instead of 642.3 Wh/kg, if it were to start as CuCl.

Thus, if the electrode starts out as $CuCl_2$, the total MTSE can be written as:

$$MTSE = 2,466.2 + 1,732.8 = 4,199 \text{ kJ/kg} \tag{12.16}$$

Or alternatively, $685.1 + 481.3 = 1,166.4$ Wh/kg.

12.7 Specific Capacity and Capacity Density in Ternary Systems

As mentioned earlier, other parameters that are often important in battery systems are the capacity per unit weight or per unit volume. In the case of ternary systems, the capacity along a constant potential plateau is determined by the length of the path of the overall composition within the corresponding triangle. This is determined by the distance along the composition line between the binary tie lines at the boundaries of the triangles.

12.8 Another Group of Examples: Metal Hydride Systems Containing Magnesium

Binary alloys are often used as negative electrodes in hydrogen-transporting electrochemical cells. When they absorb or react with hydrogen, they are generally called *metal hydrides*. Because of the presence of hydrogen as well as the two metal components, they become ternary systems.

There is a great interest in the storage of hydrogen for a number of purposes related to the desire to reduce the dependence on petroleum. The reversible hydrogen absorption in some metal hydrides is a serious competitor for this purpose.

If the kinetics of hydrogen absorption or reaction are relatively fast, and the motion of the other constituents in the crystal structure is very sluggish, so that no structural reconstitution of the metal constituents in the microstructure takes place in the time scale of interest, such metal hydride systems can be treated as *pseudobinary systems*, that is, hydrogen plus the metal alloy. This is the general assumption that is almost always found in the literature on the behavior of metal hydrides.

On the other hand, there are materials in which this is not the case, and the hydrogen–metal hydride combination should be treated as a ternary system. Experiments have shown that the reaction of hydrogen with several binary magnesium alloys provides examples of such ternary systems [1, 2].

The prior examples of the reaction of lithium with the two copper chloride phases were used to illustrate how thermodynamic information can be used to determine the phase diagram and the electrochemical properties. These hydrogen/magnesium-alloy systems will be discussed, however, *as reverse examples*, in which electrochemical methods can be used in order to determine the relevant phase diagrams and thermodynamic properties, as well as to determine the practical parameters of energy and capacity.

Metal hydride systems are typically studied by the use of gas absorption experiments, in which the hydrogen pressure and temperature are the primary external variables. Electrochemical methods can generally also be employed by the use of a suitable electrolyte and cell configuration. The variation of the cell voltage can cause a change in the difference between the effective hydrogen pressure in the two electrodes. If one electrode has a fixed hydrogen activity, the hydrogen activity in the other can be varied by the use of an applied voltage. This then causes either the absorption or desorption of hydrogen. This can be expressed by the *Nernst* relation

$$E = (RT/zF)\Delta \ln p(H_2), \tag{12.17}$$

where E is the cell voltage, R the gas constant, T the absolute temperature, z the charge carried by the transporting ion (hydrogen), and F the Faraday constant. The term $\Delta \ln p(H_2)$ is the difference in the natural logarithms of the effective partial pressures, or activities, of hydrogen at the two electrodes.

Electrochemical methods can have several advantages over the traditional pressure–temperature methods. Since no temperature change is necessary for the absorption or desorption, data can be obtained at a constant temperature. If a stable reference is used, the variation of the cell voltage determines the hydrogen activity at the surface of the alloy electrode. Large changes in hydrogen activity can be obtained by the use of relatively small differences in cell voltage. Thus, the effective pressure can be easily and rapidly changed over several orders of magnitude. The amount of hydrogen added to, or deleted from, an electrode can be readily determined from the amount of current that passes through the cell.

One of the important parameters in the selection of materials for hydrogen storage is the amount of hydrogen that can be stored per unit weight of host material, the specific capacity. This is often expressed as the ratio of the weight of hydrogen absorbed to the weight of the host material. Magnesium-based hydrides are considered to be potentially very favorable in this regard. The atomic weight of magnesium is quite low, 24.3 g per mol. MgH_2 contains 1 mol of H_2, and the ratio 2/24.3 means 8.23 w/% hydrogen. This can be readily converted to the amount of charge stored per unit weight, that is, the number of mAh/g. One Faraday is 96,500 Coulombs, or 26,800 mAh, per equivalent. The addition of two hydrogens per magnesium means that two equivalents are involved. Thus 2,204 mAh of hydrogen can be reacted per gram of magnesium to form MgH_2.

On the other hand, one is often interested in the amount of hydrogen that can be obtained by the decomposition of a metal hydride. This means that the weight to be considered is that of the metal plus the hydrogen, rather than just the metal itself. When this is done, it is found that 7.6 w/% or 2,038 mAh/g hydrogen can be obtained from MgH_2.

These values for magnesium hydride are over five times those of the materials that are commonly used as metal hydride electrodes in commercial battery systems. Thus, there is continued interest in the possibility of the development of useful alloys based upon magnesium. The practical problem is that magnesium forms a very stable oxide, which acts as a barrier to the passage of hydrogen. It is very difficult to prevent the formation of this oxide on the alloy surface in contact with the aqueous electrolytes commonly used in battery systems containing metal hydrides.

One of the strategies that has been explored is to put a material such as nickel, which is stable in these electrolytes, on the surface of the magnesium. It is known that nickel acts as a mixed conductor, allowing the passage of hydrogen into the interior of the alloy. However, this surface covering cannot be maintained over many charge/discharge cycles, with the accompanying volume changes.

A different approach is to use an electrolyte in which magnesium is stable, but its oxide is not. This was demonstrated by the use of a novel intermediate temperature alkali organo-aluminate molten salt electrolyte $NaAlEt_4$ [1]. The hydride salt NaH can be dissolved into this melt, providing hydride ions, H^-, which can transport hydrogen across the cell. This salt is stable in the presence of pure Na, which can then be used as a reference, as well as a counter, electrode. More information about this interesting electrolyte will be presented in Chap. 14.

This experimental method was used to study hydrogen storage in three ternary systems involving magnesium alloys, the H–Mg–Ni, H–Mg–Cu and H–Mg–Al systems. In order to be above the melting point of this organic anion electrolyte, these experiments were performed somewhat above 140°C.

The magnesium–nickel binary phase diagram is shown in Fig. 12.14. It shows that there are two intermediate phases, Mg_2Ni and $MgNi_2$. It is also known that magnesium forms the dihydride MgH_2. These compositions are shown on the H–Mg–Ni ternary diagram shown in Fig. 12.15. Note that the ternary diagram is drawn with hydrogen at the top in this case.

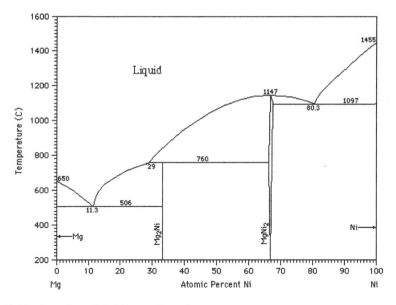

Fig. 12.14 Magnesium–nickel binary phase diagram

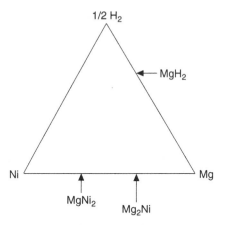

Fig. 12.15 The H–Mg–Ni ternary diagram showing only the known compositions along the binary edges

In order to explore this ternary system, an electrochemical cell was used to investigate the reaction of hydrogen with three compositions in this binary alloy system, $MgNi_2$, Mg_2Ni, and $Mg_{2.35}Ni$. Thus, the overall compositions of these materials moved along the dashed lines shown in Fig. 12.16 as hydrogen was added.

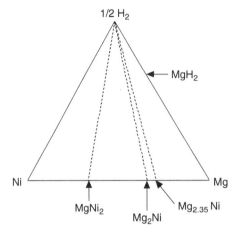

Fig. 12.16 Loci of the overall composition as hydrogen reacts with three initial Mg–Ni alloy compositions

It was found that the voltage went to zero as soon as hydrogen was added to the phase $MgNi_2$. However, in the other cases, it changed suddenly from one plateau potential to another as certain compositions were reached. These transition compositions are indicated by the circles in Fig. 12.17. The values of the voltage vs. the hydrogen potential in the different compositions regions are also shown in that figure.

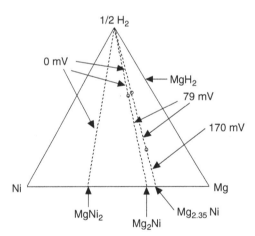

Fig. 12.17 Plateau voltages found in different composition regions

This information can be used to construct the ternary equilibrium diagram for this system. As described earlier, constant potential plateaus are found for compositions in three-phase triangles, and potential jumps occur when the composition crosses two-phase tie lines. The result is that there are no phases between $MgNi_2$

and pure hydrogen, but there must be a ternary phase with the composition Mg_2NiH_4. The resulting H–Mg–Ni ternary diagram at this temperature is shown in Fig. 12.18.

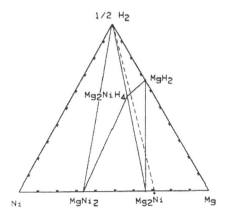

Fig. 12.18 Ternary phase stability diagram for the H–Mg–Ni system at about 140°C, derived from the compositional variation of the potential as hydrogen was reacted with three different initial binary alloy compositions

The phase Mg_2Ni reacts with four hydrogen atoms to form Mg_2NiH_4 at a constant potential of 79 mV vs. pure hydrogen. The weight of the Mg_2Ni host is 107.33 g, which is 26.83 g per mol of hydrogen atoms. This amounts to 3.73% hydrogen atoms stored per unit weight of the initial alloy. This is quite attractive and is considerably more than the specific capacity of the materials that are currently used in the negative electrodes of metal hydride/nickel batteries.

On the other hand, pure magnesium reacts to form MgH_2 at a constant potential of 107 mV vs. pure hydrogen. Because of the lighter weight of magnesium than nickel, this amounts to 8.23% hydrogen atoms per unit weight of the initial magnesium, or 7.6 w/% relative to MgH_2. Thus, the use of magnesium, and its conversion to MgH_2 are very attractive for hydrogen storage. There is a practical problem, however, due to the great sensitivity of magnesium to the presence of even small amounts of oxygen or water vapor in its environment.

If the initial composition is between Mg_2Ni and Mg, as is the case for the composition $Mg_{2.35}Ni$ that has been discussed above, there will be two potential plateaus, and their respective lengths, as well as the total amount of hydrogen stored per unit weight of the electrode, will have intermediate values, varying with the initial composition. As an example, the variation of the potential with the amount of hydrogen added to the $Mg_{2.35}Ni$ is shown in Fig. 12.19.

Similar experiments were carried out on the reaction of hydrogen with two other magnesium alloy systems, the H–Mg–Cu and H–Mg–Al systems [1]. Their ternary equilibrium diagrams were determined by using analogous methods. They are shown in Figs. 12.20 and 12.21.

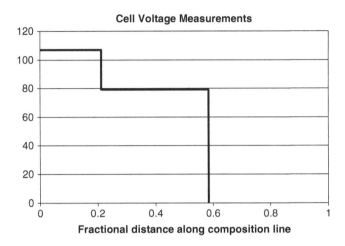

Fig. 12.19 Variation of potential as hydrogen is added to alloy with initial composition $Mg_{2.35}Ni$

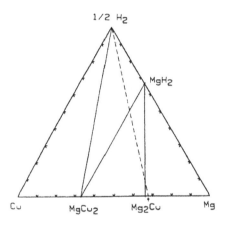

Fig. 12.20 Ternary phase stability diagram for the H–Mg–Cu system at about 140°C, derived from the compositional variation of the potential as hydrogen was reacted with different initial binary alloy compositions using organic anion molten salt electrolyte

12.9   Further Ternary Examples: Lithium-Transition Metal Oxides

These same concepts and techniques have been used to investigate several lithium-transition metal oxide systems [3,4]. They will be discussed briefly here. These examples are different from those that have been discussed thus far, for in a number of cases the initial compositions are themselves ternary phases, not just binary phases.

They further illustrate how electrochemical measurements on selected compositions can be used to determine the relevant phase diagrams. This makes it possible

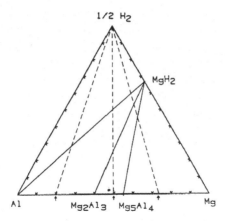

Fig. 12.21 Ternary phase stability diagram for the H–Mg–Al system at about 140°C, derived from the compositional variation of the potential as hydrogen was reacted with different initial binary alloy compositions using organic anion molten salt electrolyte

to predict the potentials and capacities of other materials within the same ternary system without having to measure them individually.

In addition, it will be seen that one can obtain a correlation between the activity of lithium, and thus the potential, and the equilibrium oxygen partial pressure, of phases and phase combinations in some cases. This provides the opportunity to predict the potentials of a number of binary and ternary materials with respect to lithium from information on the properties of relevant oxide phases alone.

Data on the ternary lithium-transition metal oxide systems that will be presented here were obtained by the use of the LiCl–KCl eutectic molten salt as electrolyte at about 400°C. They were studied at a time when there was a significant development program in the United States to develop large-scale batteries for vehicle propulsion using lithium alloys in the negative electrode and iron sulfide phases in the positive electrode. The transition metal oxides were being considered as alternatives to the sulfides.

Experiments employing this molten salt electrolyte system required the use of glove boxes that maintained both the oxygen and nitrogen concentrations at very low levels. This salt could be used for experiments to very negative potentials, limited by the evaporation of potassium. The maximum oxygen pressure that can be tolerated is limited by the formation of Li_2O. This occurs at a partial pressure of 10^{-15} atm at 400°C. This is equivalent to 1.82 V vs. lithium at that temperature. As a result, this electrolyte cannot be used to investigate materials whose potentials are above 1.82 V relative to that of pure lithium. As will be seen in Chap. 9, many of the positive electrode materials that are of interest today operate at potentials above this limit.

The first example is the lithium–cobalt oxide ternary system. Experiments were made in which lithium was added to both the binary phase CoO and the ternary phase $LiCoO_2$. The variations of the observed equilibrium potentials as lithium was added to these phases are indicated in Fig. 12.22. It is seen that there were sudden drops from 1.807 to 1.636 V, and then to zero in the case of CoO. Starting with LiCoO₂, however, only one voltage jump was observed, from 1.636 to zero. Since these jumps occur when the composition crosses binary tie lines in such diagrams, it was very easy to plot the ternary figure in this case. The result is shown in Fig. 12.23, in which the values of the potential (voltage vs. pure lithium), lithium activity, cobalt activity, and oxygen partial pressure for the two relevant compositional triangles are indicated. As mentioned earlier, it was not possible to investigate the higher potential regions that are being used in positive electrodes today.

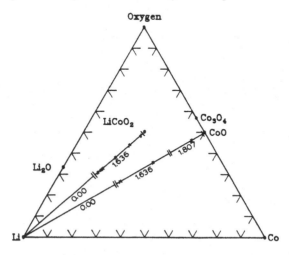

Fig. 12.22 Results of coulometric titration experiments on two compositions in the lithium–cobalt oxide system. After [4]

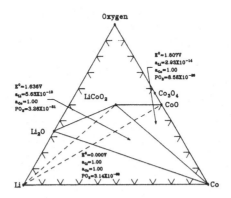

Fig. 12.23 Ternary phase stability diagram derived from the coulometric titration experiments shown in Fig. 12.22. After [4]

A more complicated case is the Li–Fe–O system. Figure 12.24 shows the variation of the equilibrium potential as lithium was added to Fe_3O_4 under near-equilibrium conditions. It is seen that this is a more complex case, for after a small initial solid solution region there are three jumps in the potential.

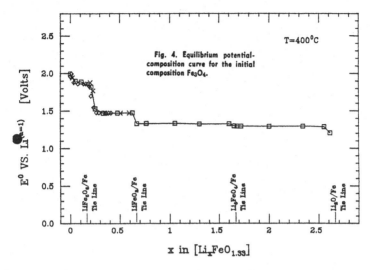

Fig. 12.24 Results of a coulometric titration experiment on a sample with an initial composition Fe_3O_4. After [4]

Similar experiments were undertaken on materials with two other initial compositions, $LiFe_5O_8$ and $LiFeO_2$. From these data, it was possible to plot the whole ternary system within the accessible potential range, as shown in Fig. 12.25.

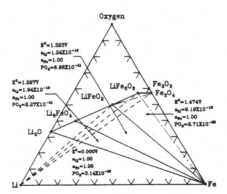

Fig. 12.25 Ternary phase stability diagram derived from coulometric titration measurements on materials in the Li–Fe–O ternary system. After [4]

Investigation of the Li–Mn–O system produced results that are somewhat different from those in the Li–Co–O and Li–Fe–O systems. The variation of the potential as lithium was added to samples with initial compositions MnO, Mn_3O_4, $LiMnO_2$ and Li_2MnO_3 is shown in Fig. 12.26. The ternary equilibrium diagram that resulted is shown in Fig. 12.27. It is seen that all of the two-phase tie lines do not go to the transition metal corner in this case. Instead, three of them lead to the composition Li_2O. Nevertheless, the principles and the experimental methods are the same.

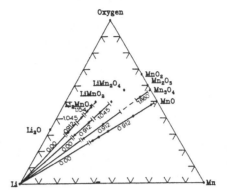

Fig. 12.26 Results of coulometric titration experiments on several phases in the Li–Mn–O ternary system. After [4]

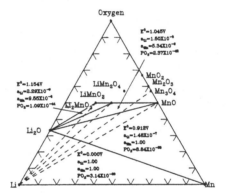

Fig. 12.27 Ternary phase stability diagram that resulted from the Coulometric titration results shown in Fig. 12.26. After [4]

It will be shown later, in Chap. 19 that some materials of this type behave quite differently at ambient temperature and higher potentials. In some cases, lithium can be extracted from individual phases, which then act as insertion–extraction electrodes, with potentials that vary with the stoichiometry of individual phases. The principles involved in insertion–extraction reactions will be discussed later, in Chap. 13.

12.10 Ternary Systems Composed of Two Binary Metal Alloys

In addition to the ternary systems that involve a non-metal component that have been discussed thus far in this chapter, it is also possible to have ternaries in which all three components are metals. Such materials are possible candidates for use as reactants in the negative electrode of lithium battery systems.

One example will be briefly mentioned here, the Li–Cd–Sn system, which is composed of two binary lithium alloy systems, Li–Cd and Li–Sn. As will be described in Chap. 18, these, as well as a number of other binary metal alloy systems, have been investigated at ambient temperature. Their kinetic behavior is sufficiently fast that they can be used at these low temperatures. This system, as well as others, will be discussed there in connection with the important mixed-conductor matrix concept.

12.10.1   An Example, The Li–Cd–Sn System at Ambient Temperature

If the two binary phase diagrams and their related thermodynamic information are known, it is possible to predict the related ternary phase stability diagram, assuming that no intermediate phases are stable. This assumption can be checked by making a relatively few experiments to measure the voltages of selected compositions. If they correspond to the predictions from the binary systems, there must be no additional internal phases. The value of this approach is that it gives a quick picture of what would happen if a third element were to be added as a dopant to a binary alloy.

As an example, the ternary phase stability diagram that shows the potentials of all possible alloys in the Li–Cd and Li–Sn system [5] is shown in Fig. 12.28.

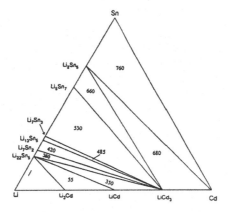

Fig. 12.28 Ternary phase stability diagram for the Li–Cd–Sn system. The numbers are the values of the voltage of all compositions in the various sub-triangles relative to pure lithium. After [5].

12.11   What About the Presence of Additional Components?

Practical materials often include additional elements, either as deliberately added dopants, or as impurities. If these elements are in solid solution in the major phases present in the ternary system, they can generally be considered to cause only minor deviations from the properties of the basic ternary system. Thus, it is not generally necessary to consider systems with more than three components.

12.12 Summary

This rather long chapter has shown that the ideal electrochemical behavior of ternary systems, in which the components can be solids, liquids or gases, can be understood by the use of phase stability diagrams and theoretical electrochemical titration curves. The characteristics of phase stability diagrams can be determined from thermodynamic information, and from them the related theoretical electrochemical titration curves can be determined. Important properties, such as the maximum theoretical specific energy, can then be calculated from this information. A number of examples have been discussed that illustrate the range of application of this powerful method.

References

1. C.M. Luedecke, G. Deublein and R.A. Huggins, *Hydrogen Energy Progress V,* ed. by T.N. Veziroglu and J.B. Taylor, Pergamon Press (1984), p. 1421.
2. C.M. Luedecke, G. Deublein and R.A. Huggins, J. Electrochem.Soc. *132*, 52 (1985)
3. N.A. Godshall, I.D. Raistrick and R.A. Huggins, Mat. Res. Bull. *15*, 561 (1980)
4. N.A. Godshall, I.D. Raistrick and R.A. Huggins, J. Electrochem. Soc. *131*, 543 (1984)
5. A. A. Anani, S. Crouch-Baker and R. A. Huggins, J. Electrochem. Soc. *135*, 2103 (1988).

Chapter 13
Insertion Reaction Electrodes

13.1 Introduction

The topic of *insertion reaction electrodes* did not even appear in the discussions of batteries and related phenomena just a few years ago, but is a major feature of some of the most important modern battery systems today. Instead of reactions occurring on the surface of solid electrodes, as in traditional electrochemical systems, what happens *inside* the electrodes is now recognized to be of critical importance.

A few years after the surprise discovery that ions can move surprisingly fast inside certain solids, enabling their use as solid electrolytes, it was recognized that some ions can move rapidly into and out of some other (electrically conducting) materials. The first use of insertion reaction materials was for non-blocking electrodes to assist the investigation of the ionic conductivity of the (then) newly discovered ambient temperature solid electrolyte, sodium beta alumina [1–3]. Their very important use as charge-storing electrodes began to appear shortly thereafter.

This phenomenon is a key feature of the electrodes in many of the most important battery systems today, such as the lithium-ion cells. Specific examples will be discussed in later chapters.

Many examples are now known in which a mobile guest species can be *inserted into*, or *removed (extracted) from*, a stable host crystal structure. This phenomenon is an example of both *soft chemistry* and *selective equilibrium*, in which the mobile species can readily come to equilibrium, but this may not be true of the host, or of the overall composition. The mobile species can be atoms, ions, or molecules, and their concentration is typically determined by equilibrium with the thermodynamic conditions imposed on the surface of the solid phase.

In the simplest cases, there is little, if any, change in the structure of the host. There may be modest changes in the volume, related to bond distances, and possibly directions, but the general character of the host is preserved. In many cases, the *insertion* of guest species is reversible, and they can also be *extracted*, or *deleted*, returning the host material to its prior structure.

R.A. Huggins, *Energy Storage*,
DOI 10.1007/978-1-4419-1024-0_13, © Springer Science+Business Media, LLC 2010

The terms "*intercalation*" and "*de-intercalation*" are often used for reactions involving the insertion and extraction of guest species for the specific case of host materials that have *layer-type crystal structures*. On the other hand, "*insertion*" and "*extraction*" are more general terms. Reactions of this type are most likely to occur when the host has an open-framework or a layered type of crystal structure, so that there is space available for the presence of additional small ionic species. Since such reactions involve a change in the chemical composition of the host material, they can also be called *solid solution reactions*.

Insertion reactions are generally *topotactic*, with the guest species moving into, and residing in specific sites within the host lattice structure. These sites can often be thought of as *interstitial sites* in the host crystal lattice that are otherwise empty. The occurrence of a *topotactic* reaction implies some three-dimensional correspondence between the crystal structures of the parent and the product. On the other hand, the term *epitaxy* relates to a correspondence that is only two-dimensional, such as on a surface.

It has been known for a long time that large quantities of hydrogen can be inserted into, and extracted from palladium and some of its alloys. Palladium–silver alloys are commonly used as hydrogen-pass filters. Several types of materials with layer structures, including graphite and some clays, are also often used to remove contaminants from water by absorbing them into their crystal structures.

The most common examples of interest in connection with electrochemical phenomena involve the insertion or extraction of relatively small guest cationic species, such as H^+, Li^+ and Na^+. However, it will be shown later that there are materials in which anionic species can also be inserted into a host structure.

It should be remembered that electrostatic energy considerations dictate that only neutral species, or neutral combinations of species, can be added to, or deleted from, solids. Thus, the addition of cations requires the concurrent addition of electrons, and the extraction of cations is accompanied by either the insertion of holes or the deletion of electrons. Thus, this phenomenon almost always involves materials that have at least some modicum of electronic conductivity.

The term "*soft chemistry*", or *chimie douce* in French, where much of the early work took place [4, 5], is sometimes used to describe reactions or chemical changes that involve only the relatively mobile components of the crystal structure, whereas the balance of the structure remains relatively unchanged.

Such reactions are often highly reversible, but in some cases, the insertion or extraction of mobile atomic or ionic species causes irreversible changes in the structure of the host material, and the reversal of this process does not return the host to its prior structure. In extreme cases, the structure may be so distorted that it becomes *amorphous*. These matters will be discussed below.

Insertion reactions are much more prevalent at lower temperatures than at high temperatures. The mobility of the component species in the host structure generally increases rapidly with temperature. This allows much more significant changes in the overall structure to occur, leading to *reconstitution reactions*, with substantial structural changes, rather than only the motion of the more mobile species at elevated temperatures. Reconstitution reactions typically can be thought of as involving bond breakage, atomic reorganization, and the formation of new bonds.

13.2 Examples of the Insertion of Guest Species into Layer Structures

Many materials have crystal structures that can be characterized as being composed of rather stiff *covalently bonded slabs* containing several layers of atoms. These slabs are held together by relatively strong, for example, covalent bonds. But adjacent slabs are bound to each other by relatively weak *van der Waals* forces. The space between the tightly bound slabs is called the *gallery space*, and additional species preferably reside there. Depending upon the identity, size, and charge of any inserted species present, the inter-slab dimensions can be varied.

Materials with the CdI_2 structure represent a simple example. They have a basic stoichiometry MX_2, and can be viewed as consisting of close-packed layers of negatively charged X ions held together by strong covalent bonding to positive M cations. In this case, the cations are octahedrally coordinated by six X neighbors, and the stacking of the X layers is hexagonal, with alternate layers directly above and below each other. This is generally described as ABABAB stacking.

This structure can be depicted as shown schematically in Fig. 13.1. Examples of materials with this type of crystal structure are CdI_2, $Mg(OH)_2$, $Fe(OH)_2$, $Ni(OH)_2$, and TiS_2. Another simpler way to depict these structures is illustrated in Fig. 13.2 for the case of TiS_2.

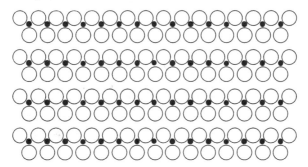

Fig. 13.1 Simple schematic model of a layer-type crystal structure with hexagonal ABABAB stacking. The empty areas between the covalently bonded slabs are called galleries

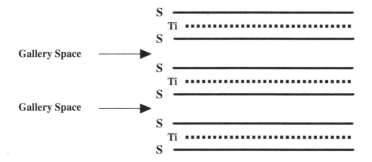

Fig. 13.2 Another type of model of a layer-type crystal structure. The example is TiS_2

13.3   Floating and Pillared Layer Structures

In many cases, the mobile species move into and through sites in the previously empty gallery space between slabs of host material that are held together only by relatively weak van der Waals forces. The slabs can then be described as floating, and the presence of guest species often results in a significant change in the inter-slab spacing.

However, in other cases, the slabs are already rigidly connected by *pillars*, which partially fill the galleries through which the mobile species move. The pillar species are typically immobile and thus are different from the mobile guest species. Because of the presence of the static pillars, the mobile species move through a two-dimensional network of interconnected tunnels, instead of through a sheet of available sites.

The presence of pillars acts to determine the spacing between the slabs of the host material, and thus the dimensions of the space through which mobile guest species can move. Examples of this kind will be discussed later. A simple schematic model of a pillared layer structure is shown in Fig. 13.3.

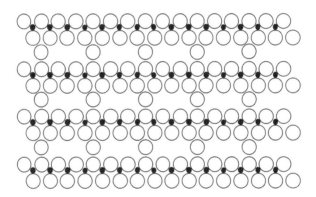

Fig. 13.3 Schematic model of pillared layer structure

13.4   More on Terminology Related to the Insertion of Species into Solids

Sheets – Single layers of atoms or ions. In the case of graphite, individual sheets are called *graphene* layers.
Stacks – Parallel sheets of chemically identical species.
Slabs or Blocks – Multi-layer structures tightly bound together, but separated from other structural features.
Example: covalently bonded MX_2 slabs such those shown above in the CdI_2 structure.

Galleries – The spaces between slabs in which the bonding is relatively weak, and in which guest species typically reside.

Pillars – Immobile species within the galleries that serve to support the adjacent slabs and to hold them together.

Tunnels – Connected interstitial space within the host structure in which the guest species can move and reside. Tunnels can be empty, partly occupied, or fully occupied by guest species.

Cavities – Empty space larger than the size of a single atom vacancy.

Windows – Locations within the host structure through which the guest species have to move in order to go from one site to another. Windows are typically defined by structural units of the host structure.

13.5   Types of Inserted Guest Species Configurations

There are several types of insertion reactions. In one case, the mobile guest species randomly occupy sites within all of the galleries, gradually filling them all up as the guest population increases. When this is the case, the variation of the electric potential with composition indicates a single-phase solid solution reaction, and there can be transient composition gradients within the gallery space.

If, however, the presence of the guest species causes a modification of the host structure, the insertion process will likely occur by the motion of an interface that separates the region into which the guest species have moved from the area in which there are no, or fewer, guest species. Thermodynamically, this has the characteristics of a polyphase reconstitution reaction, and occurs at a constant potential.

Alternatively, there can be two or more types of sites in the gallery space, with different energies, and the guest species can occupy an ordered array of sites, rather than all of them. When this is the case, changes in the overall concentration of mobile species require the translation of the interface separating the occupied regions from those that are not occupied, again characteristic of a constant-potential reconstitution reaction. These moving interfaces can remain planar, or they can develop geometrical roughness. Several possibilities are illustrated schematically in Figs. 13.4–13.6.

13.6   Sequential Insertion Reactions

If there are several different types of sites with different energies, insertion generally occurs on one type of site first, followed by the occupation of the other type of site. Figure 13.7 shows the potential as a function of composition during the insertion of lithium into $NiPS_3$, in which there are two types of sites available. They are occupied in sequence, with random occupation in both cases.

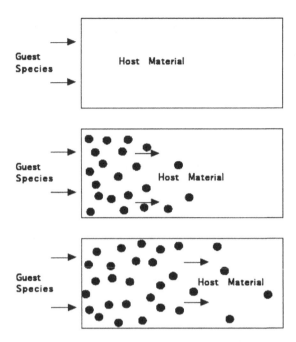

Fig. 13.4 Random diffusion of guest species into gallery space

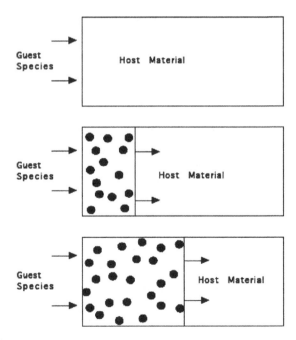

Fig. 13.5 Motion of two-phase interface when guest species is not ordered upon possible sites

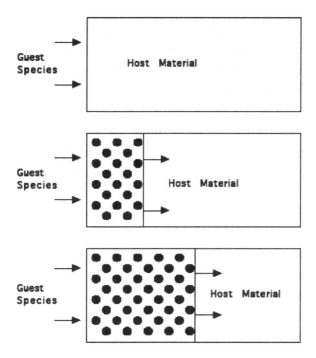

Fig. 13.6 Motion of two-phase interface when guest species is ordered upon possible sites in the gallery space

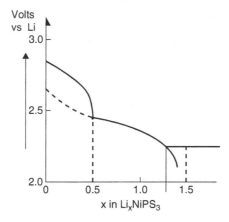

Fig. 13.7 Coulometric titration curve related to the insertion of lithium into $NiPS_3$. There is random filling of the first two types of sites. A reconstitution reaction occurs above about 1.4 Li

Another example in which there are also different types of sites available for the insertion of Li ions involves the $K_xV_2O_5$ structure. The host crystal structure illustrating the several different types of sites for guest ions is shown schematically in Fig. 13.8 [6].

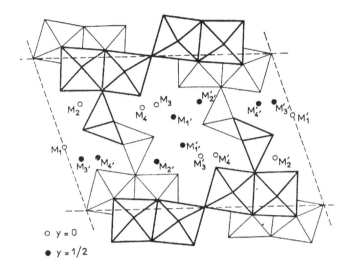

Fig. 13.8 (010) Projection of the $K_xV_2O_5$ structure, showing the different types of sites for the guest species. After [6]

The experimentally measured coulometric titration curve for the insertion of Li ions into a member of this group of materials is shown in Fig. 13.9 [7]. It shows that the reaction involves three sequential steps. Up to about 0.4 Li can be incorporated into the first set of sites randomly. This is followed by the insertion of another 0.4 Li into another set of sites in an ordered arrangement. This means that there are two different lithium arrangements, with a moving interface between them. Thus there are two phases present, so this corresponds to a reconstitution reaction. This is then followed by another reconstitution reaction, the insertion of about one additional Li into another ordered structure.

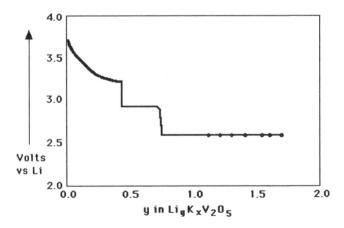

Fig. 13.9 Coulometric titration curve for the insertion of Li into $K_{0.27}V_2O_5$. After [6]

A different type of ordered reaction involves the selective occupation of particular galleries, and not others, in a material with a layered crystal structure. This phenomenon is described as "staging". If alternate galleries are occupied and intervening ones are not, the material is described as having a "second stage" structure. If every third gallery is occupied, the structure is "third stage" and so forth. A simple model depicting staging is shown in Fig. 13.10.

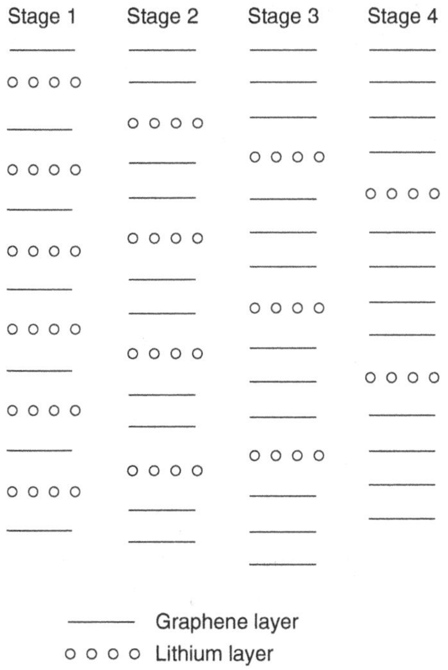

Fig. 13.10 Simple model depicting staging when lithium is inserted in the galleries of graphite

13.7   Co-insertion of Solvent Species

In some cases it is found that species from the electrolyte can also move into the gallery space. This tends to be the case when the electrolyte solvent molecules are relatively small, so that they can enter without causing a major disruption of the host structure. This is found to occur in some organic solvent systems, and also some aqueous electrolyte systems where the electroactive ion is surrounded by a hydration sheath. This is a matter of major concern in the case of negative electrodes in lithium systems, and will be discussed at much greater length in a later chapter.

13.8   Insertion into Materials with Parallel Linear Tunnels

The existence of staging indicates that, at least in some materials, the presence of inserted species in one part of the structure is "seen" in other parts of the structure. An interesting example of this involves the presence of mobile guest species in the material Hollandite, which has a crystal structure with parallel linear tunnels, rather than slabs.

A drawing of this structure is shown in Fig. 13.11. At low temperatures, the interstitial ions within the tunnels are in an ordered arrangement upon the available sites. In addition, there is coordination between the arrangement in one tunnel with that of other nearby tunnels. Thus, there is three-dimensional ordering of the guest species.

As the temperature is raised somewhat, increased thermal energy causes the ordered interaction between the mobile ion distributions in nearby tunnels to relax, although the ordering within tunnels is maintained.

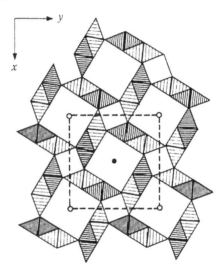

Fig. 13.11 Hollandite structure. Viewed along the c-axis

At even higher temperatures the in-tunnel ordering breaks down, so that the species are distributed randomly inside the tunnels, as well. The influence of temperature is illustrated schematically in Fig. 13.12.

13.9   Changes in the Host Structure Induced by Guest Insertion or Extraction

It was mentioned earlier that the insertion or extraction of mobile guest species can cause changes in the host structure. There are several types of such structural changes that can occur. They will be briefly discussed in the next sections.

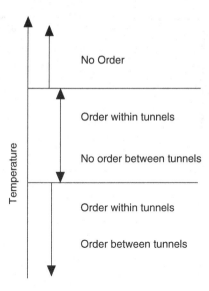

Fig. 13.12 Influence of temperature upon various types of order in structure with parallel tunnels

13.9.1 Conversion of the Host Structure from Crystalline to Amorphous

There are a number of examples in which an initially crystalline material becomes amorphous as the result of the insertion of guest species, and the corresponding mechanical strains in the lattice. This often occurs gradually as the insertion/extraction reaction is repeated; for example, upon electrochemical cycling. One example of this, the $Li_xV_6O_{13}$ binary system, is shown in Figs. 13.13 and 13.14 [8].

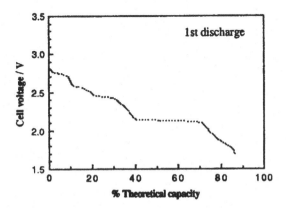

Fig. 13.13 Discharge curve observed during the initial insertion of lithium into a material that was initially V_6O_{13}. After [8]

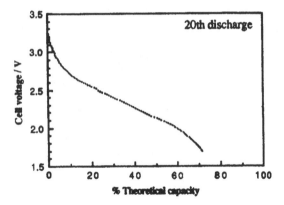

Fig. 13.14 Discharge curve observed during the 20th insertion of lithium into a material that was initially V_6O_{13}. After [8]

In this case, the shape of the potential curve during the first insertion of lithium into crystalline V_6O_{13} shows that a sequence of reconstitution reactions take place that give rise to a series of different structures, and a discharge curve with well-defined features.

After a number of cycles, however, the discharge curve changes, with a simple monotonous decrease in potential, indicative of a single-phase insertion reaction. X-ray diffraction experiments confirmed that the structure of the material had become amorphous.

Another example of changes resulting from an insertion reaction is shown in Fig. 13.15. In this case, lithium was inserted into a material that was initially V_2O_5 [9]. The result is similar to the V_6O_{13} case, with clear evidence of the formation of a

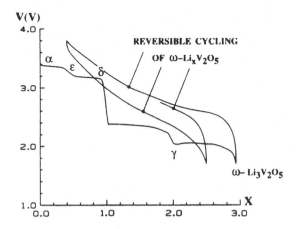

Fig. 13.15 The variation of the potential as lithium is added to V_2O_5. When the composition reached $Li_3V_2O_5$ an amorphous phase was formed. After [9]

series of different phases as lithium was added. It was found that the insertion reaction was reversible, forming the ε and δ structures, if only up to about 1 Li was inserted into α V_2O_5. The addition of more lithium resulted in the formation of different structural modifications, called the γ and ω structures, which have nominal compositions of $Li_2V_2O_5$ and $Li_3V_2O_5$, respectively. These two reactions are not reversible, however.

When lithium was extracted from the ω phase, its charge–discharge curve became very different, exhibiting the characteristics of a single phase with a wide range of solid solution. The amount of lithium that could be extracted from this phase was quite large, down to a composition of about $Li_{0.4}V_2O_5$. Upon the re-insertion of lithium, the discharge curve maintained the same general form, indicating that a reversible amorphous structure had been formed during the first insertion process.

13.9.2   Dependence of the Product upon the Potential

It has been found that displacement reactions can occur in a number of materials containing silicon when they are reacted with lithium to a low potential (high lithium activity). An irreversible reaction occurs that results in the formation of fine particles of amorphous silicon in an inert matrix of a residual phase that is related to the precursor material [10, 11]. Upon cycling, the amorphous Li–Si structure shows both good capacity and reversibility.

However, it has also been shown [12] that if further lithium is inserted, going to a potential below 50 mV, a crystalline Li–Si phase forms that replaces of the amorphous one.

Because of the light weight of silicon, the large amount of lithium that can react with it, and the attractive potential range, it appears as though silicon or its alloys may play an important role as a negative electrode reactant in lithium batteries in the future.

13.9.3   Changes upon the Initial Extraction of the Mobile Species

Similar phenomena can also occur during the initial extraction of a mobile species that is already present in a solid. This is shown in Fig. 13.16 for the case of a material with an initial composition of about $Li_{0.6}V_2O_4$ [13]. It can be seen that the potential starts between 3.0 and 3.5 V vs. pure Li, as is generally found for materials that have come into equilibrium with air. The reason for this will be discussed later.

The initial lithium could be essentially completely deleted from the structure, causing the potential to rise to over 4 V vs. pure lithium. When lithium was subsequently re-inserted, the discharge curve had a quite different shape, indicating the presence of a reconstitution reaction resulting in the formation of an intermediate phase.

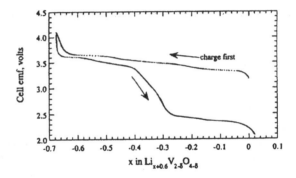

Fig. 13.16 Initial charging and discharge curves of a material with a composition of $Li_{0.6}V_2O_4$. After [13]

13.10 The Variation of the Potential with Composition in Insertion Reaction Electrodes

13.10.1 Introduction

The externally measured electrical potential of an electrode is determined by the electrochemical potential of the electrons within it, η_{e^-}. This is often called the *Fermi level*, E_F. Since potentials do not have absolute values, they are always measured as differences. The voltage of an electrochemical cell is the electrically measured difference between the Fermi levels of the two electrodes.

$$\Delta E = \Delta \eta_{e^-} \tag{13.1}$$

As has been demonstrated many times in this text already, the measured potential of an electrode often varies with its composition, for example, as guests species are added to, or deleted from, a host material. The relevant thermodynamic parameter is the chemical potential of the electrically neutral electroactive species. If this species exists as a cation M^+ within the electrode, the important parameter is the chemical potential of neutral M, μ_M. This is related to the electrochemical potentials of the ions and the electrons by

$$\mu_M = \eta_{M^+} + \eta_{e^-} \tag{13.2}$$

Under open circuit conditions, there is no flux of ions through the cell. Since the driving force for the ionic flux through the electrolyte is the gradient in the electrochemical potential of the ions, for open circuit

$$\frac{d\eta_{M^+}}{dx} = \Delta \eta_{M^+} = 0 \tag{13.3}$$

Therefore, the measured voltage across the cell is simply related to the difference in the chemical potential of the neutral electroactive species in the two electrodes by

$$\Delta E = \Delta \eta_{e^-} = \frac{-1}{z_{M^+} q} \Delta \mu_M \tag{13.4}$$

The common convention is to express both the difference in the electrical potential (the voltage) and the difference in chemical potential as the values in the right hand (positive) electrode less those in the left hand electrode.

A general approach that is often used is to understand the potentials of electrons is based upon the *electron energy band model*. The critical features are the variation of the density of available states with the energy of the electrons, and the filling of those states up to a maximum value that is determined by the chemical composition. The energy at this maximum value is the *Fermi level*.

In the case of metals, the variation of the potential of the available states is a continuous function of the composition, and the *free electron theory* can be used to express this relationship.

In non-metals, semiconductors, and insulators, the density of states is not a continuous function of the chemical composition. Instead, there are potential ranges in which there are no available states that can be occupied by electrons. In the case of the simple semiconductors such as silicon or gallium arsenide, one speaks of a *valence band*, in which the states are generally fully occupied, an *energy gap* within which there are no available states, and a *conduction band* with normally empty states. The concentration of electrons in these bands varies with the temperature due to *thermal excitation*, and can also be modified by the presence of *aliovalent species*, generally called *dopants*. *Optical excitation* has an effect similar to that of *thermal excitation*.

In a number of materials, particularly those in which the electronic conductivity is relatively low, it is convenient to think of the relation between the occupation of energy states and a change in the formal valence, or charge, upon particular species within the host structure. For example, the addition of an extra electron could result in a change of the formal charge of W^{6+} to W^{5+}, Ti^{4+} to Ti^{3+}, Mn^{4+} to Mn^{3+}, or Fe^{3+} to Fe^{2+} in a transition metal oxide. Such phenomena are called *redox* reactions.

These different cases will be discussed below, and it will be seen that there is a clear relationship between them.

13.10.2 The Variation of the Electrical Potential with Composition in Simple Metallic Solid Solutions

There are a number of metals in which the insertion of mobile guest species can occur. As mentioned already, this can be described as a solid solution of the guest species in the host crystal structure. The important quantity controlling the potential is the variation of the chemical potential of the neutral guest species as a function of its concentration. This can be formally divided into two components, the influence of the change in the electron concentration in the host material, and the effect due to a change in the concentration of the ionic guest species, M^+.

In the case of a random solid solution in a material with a high electronic conductivity, the two major contributions are the contribution from the composition dependence of the Fermi level of the degenerate electron gas that is characteristic of such mixed conductors, and that due to the composition dependence of the enthalpy and configurational entropy of the guest ions in the host crystal lattice [14].

13.10.3 Configurational Entropy of the Guest Ions

If the guest ions are highly mobile and can move rapidly through the host crystal structure, it may be assumed that all of the identical crystallographic sites are equally accessible. There will be a contribution to the total free energy due to the *configurational entropy* S_c related to the random distribution of the guest atoms over the available sites. This is given by

$$ S_c = -k \left(\ln \frac{x}{x_0 - x} \right) \tag{13.5} $$

where x is the concentration of guest ions, x_0 the concentration of identical available sites, and k is Boltzmann's constant. This configurational entropy contribution to the potential is the product of the absolute temperature and the entropy. This is plotted in Fig. 13.17.

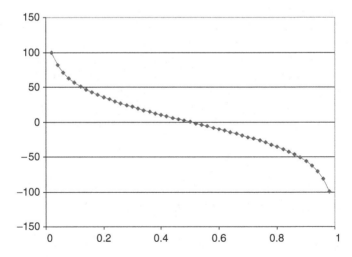

Fig. 13.17 Contribution to the potential due to the configurational entropy of a random distribution of the guest ions upon the available identical positions in a host crystal structure. The values on the abscissa are the fractional site occupation, and those on the ordinate are mV

This model assumes that there is no appreciable interaction between nearby guest species, and that there is only one type of site available for occupation.

13.10.4   The Concentration Dependence of the Chemical Potential of the Electrons in a Metallic Solid Solution

In a simple metal, the electron concentration is typically sufficiently high so that at normal temperatures, the electron gas can be considered to be completely degenerate. Under those conditions, the electrochemical potential of the electrons can be approximated by the energy of the Fermi level E_F.

In the free electron model, this can be expressed as

$$E_F = \frac{h^2}{8m\pi^2}\left(\frac{3\pi^2 N_A}{V_m}\right)^{2/3}N^{2/3} \tag{13.6}$$

where m is the electron mass, N_A is Avogadro's number, V_m is the molar volume, and E_F is calculated from the bottom of the conduction band.

Thus, the electronic contribution to the total chemical potential is proportional to the 2/3 power of the guest species concentration if the simple free electron model is valid. More generally, however, the electron mass is replaced by an effective mass m^*. This takes into account other effects, such as the non-parabolicity of the conduction band.

If one can assume that the electrons can be treated as fully degenerate, the chemical potential of the electrons is directly related to the Fermi level, E_F, which can be written as

$$E_F = (\text{Constant})\left(\frac{x^{2/3}}{m^*}\right) \tag{13.7}$$

where x is the guest ion concentration and m^* is the effective mass of the electrons.

13.10.5   Sum of the Effect of These Two Components upon the Electrical Potential of a Metallic Solid Solution

Thus, the composition dependence of the electrode potential in a metallic solid solution can be written as the sum of the influence of composition upon the configurational entropy of the guest ions, and the composition dependence of the Fermi level of the electrons. This can be simply expressed as

$$E = (\text{Constant})\left(\frac{x^{2/3}}{m^*}\right) - \left(\frac{RT}{zF}\right)\ln\left(\frac{x}{1-x}\right) \tag{13.8}$$

This relationship is illustrated in Fig. 13.18 for several values of the electron effective mass. It also shows the influence of the value of the electron effective mass upon the general slope of the curve. From (13.8), it can be seen that smaller effective masses make the first term larger, and thus result in the potential being more composition-dependent.

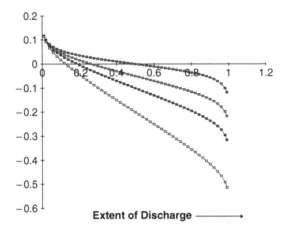

Fig. 13.18 Calculated influence of the value of the electronic effective mass upon the composition-dependence of the potential in an insertion reaction in a simple metal

An example showing experimental data [14] that illustrate the general features of this model is shown in Fig. 13.19. Although the host material in this case was an oxide, it is an example of a "tungsten bronze". In this family of oxides, the electron energy spectrum approximates that of a free electron metal.

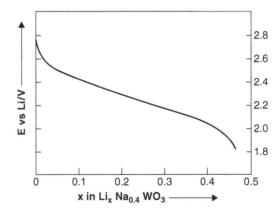

Fig. 13.19 Variation of the potential as a function of lithium concentration in $Li_xNa_{0.4}WO_3$. After [14]

It should be remembered that although the band diagrams commonly used in discussing semiconductors are plotted with greater energy values higher, the scale of the electrical potential is in the opposite direction. This is because the energy of a charged species is the product of its charge and the electrical potential, and the charge on electrons is negative.

13.10.6   The Composition-Dependence of the Potential in the Case of Insertion Reactions That Involve a Two-Phase Reconstitution Reaction

The earlier discussion of the influence of the Gibbs Phase Rule upon the compositional variation of the potentials in electrodes pointed out that when there are two phases present in a two-component system, the potential will have a fixed, or constant, value, independent of the composition. This will also be the case for materials that act as pseudo-binaries, regardless of how many different species are actually present. A number of insertion reaction materials are of this type, with one relatively mobile species inside a relatively stable host structure. If, in the time span of interest, the host structure does not undergo any changes it can be considered to be a single component thermodynamically. This is what is found in a number of materials in which the host structure is a transition metal oxide.

One example in which the potential is composition-independent involves the insertion and extraction of lithium in materials with the composition $Li_4Ti_5O_{12}$, which has a defect spinel structure [15]. The general composition of spinel structure materials can be described as AB_2O_4, where the A species reside on tetrahedral sites, and the B species on octahedral sites within a close-packed face-centered cubic oxygen lattice. If one assumes this general stoichiometry, one of the four lithium ions would share the octahedral sites with the titanium ions, and the other ones would reside on tetrahedral sites. Thus, the composition can be written as $Li_3[LiTi_5]O_{12}$, or alternatively, $Li[Li_{1/3}Ti_{5/3}]O_4$.

It has been found that an additional lithium ion can react with this material, and this can be written as

$$Li + Li[Li_{1/3}Ti_{5/3}]O_4 = Li_2[Li_{1/3}Ti_{5/3}]O_4 \tag{13.9}$$

X-Ray diffraction data have indicated that the lithium ions now occupy octahedral sites instead of tetrahedral sites. Since there are only as many octahedral sites available as oxide ions in this structure, they must now be all filled. This is likely why the capacity of this electrode material is limited at this composition.

These materials were prepared in air, and were white in color. As with all materials prepared in air, their potential was initially near 3 V versus lithium. When lithium was added by transfer from the negative electrode, lithium in carbon, the potential went rapidly down to 1.55 V, and remained there until the reaction was complete. Thus, this insertion reaction has the characteristics of a moving-interface reconstitution reaction.

Upon deletion of the inserted lithium, the potential retraced the discharge curve closely, with very little hysteresis. This is illustrated in Fig. 13.20 [15]. Because of the small volume change, negligible hysteresis, and rapid kinetics, this material acts as a very attractive electrode in lithium cells. The one disadvantage is that its potential is unfortunately about half way between the negative and positive electrode potentials in most lithium batteries.

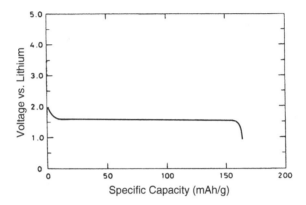

Fig. 13.20 Discharge curve of $Li_4Ti_5O_{12}$ cell. After [15]

As will be discussed later, hysteresis, which leads to a difference in the composition-dependence of the potential when charging and discharging, is often related to mechanical strain energy caused by dislocation generation and motion, as a consequence of volume changes that occur due to the insertion and extraction of the guest ions.

Another example of an insertion-driven reconstitution reaction is the reaction of lithium with $FePO_4$ to form $LiFePO_4$, which also happens readily at ambient temperature. This also has a very flat reaction potential, as shown in Fig. 13.21 [16].

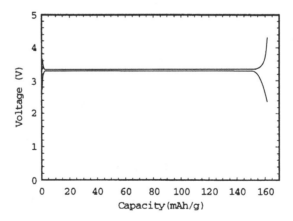

Fig. 13.21 Charge and discharge curves for the reaction of lithium with $FePO_4$ to form $LiFePO_4$. After [16]

In this case the material is prepared (in air) as $LiFePO_4$, and the initial reaction within the cell involves charging, that is, deleting lithium from it. This lithium goes across the electrochemical cell and into the carbon material in the negative electrode. The reaction that occurs at the operating potential during the initial charge can be simply written as

$$LiFePO_4 = Li + FePO_4 \qquad (13.10)$$

Upon discharge of the cell, the reaction goes in the opposite direction, of course.

This material is now one of the most important positive electrode reactants in lithium batteries, and will be discussed further in a later chapter.

13.11   Final Comments

This chapter has been intended to be only a general introduction to the scope of insertion reactions in electrode materials. This is a very important topic and will be addressed further in the discussions of specific materials in later chapters.

References

1. M.S. Whittingham and R.A. Huggins, J. Electrochem. Soc. *118*, 1 (1971)
2. M.S. Whittingham and R.A. Huggins, J. Chem. Phys. *54*, 414 (1971)
3. M.S. Whittingham and R.A. Huggins, in *Solid State Chemistry,* ed. by R.S. Roth and S.J. Schneider, Nat. Bur. of Stand. Special Publication 364 (1972), p. 139
4. J. Livage, *Le Monde*, October 26, 1977
5. J. Rouxel, Materials Science Forum (1994), p. 152
6. J. Goodenough, in Annual Review of Matls Sci., Vol. 1, ed. by R.A. Huggins (1970), p. 101
7. I.D. Raistrick and R.A. Huggins, Mat. Res. Bull. *18*, 337 (1983)
8. W.J. Macklin, R.J. Neat and S.S. Sandhu, Electrochim. Acta *37*, 1715 (1992)
9. C. Delmas, H. Cognac-Auradou, J.M. Cocciantelli, M. Menetrier and J.P. Doumerc, Solid State Ionics *69*, 257 (1994)
10. A. Netz, R.A. Huggins and W. Weppner, J. Power Sources *119–121*, 95 (2003)
11. A. Netz and R.A. Huggins, Solid State Ionics *175*, 215 (2004)
12. M.N. Obrovac and L. Christensen, Electrochem. Solid State Lett. *7*, A93 (2004)
13. T.A. Chirayil, P.Y. Zavalij and M.S. Whittingham, J. Electrochem. Soc. *143*, L193 (1996)
14. I.D. Raistrick, A.J. Mark and R.A. Huggins, Solid State Ionics *5*, 351 (1981)
15. T. Ohzuku, A. Ueda and N. Yamamoto, J. Electrochem. Soc. *142*, 1431 (1995)
16. A. Yamada, S.C. Chung and K. Hinokuma, J. Electrochem. Soc. *148*, A224 (2001)

Chapter 14
Electrode Reactions that Deviate from Complete Equilibrium

14.1 Introduction

The example that was discussed earlier, the reaction of lithium with iodine to form LiI, dealt with elements and thermodynamically stable phases. By knowing a simple parameter, the standard Gibbs free energy of formation of the reaction product, the cell voltage under equilibrium and near-equilibrium conditions can be calculated for this reaction. If the cell operates under a fixed pressure of iodine at the positive electrode and at a stable temperature, the Gibbs phase rule indicates that the number of the residual degrees of freedom F in both the negative and positive electrodes is zero. Thus, the voltage is independent of the extent of the cell reaction.

This is a case in which the reaction involves species that are *absolutely stable*. The description of a phase as absolutely stable means that it is in the thermo-dynamic state, e.g., crystal structure, with the lowest possible value of the Gibbs free energy for its chemical composition.

14.2 Stable and Metastable Equilibrium

On the other hand, there could be several versions of a phase with different structures that might be stable in the sense that they have lower values of the Gibbs free energy than would be the case with minor changes. Such a situation, in which a phase is stable against small perturbations, is described by the term *metastable*. On the other hand, it may be less stable than the *absolutely stable* modification. This can be illustrated schematically by the use of a simple mechanical model, as is illustrated in Fig. 14.1.

This situation can also be described in terms of the changes in the potential energy of a simple block. If the block sits on its end, it is in a metastable state, and if it is tipped a small amount, its potential energy will be increased, but it will tend to

R.A. Huggins, *Energy Storage*,
DOI 10.1007/978-1-4419-1024-0_14, © Springer Science+Business Media, LLC 2010

revert back to its initial metastable condition. But a larger perturbation will get it over this potential energy hump so that it will tip over and land in the flat position, the absolutely stable state.

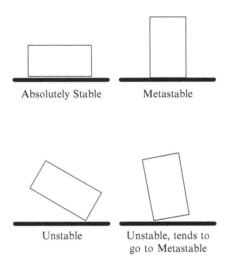

Fig. 14.1 Simple mechanical model illustrating metastable and absolutely stable states

This situation can also be illustrated by the use of a reaction coordinate diagram of the type often used in discussions of chemical reaction kinetics, as shown in Fig. 14.2.

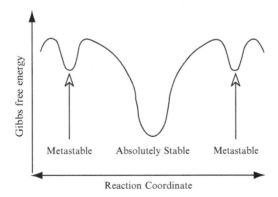

Fig. 14.2 Reaction coordinate representation of a system with metastable and absolutely stable states

This discussion does not only apply to single phases, for it is possible to have a situation in which a material has a microstructure that consists of a metastable single phase, whereas the absolutely stable situation involves the presence of two, or perhaps more, phases. In order for the system to go from the metastable single-phase situation to the more stable *polyphase* structure, it is necessary to *nucleate* the additional phase or phases as well as to change the composition of the initial metastable phase. This may be kinetically very difficult.

In the case of the Li/I system, where there is only one realistic structure for the reactant and product phases, only the absolutely stable situation has to be considered.

However, in other materials systems, the situation is often different at lower temperatures from that at high temperatures, where absolutely stable phases are generally present. As will be discussed later, metastable phases and metastable crystal structures often play significant roles at ambient temperatures.

14.3   Selective Equilibrium

There is also the possibility that a material may attain equilibrium in some respects, but not in others. Some of the most interesting and important ambient temperature materials fall into this category.

A number of the reactants in ambient temperature battery systems that undergo insertion reactions have crystal structures that can be described as a composite consisting of a highly mobile ionic species within a relatively stable host structure.

Such structures are sometimes described as having two different *sub-lattices*, one of which has a high degree of mobility, and the other is highly stable, for its structural components are rigidly bound. As discussed earlier, the guest species with high mobilities are typically rather small and move about through interstitial tunnels in the surrounding rigid host structure. The species in the mobile sub-lattice can readily come to equilibrium with the thermodynamic forces upon them, whereas the more tightly bound parts of the host structure cannot. The term soft chemistry, or chimie douce in French, was introduced in Chap. 13 for this situation.

The term *selective equilibrium* can be used for this situation. Under these conditions, the stable part of the crystal structure can be treated as a single component when considering the applicability of the Gibbs phase rule. An example that was discussed earlier is the phase Li_xTiS_2. The structure of this material can be thought of as consisting of rigid planar slabs of covalently bonded TiS_2, with mobile lithium ions in the space between them. The lithium species readily attain equilibrium with the external environment at ambient temperatures, whereas the TiS_2 part of the structure is relatively inert and thus can be considered to be a single component. Thus, at a fixed temperature and total pressure, the number of residual degrees of freedom is 1. This means that the value of one additional thermodynamic parameter will determine all of the intensive variables. As an example, a change in the electrical potential is related to a change in the equilibrium amount of lithium in the structure, the value of x in Li_xTiS_2.

14.4   Formation of Amorphous Vs. Crystalline Structures

An amorphous structure can result when a phase is formed under conditions in which complete equilibrium and the expected crystalline structure cannot be attained. Although they may have some localized ordered arrangements, amorphous structures do not have regular long-range arrangements of their constituent atoms or ions. Amorphous structures are always less stable than the crystalline structure with the same composition. Thus they have less negative values of the Gibbs free energy of formation than their crystalline cousins.

If the phase LiM can be electrochemically synthesized by the reaction of lithium with species M, a type of reconstitution reaction, there will be a corresponding constant voltage two-phase plateau related to that reaction. The magnitude of the plateau voltage is determined by the Gibbs free energy of the product phase, as described earlier. Because of its less negative Gibbs free energy of formation, the potential of the plateau related to the formation of an amorphous LiM phase must always be lower than that of the corresponding crystalline version of LiM. This is illustrated schematically in Fig. 14.3.

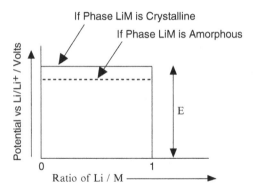

Fig. 14.3 Schematic drawing of the voltage of galvanic cell as a function of overall composition for a simple formation reaction Li + M = LiM for two cases, one in which the LiM product is crystalline, and the other in which it is amorphous

This has interesting consequences for the case in which two intermediate phases can be formed. As an example, assume that lithium can react with material M to form two phases in sequence, LiM and Li_2M. The reaction for the formation of the first phase, LiM is

$$Li + M = LiM \qquad (14.1)$$

and if the phase LiM has a very narrow range of composition, the equilibrium titration curve, a plot of potential E versus composition, will look like that shown in Fig. 14.3.

The plateau voltage is given by:

$$E = -\Delta G_f{}^0(\text{LiM})/F \qquad (14.2)$$

If additional lithium can react with LiM to form the phase Li_2M, there will be an additional voltage plateau whose potential is determined by the reaction

$$\text{Li} + \text{LiM} = \text{Li}_2\text{M} \qquad (14.3)$$

This is shown schematically in Fig. 14.4.

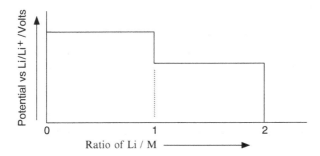

Fig. 14.4 Schematic titration curve for a sequence of two reactions of Li with M, first forming LiM, and then forming Li_2M

The voltage of the second plateau is lower than that of the first, and is given by:

$$E = -[\Delta G_f{}^0(\text{Li}_2\text{M}) - \Delta G_f{}^0(\text{LiM})]/F \qquad (14.4)$$

But what if the first phase, LiM, is amorphous, rather than crystalline? As mentioned above, this means that Gibbs free energy of formation of that phase is smaller and the voltage of the first plateau is reduced.

The total Gibbs free energy of the two reactions is determined, however, by the Gibbs free energy of formation of the final phase, Li_2M. This is not changed by the formation of the intermediate phase LiM. The total area under the curve is thus a constant. The interesting result is that if the voltage of the first plateau is reduced, the voltage of the second one must be correspondingly increased. This can be depicted as in Fig. 14.5.

Thus, the reduced stability of the intermediate phase reduces the magnitude of the step in the titration curve. Therefore, the overall behavior approaches what it would be if the intermediate phase did not form at all and there would only be one reaction, the direct formation of phase Li_2M.

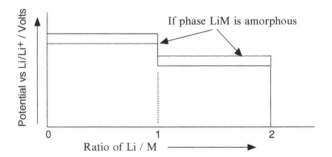

Fig. 14.5 Change in the schematic titration curve if the first product, LiM, is amorphous. The voltage of the second plateau must be higher to compensate for the reduced voltage of the first plateau

There are a number of cases in which it has been found that the insertion of guest species into host crystal structures can cause them to become amorphous. This was discussed in Chap. 13.

14.5 Deviations from Equilibrium for Kinetic Reasons

The observed potentials and capacities of electrodes are often displaced from those that would be expected from equilibrium thermodynamic considerations because of kinetic limitations. There may not be sufficient time to attain compositional and/or

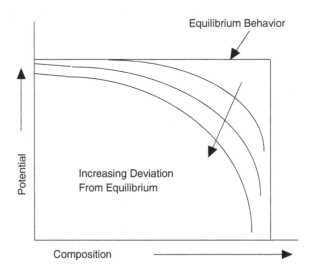

Fig. 14.6 Schematic representation of the influence of kinetic limitations upon both the potential and capacity of an electrode reaction

structural changes that should, in principle, occur. This is more likely to occur at lower temperatures, and under higher current conditions.

The influence of increasing deviations from equilibrium conditions upon the behavior of a simple reconstitution reaction is shown schematically in Fig. 14.6. It is seen that both the potential and the apparent capacity can deviate significantly from equilibrium values.

The kinetics of electrode reactions, and methods that can be used to evaluate them, will be discussed in subsequent chapters.

Chapter 15
Lead-Acid Batteries

15.1 Introduction

Over many years, the most common use of the word "battery" was in connection with the rechargeable energy source that was used to start automobiles. These were almost always what are generally called *Pb-acid batteries*, and were often a source of aggravation. A considerable amount of progress has been made in recent years, so that the SLI (starting-lighting-ignition) batteries now used in autos are actually quite reliable, assuming that they are not abused. Different types of Pb-acid batteries are used for a number of other applications, both mobile and stationary, ranging from quite small to very large, and the greatest fraction of the total battery market world-wide is now based upon this technology.

There are several reasons for the widespread use of lead-acid batteries, such as their relatively low cost, ease of manufacture, and favorable electrochemical characteristics, such as rapid kinetics and good cycle life under controlled conditions.

Pb-acid cells were first introduced by G. Planté in 1860 [1], who constructed them using coiled lead strips separated by linen cloth and immersed in sulfuric acid. By initially passing a dc current between the two lead strips, an oxide grew on the one on the positive side, forming a layer of lead dioxide. This caused the development of a voltage between them, and it was soon found that charge could be passed reversibly through this configuration, so that it could act to store electrical energy.

Significant improvements have been made over the years. One of the most important was the invention of the pasted plate electrode by C. Fauré in 1881 [2]. This involved the replacement of solid metal negative electrodes by a paste of fine particles held in a lead or lead alloy grid. By doing this, the reaction surface area is greatly increased.

Another significant improvement has been the development of sealed cells during the last several decades. This is sometimes called *valve-regulated lead-acid* technology. These matters will be discussed in the following sections.

R.A. Huggins, *Energy Storage*,
DOI 10.1007/978-1-4419-1024-0_15, © Springer Science+Business Media, LLC 2010

There are two general types of applications that are commonly considered for lead-acid cells, and they impose quite different requirements. One type involves keeping the cell essentially fully charged so that it maintains a constant output voltage. This is sometimes called *float charging*. Such cells are generally stationary, and are expected to have high reliability, long life, a low self-discharge rate, and a good cycling efficiency with low loss under cycling and overcharge. They are often used in telecommunication and large computer systems, railroad signaling systems, and to supply standby power as uninterruptible power sources (UPS). They are generally not optimized for energy density or specific energy, but are attractive because of their low cost.

The other general type is targeted toward applications that may involve deep discharging, such as in load leveling systems and traction applications. In these cases, the specific energy and/or specific power can be very important, perhaps as the cost or cycle efficiency and lifetime. Periodic, rather than continuous, charging is more common.

Although hydride/"nickel" and lithium-ion cells are generally used in smaller portable applications, some sealed Pb-acid cells are now used for such applications where the lowest cost is particularly important.

15.2   Basic Chemistry of the Pb-Acid System

The lead-acid cell is often described as having a negative electrode of finely divided elemental lead, and a positive electrode of powdered lead dioxide in an aqueous electrolyte. If this were strictly true and there were no other important species present, the cell reaction would simply involve the formation of lead dioxide from lead and oxygen.

$$Pb + O_2 = PbO_2 \tag{15.1}$$

The open circuit voltage of such a cell would be determined from the Gibbs free energy of formation of PbO_2, $\Delta G_f^0(PbO_2)$, by

$$E = -\left(\frac{\Delta G_f^0(PbO_2)}{zF}\right) \tag{15.2}$$

in which z is 4, the number of charges involved in reaction (15.1), and F is the Faraday constant. The value of $\Delta G_f^0(PbO_2)$ can be calculated to be -215.4 kJ/mol [3]. Thus, the cell voltage would be 0.56 V. However, this is far from what is actually observed, so that the reaction that determines the cell voltage must be quite different from (15.1).

Instead of (15.1), the overall chemical process involved in the discharge of Pb-acid cells is generally described [4], in accordance with the *double sulfate theory* [5–8] as

$$Pb + PbO_2 + 2H_2SO_4 = 2PbSO_4 + 2H_2O \qquad (15.3)$$

This relation goes to the right hand side during discharge and toward the left side when the cell is recharged. This has been demonstrated by observations of morphological changes in both the negative [9, 10] and positive electrodes [11]. The mechanism whereby this takes place, and the thermodynamic basis for the cell voltage will be discussed later.

Using the values of the standard Gibbs free energy of formation, ΔG_f^0, of the phases in this reaction, it has been shown that the equilibrium voltage of this reaction under standard conditions is 2.041 [12].

15.2.1 Calculation of the MTSE

It is interesting to calculate the maximum theoretical specific energy of Pb-acid cells. As discussed in Chap. 9, this can be expressed as

$$MTSE = 26,805 \left(\frac{xE}{W} \right) \qquad (15.4)$$

in which x is the number of elementary charges, E the average cell voltage, and W the sum of the atomic weights of either the reactants or the products. In this case, x is 2, E is 2.05 V, and W is 642.52 g. Inserting these values, the maximum theoretical specific energy, calculated from these reactions, is 171 Wh/kg. This is fallacious, however, for it is necessary to have additional water present for the cell to operate. This increases the weight, and thus reduces the specific energy. But in addition, other passive components add significant amounts of weight, as is always the case in practical batteries. Values of the practical specific energy of lead-acid batteries are currently in the range of 25–40 Wh/kg. Higher values are typical for those optimized for energy, and lower values for those designed to provide more power.

15.2.2 Variation of the Cell Voltage with the State of Charge

From (15.3), it is obvious that the electrolyte changes, the amount of sulfuric acid decreases, and the amount of water present increases, as the cell becomes discharged. This causes a change in the electrolyte density. It is about 40% by weight H_2SO_4 at full charge, but only 16% when the cell is fully discharged. The corresponding values of equilibrium open circuit voltage are 2.15 V and 1.98 V at 25°C. These density and voltage variations are illustrated in Fig. 15.1. Whereas it may take some time to reach the equilibrium voltage because of temporary structural and compositional inhomogenieties in the electrodes, the electrolyte density can be readily measured, and is often used to indicate the state of charge of the cell.

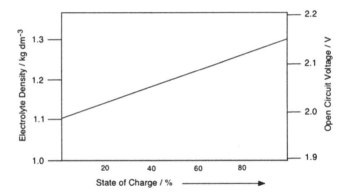

Fig. 15.1 Variations of the electrolyte density and open circuit voltage in Pb-acid cells as functions of the state of charge

15.3   Potentials of the Individual Electrodes

It is clear that the cell voltage in lead-acid cells is significantly greater than the theoretical stability range of water, which is 1.23 V under standard state equilibrium conditions. This disparity is often attributed to (unspecified) kinetic factors in the literature.

However, it means that the electrolytic stability window is extended by the presence of at least one additional ionically-conducting, but electronically-insulating (electroyte) phase in series with the aqueous electrolyte. When this is the case the voltage across both electrolytes can exceed the stability limit of either one [13].

Whereas it is easy to measure the cell voltage with a voltmeter, such a measurement does not give any information about the potentials of either electrode, just the difference. To get information about the individual potentials, it is necessary to use reference electrodes.

This was done by Ruetschi [12, 14, 15], who used this information to determine the potential-determining microstructure in each electrode. He found that the surface of the lead in the negative electrode reactant is covered by a completely formed membrane layer of $PbSO_4$. This phase can thus be considered to be an ionic conductor that extends the electrolytic stability window of the system. The negative electrode potential is determined by the Pb, $PbSO_4$ equilibrium, which he found to be -0.97 V relative to the Hg/Hg_2SO_4 reference electrode potential. This reference potential is $+0.65$ V relative to the standard hydrogen reference potential, the SHE. PbO cannot play a role in this electrode potential, for it is only stable at potentials above -0.40 V versus the Hg/Hg_2SO_4 reference.

Likewise, Ruetschi found that the positive electrode microstructure consists of three phases. $PbSO_4$ and PbO_2 are on top of the underlying lead structure. Again, there is a perm-selective layer of $PbSO_4$ on top of the PbO_2, which is an electronic conductor. Thus, the potential is determined by the $PbSO_4$, PbO_2 equilibrium in that case.

On the basis of this quantitative work on their potentials and local corrosion films, the potential-determining parts of the Pb-acid battery system can be understood by considering the compositions of the two electrodes, as schematically illustrated in Fig. 15.2.

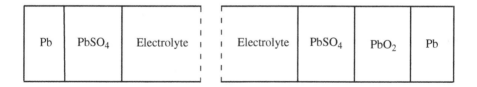

Fig. 15.2 Schematic structures of the potential-determining portions of the electrodes in Pb-acid cells

It was found that the PbO_2, $PbSO_4$ positive electrode potential depends upon the acid concentration in the same way as the voltage of the Pb-acid cell. This is consistent with the Gibbs phase rule discussed earlier, it can be written as

$$F = C - P + 2 \qquad (15.5)$$

For a fixed temperature and total pressure, it becomes simply

$$F = C - P \qquad (15.6)$$

And since the PbO_2, $PbSO_4$ system has three components, Pb, S, and O, and only two phases, there is one degree of freedom left. Thus, the potential is composition-dependent.

Table 15.1 shows the values of the relevant potentials in the lead-acid system, including two different liquid electrolyte compositions.

Table 15.1 Potentials relevant to the Pb-acid battery

Equilibrium	Potential vs. Hg/Hg_2SO_4 (V)	Potential vs. SHE (V)
Hg/Hg_2SO_4	0.00	+0.65
$PbSO_4/PbO_2$ (unit activities)	+1.03	+1.68
$PbSO_4/PbO_2$ (5 m H_2SO_4)	+1.08	+1.73
$PbSO_4/PbO_2$ (1 m H_2SO_4)	+0.91	+1.56
Ag/Ag_2SO_4	+0.04	+0.69
$Pb/PbSO_4$	−1.01	−0.36
Lowest potential at which PbO is stable	−0.40	+0.25

15.4   Relation to the Mechanism of the Electrochemical Reactions in the Electrodes

The electrochemical reaction at the negative electrode is generally expressed in the battery literature as

$$Pb + HSO_4^- = PbSO_4 + H^+ + 2e^- \tag{15.7}$$

and that at the positive electrode as

$$PbO_2 + 3H^+ + HSO_4^- + 2e^- = PbSO_4 + 2H_2O \tag{15.8}$$

It can be seen that both of these reactions involve H^+ ions. As they move into and out of the local aqueous environment, they cause its pH to vary, changing the solubility of the $PbSO_4$ in solution in the negative electrode structure, and that of both PbO_2 and $PbSO_4$ in the positive electrode structure. These solid phases are caused to precipitate and/or dissolve. This is generally called a *dissolution–precipitation mechanism*.

Thus, the overall reaction in this type of battery is a composite of ionic transport of protons through a dense solid electrolyte layer of $PbSO_4$ that causes changes in the local pH and thus of the solubility of $PbSO_4$ and PbO_2 in the adjacent liquid electrolyte. This results in their dissolution or precipitation within the multiphase electrode structure. Whereas the electrode potentials are determined by the 2-phase equilibria under the $PbSO_4$ layer, the electrode capacity is determined by the amounts of the precipitate phases that react within the liquid electrolyte portion of the electrodes.

15.5   Construction of the Electrodes

Although descriptions of Pb-acid cells always say that the negative electrodes are primarily lead, and the positive electrodes primarily PbO_2, they are both initially made from the same material, a paste consisting of a mixture of PbO and Pb_3O_4 [16]. It can be considered to be lead powder that is 70–85% oxidized, and is traditionally called "leady oxide". Measured amounts of water and an H_2SO_4 solution are added, along with small polymer fibers to influence the mechanical properties, under carefully controlled temperature conditions. This results in the formation of basic lead sulfates, $3 PbO \cdot PbSO_4 \cdot H_2O$ or $4 PbO \cdot PbSO_4$. Various other materials are sometimes added to this mix. An example is the use of lignin, a component of wood, as a spacer, or "expander", in the paste that is used in negative electrodes [17]. Its presence reduces the tendency to form large Pb_2SO_4 crystals upon cycling those electrodes.

The paste is inserted into the electrodes by spreading into an open grid structure. Besides mechanically holding the paste material in place, the grid, which has better electronic conductivity than the fine-particle paste, also serves to carry the current throughout the total electrode structure.

Following this process, during which the paste is inserted into both electrode structures, generally before their final insertion into the battery case, a process called *formation* is undertaken. This process converts the materials in the two electrodes into the different compositions and structures required for the fully charged state of the cell.

This forming process is equivalent to an initial electrochemical charge. It is carried out under carefully controlled conditions, typically at a very low current density, so that the total structure in the two sets of electrodes is converted into the desired chemical species without any disruption of the physical state of the paste-impregnated electrodes. This is essentially what Plante [1] did, for he made his battery using sheet lead electrodes and cycling them in dilute H_2SO_4.

The variation in the potentials of the two electrodes during the formation process is shown in Fig. 15.3.

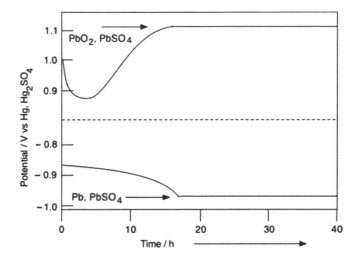

Fig. 15.3 Variation of the potentials of positive (*top*) and negative (*bottom*) electrodes during low rate formation process. After [12]

In an increasing number of cases, formation is followed by a carefully controlled drying process, and the batteries are supplied to the user in the dry state. The acid electrolyte is inserted into the cell at the time of the first use.

15.5.1 Volume Changes and Shedding

As is the case with a number of other battery systems, significant volume changes can occur in the electrodes as the result of the reactions that take place during charging and discharging. $PbSO_4$ has a substantially greater volume than both PbO_2 and elemental lead. The conversion of PbO_2 to $PbSO_4$ results in an increase of 92%, whereas the change in volume from Pb to $PbSO_4$ is 164%.

These volume changes can cause electronically-conducting material to fall to the bottom of the cell, sometimes causing electrical shorting between adjacent electrode plates. In the past, when battery construction was different, it was sometimes found that apparently dead batteries could be "cured" by simply extracting this material from the bottom of the cells.

Whereas the electrodes in these batteries are generally flat, and assembled in stacks, a different configuration is also sometimes employed, in which the active electrode reactants are enclosed in tubes of an inert material, which serve to contain the active material and reduce the shedding problem. Some manufacturers now encase each electrode in a porous plastic bag to prevent the results of shedding from shorting out the electrodes.

15.6   Alloys Used in Electrode Grids

The grid is generally considered to be the most critical passive component of lead-acid cells. It has two functions. One is to physically contain the active materials in the electrodes, and the other is to conduct electrons to and from the active materials. Both (relatively) pure lead and several lead alloys have been used in the manufacture of the grids in lead-acid batteries. There are two basic considerations, their mechanical, and their corrosion, properties.

Pure lead is quite soft, and although this might be an advantage in a mechanical manufacturing process, most grids are currently manufactured by casting liquid lead alloys.

Several lead alloys were developed in order to increase the mechanical strength of grids without significantly changing their electrochemical properties. Lead–antimony alloys were initially preferred. The phase diagram for this system is shown in Fig. 15.4. Compositions not far from the eutectic, which is at 17% antimony, are quite fluid, making it relatively easy to cast grids with rather complex shapes. After freezing, the solid contains two phases, relatively pure lead containing a precipitate of finely divided antimony particles.

Partly because of concern regarding the health issues related to the use of antimony – SbH_3 gas, which can form in the presence of moisture, is poisonous – attention was given to reductions in the antimony content from up to 11 wt percent down to some 6–9 wt percent. But with less Sb, these alloys are not so readily cast, have reduced mechanical strength and are less resistant to corrosion – especially if less than 6%.

Several other alloys were introduced subsequently. One direction that was followed by a number of manufacturers was the use of lead–calcium alloys, particularly in float applications.

The phase diagram for this alloy system is quite different. Instead of having a eutectic reaction and appreciable solid solubility, as in the lead–antimony system,

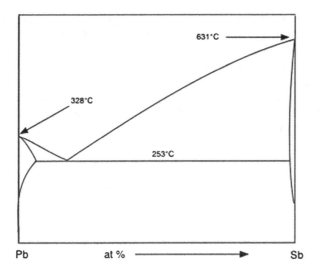

Fig. 15.4 Lead–antimony phase diagram

there is a peritectic reaction at low calcium contents at a temperature quite close to the melting point of pure lead. This is shown in Fig. 15.5. Since the solid solubility of calcium in lead is quite small and rather temperature dependent, it is possible to form fine precipitate particles of the phase $CaPb_3$. The microstructure of these alloys changes gradually after they are cast, resulting in age hardening, which increases the resistance to mechanical deformation by creep.

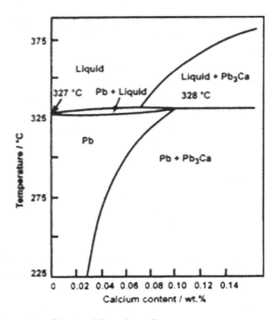

Fig. 15.5 Lead-rich region of lead–calcium phase diagram

In addition, alloys in this system have good corrosion resistance at the potentials of the positive electrode. Their use in negative electrodes is generally thought to result in reduced hydrogen gas evolution.

This has led to the investigation of ternary alloys. The presence of a small (0.5%) amount of As increases the rate of age hardening and provides better creep resistance, which is important for positive plates during deep discharging. Small amounts of Sn (2.5%) increase the fluidity, making the grids easier to cast, and also give better cycle life.

15.7 Alternative Grid Materials and Designs

There have been several other approaches to the design and materials used in electrode grids. One of these involves the strengthening the lead by the inclusion of glass fibers, or polymers.

An alternative that has also been explored somewhat is the use of partially reduced titanium oxides for the construction of grids for positive electrodes [18]. This material, which is primarily TiO_x, where x is between 1.75 and 1.8, is called by the trade name *Ebonex*.

15.8 Development of Sealed Pb-Acid Batteries

For more than 100 years, lead-acid batteries were designed as "flooded" open cells, so that the hydrogen and oxygen products that are developed upon overcharge could escape into the atmosphere. To compensate for these losses, water (preferably distilled) had to be periodically added to the electrolyte.

The technology has now changed significantly, and the most common batteries do not require water replenishment. In addition, the electrolyte is immobilized, so that these products are essentially "spill-proof", and can be used in any physical orientation, upright, on the side, or even upside-down [16].

There are two general approaches that are used. One of these is generally called the "*gel*" *technology*, and was first developed by Sonnenschein in Germany [19], and the other is known as the "*glass mat*" *technology* and was initially developed by Gates Energy Products in the United States [20]. Both are now described by the general term: *valve-regulated lead-acid (VRLA) cells*. This name is related to the fact that a small pressure valve must be present in such sealed cells for security purposes. It reversibly opens if the interior gas pressure exceeds about 1.5 bar.

In the case of the gel technology, the addition of fumed silica, a very fine amorphous form of silicon dioxide that has a very high surface area, to the sulfuric acid electrolyte causes it to thicken, or harden, into a gel. Upon the loss of some water, this mechanically stable structure develops cracks and fissures that can allow the passage of gaseous oxygen across the cell from the positive electrode to the negative electrode upon overcharge.

The other approach involves the use of a highly porous microfiber glass mat between the electrodes. This mat functions as a mechanical separator, and also as a container for the electrolyte, which adsorbs on the surface of the very fine – for example, 1–2 μm diameter–glass fibers. If the mat is only partially filled with the liquid electrolyte, there is also space in this structure for gas to move between the positive and negative electrodes.

In both cases, the cells are designed to be positive electrode limited. This means that the capacity of the positive electrode is less than that of the negative electrode. The cells operate by means of an *internal oxygen cycle*, or *oxygen recombination cycle*. When the positive electrode reaches the limit of its capacity, further charging causes the decomposition of water and the formation of neutral oxygen gas.

$$H_2O = 2H^+ + \frac{1}{2}O_2 + 2e^- \tag{15.9}$$

This gas travels through the gel or glass mat electrolyte to the negative electrode, where it reacts with hydrogen in the negative electrode to again form water.

$$O_2 + 4H^+ + 4e^- = 2H_2O \tag{15.10}$$

The result is that the cell suffers self-discharge as the negative electrode loses capacity.

This latter reaction is exothermic, whereas the oxygen formation reaction is endothermic.

Upon charging, part of the electrical energy is consumed by the oxygen-recombination cycle and converted into heat.

15.9 Additional Design Variations

There have been several new approaches to the design of lead-acid cells in addition to the standard parallel flat plate and tubular configurations. One of these is the bipolar concept, which involves the construction of a stack of cells that are connected in series. To do this, it is necessary to have an electronically conducting bipolar plate that acts as a separator between the electrodes in adjacent cells. The negative electrode of one cell is on one side of the bipolar plate, and the positive electrode of the adjacent cell is on the other side. An example of such a configuration is shown in Fig. 15.6.

It is necessary to have seals to separate the electrolytes in adjacent cells in order to prevent current flow between them. It would also be advantageous to get rid of one of the current collectors, with one layer serving as the positive electrode for one cell and the negative electrode for the other.

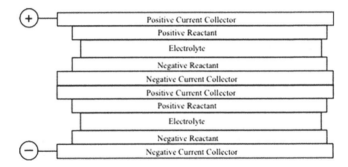

Fig. 15.6 Simple bipolar arrangement

The simplest case is to have a metal sheet or foil act in this way. However, this requires that this metal be stable in contact with these two electrode materials and the potentials at which they both exist, reducing on one side and oxidizing on the other side. Alternatively, one could have an electronically conducting nonmetal serve this function. One example could be graphite. Other materials might also be considered, such as oxides, nitrides, borides, etc., but they also have to meet the same requirements.

Another approach would be to use a bimetallic sheet, fabricated with one material on one side, and a different one on the other side. Such double-layer sheets could be produced by electrodeposition, sputtering, or other such processes. Simply rolling the two materials together might well be the best, and least expensive, method for modest to large scale applications.

A further approach would be to put metallically conducting layers on both sides of a mechanical support material – perhaps a polymer or ceramic. These two conducting layers could be electronically connected by the use of holes through the support material. This might be represented schematically as shown below. In this case, one conductor is both on the top and in the holes, and the other is on the other side. This is shown schematically in Fig. 15.7.

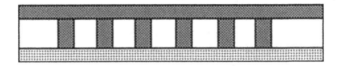

Fig. 15.7 Simple model of mechanically-supported three-layer bipolar plate

Another design variant would be to enhance the power output, rather than the voltage. One approach to this was developed by Bolder Technology [21]. It involves the use of a spiral-wound thin layer concept that essentially places a large number of local cells in parallel. This is represented schematically in Fig. 15.8. But the construction is unique. Thin film electrodes, separated by a separator are wound into a spiral. They protrude out of opposite ends of a cylindrical can and are electrically connected by melting and freezing caps on the two ends of the container. This provides a very large contact area, and produces a configuration with very low internal impedance. Such a design does not store much energy, but can operate at very high power for short times. In one application, starting power is supplied for internal combustion engines by a cell that is only one-eighth to one-tenth times the size and weight of conventional lead-acid designs.

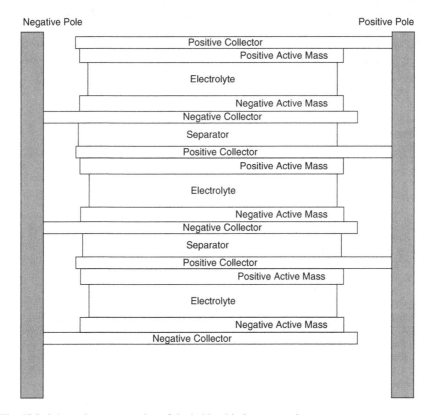

Fig. 15.8 Schematic representation of the bolder thin layer wound system

15.9.1 Other Improvements

In addition to these innovative design changes, a number of improvements have been made in lead-acid cell components. One area involves the enhancement of the mechanical properties, or the reduction in weight, of grid materials. An example is the use of mechanically expanded metal, rather than cast, grids. Another has been the development of extruded lead-covered glass fibers for grid structures.

Both polymer fibers and graphite particles have been introduced into the active materials in some cases to increase their mechanical strength or electronic conductivity.

15.10 Rapid Diffusion of Hydrogen in PbO$_2$

Experiments [22, 23] have shown that the chemical diffusion of hydrogen in PbO$_2$ is very fast, in the range of 0.4 to about 5×10^{-7} cm^2 s^{-1}, varying with the potential and thus with the hydrogen content. These values are some six orders of magnitude greater than hydrogen diffusion in MnO$_2$, which is the positive electrode reactant in the common alkaline Zn/MnO$_2$ cells. This very rapid diffusion explains why Pb-acid cells can provide such very high values of initial current, which is useful when they are used as starter batteries. The quantity of this proton-related charge is relatively small, however, only about 1% of the total capacity of the positive electrode. Thus, this effect does not last very long.

References

1. G. Planté, C. R. Acad. Sci. Paris 50, 640 (1860)
2. C.A. Faure, C. R. Acad. Sci. Paris 92, 951 (1881)
3. I. Barin, Thermochemical Data of Pure Substances, 3 rd Edition, VCH (1995)
4. P. Ruetschi, J. Power Sources 2, 3 (1977/78)
5. J.H. Gladstone and A. Tribe, Nature 25, 221 (1882)
6. F. Dolezalek, Die Theorie des Bleiakkumulators, Halle, Paris (1901)
7. W.H. Beck and W.F.K. Wynne-Jones, Trans. Faraday Soc. 50, 136, 147, 927 (1954)
8. J.A. Duis and W.F. Giauque, J. Phys. Chem. 72, 562 (1968)
9. J.L. Weininger, J. Electrochem. Soc. 121, 1454 (1974)
10. J.L. Weininger and F.W. Secor, J. Electrochem. Soc. 121, 1541 (1974)
11. K.J. Euler, Bull. Schweiz. Elektrotech. Ver. 61, 1054 (1970)
12. P. Ruetschi, J. Electrochem. Soc. 120, 331 (1973)
13. R.A. Huggins, Advanced Batteries: Materials Science Aspects, Springer (2009), Chapter 16
14. P. Ruetschi and R.T. Angstadt, J. Electrochem. Soc. 111, 1323 (1964)
15. P. Ruetschi, J. Power Sources 113, 363 (2003)
16. D.A.J. Rand, P.T. Moseley, J. Garche and C.D. Parker, eds. Valve-Regulated Lead–Acid Batteries, Elsevier (2004)
17. T.A. Willard, US Patents 1,432,508 and 1.505,990 (1920)
18. K. Ellis, A. Hill, J. Hill, A. Loyns and T. Partington, J. Power Sources 136, 366, (2004)
19. O. Jache, U.S. Patent 3,257,237 (1966)
20. D.H. McClelland and J.L. Devitt, U.S. Patent 3,862,861 (1975)
21. http://www.boldertmf.com
22. R. Münzberg and J.P. Pohl, Z. Phys. Chem. 146, 97 (1985)
23. G.P. Papazov, J.P. Pohl and H. Rickert, Power Sources 7, 37 (1978)

Chapter 16
Negative Electrodes in Other Rechargeable Aqueous Systems

16.1 Introduction

This chapter will discuss three examples of negative electrodes that are used in several aqueous electrolyte battery systems, the zinc electrode, the "cadmium" electrode and metal hydride electrodes.

It will be seen that these operate in quite different ways. In the first case, there is an exchange between solid zinc and zinc (zincate) ions in the electrolyte. In the second, the "cadmium" electrode is actually a two-phase system, with elemental cadmium in equilibrium with another solid, its hydroxide. And in the third, hydrogen is exchanged between a solid metal hydride and hydrogen-containing ionic species in the electrolyte.

16.2 The Zinc Electrode in Aqueous Systems

16.2.1 Introduction

Elemental zinc is used as the negative electrode in a number of aqueous electrolyte batteries. The most prominent example is the very common Zn/MnO_2 primary "alkaline cell" that is used in a wide variety of small electronic devices. As will be discussed later, the positive electrode reaction in this case involves the insertion of hydrogen into the MnO_2 crystal structure. A discussion of Zn/MnO_2 technology can be found in [1].

The initial open circuit voltage of these cells is in the range of 1.5–1.6 V. This is greater than the decomposition voltage of water, which can be calculated from its Gibbs free energy of formation, 237.1 kJ/mol, to be 1.23 V at ambient temperatures from

$$\Delta E = -\left(\frac{\Delta G_f^0(H_2O)}{2F}\right) \tag{16.1}$$

R.A. Huggins, *Energy Storage*,
DOI 10.1007/978-1-4419-1024-0_16, © Springer Science+Business Media, LLC 2010

It will be shown here that this is possible because the zinc negative electrode is covered by a thin layer of ionically-conducting ZnO, the thermodynamic result is that its potential is several hundred millivolts lower than the potential at which gaseous hydrogen is normally expected to evolve if an unoxidized metal electrode were to be in contact with water.

16.2.2 Thermodynamic Relationships in the H–Zn–O System

The potential and stability of the zinc electrode can be understood by consideration of the thermodynamics of the ternary H–Zn–O system and its representation in a ternary phase stability diagram.

In addition to the elements and water, the only other relevant phase in this system is ZnO, and the value of its Gibbs free energy of formation is -320.5 kJ/mol at $25°C$.

As discussed earlier, one can use the values of the Gibbs free energy of formation of the different phases to determine which tie lines are stable in a ternary phase stability diagram. In this case, the only possibilities would be either a line between Zn and H_2O or a line between ZnO and hydrogen. Because the standard Gibbs free energy of formation of ZnO is more negative than that of water, the second of these possible tie lines must be the more stable. The simple result in this case is shown in Fig. 16.1. It shows that a subtriangle is formed that has Zn, ZnO, and H_2 at its corners. Another has water, ZnO, and hydrogen at its corners. The potentials of all compositions in the first triangle with respect to oxygen can be calculated from the Gibbs free energy change related to the simple binary reaction along its edge,

$$Zn + \frac{1}{2}O_2 = ZnO \tag{16.2}$$

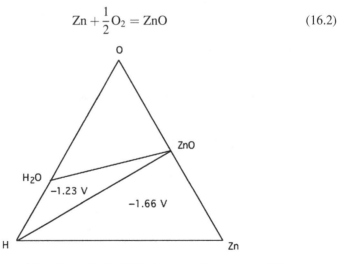

Fig. 16.1 Ternary phase stability diagram for the H–Zn–O system. The numbers within the ternary subtriangles are the potentials relative to pure oxygen

and the result is -1.66 V. The potential of all compositions in the second triangle is likewise related to the standard Gibbs free energy of formation of water, or -1.23 V relative to pure oxygen. That means that zinc has a potential that is 0.43 V more negative than the potential of pure hydrogen in aqueous electrolytes.

Because of the presence of the thin ionically conducting, but electronically insulating, layer of ZnO, water is not present at the *electrochemical interface*, the location of the transition between ionic conduction and electronic conduction, and hydrogen gas is not formed on the zinc. Thus, the effective stability range of the electrolyte is extended, as a topic that will be more fully discussed later.

16.2.3   Problems with the Zinc Electrode

Whereas its low potential is very attractive, there are two negative features of the use of zinc electrodes in aqueous systems. Both relate to its rechargeability in basic aqueous electrolytes.

One of these is that ZnO dissolves in KOH electrolytes, producing an appreciable concentration of zincate ions, $Zn(OH)_4^{2-}$, in which the Zn^{2+} cations are tetrahedrally surrounded by four OH^- groups. Nonuniform zincate composition gradients during recharging, as well as the ZnO on the surface, can lead to the formation of *protrusions*, *filaments,* and *dendrites* during the re-deposition of zinc from the electrolyte at appreciable currents.

The other is that the zinc has a tendency to not redeposit upon the electrode at the same locations during charging of the cell as those from which it was removed during discharge. Gravitational demixing causes the concentration of zincate ions to increase at lower locations, leading to slight differences in the electrolyte conductivity. The result is that there is a gradual redistribution of the zinc, so that the lower portions of the electrode become somewhat thicker or denser as it is discharged and recharged. This effect is often called *shape-change*.

16.3   The "Cadmium" Electrode

16.3.1   Introduction

Cadmium/"nickel" cells have been important products for many years. They have alkaline electrolytes and use the "nickel" positive electrode, H_xNiO_2, which is discussed in the next chapter. Because they have both higher capacity and a reduced problem with environmental pollution – cadmium is considered to be environmentally hazardous – batteries with metal hydride, rather than cadmium, negative electrodes are gradually taking a larger part of the market. They are discussed later in this chapter.

16.3.2 Thermodynamic Relationships in the H–Cd–O System

As in the case of the H–Zn–O system described in the last chapter, the first step in understanding what determines the potential of the cadmium electrode involves the use of available thermodynamic data to determine the relevant ternary phase stability diagram, for the driving forces of electrochemical reactions are the related reactions between electrically neutral species.

In this case, the key issue is the value of the standard Gibbs free energy of formation of CdO, which has been found to be -229.3 kJ/mol at $25°C$. This is less than the value for the formation of water, so a tie line between water and cadmium must be more stable than a tie line between CdO and hydrogen. Also, $Cd(OH)_2$ is a stable phase between water and CdO, because its Gibbs free energy of formation at $25°C$ is -473.8 kJ/mol, whereas the sum of the others is -466.4 kJ/mol. From these data, the ternary phase stability diagram shown in Fig. 16.2 can be drawn. It is clear that it is different from the one for the H–Zn–O system discussed earlier.

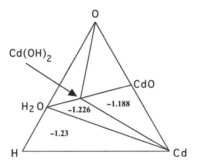

Fig. 16.2 The H–Cd–O ternary phase stability diagram, showing the potentials of the compositions in the subtriangles versus pure oxygen, in volts

It is seen that $Cd(OH)_2$ is also stable when Cd is in contact with water. The potential of the cadmium electrode is determined by the potential of the subtriangle that has water, $Cd(OH)_2$, and cadmium at its corners.

Since there are three phases as well as three components, Cd, hydrogen, and oxygen, present, there are no degrees of freedom, according to the Gibbs phase rule, as discussed earlier. Therefore, the cadmium reaction should occur at a constant potential, independent of the state of charge. This is what is experimentally found.

The potential of all compositions in this triangle is determined by the reaction

$$\frac{1}{2}O_2 + Cd + H_2O = Cd(OH)_2 \tag{16.3}$$

and from the Gibbs free energies of formation of the relevant phases, it is found that its value is -1.226 V relative to that of pure oxygen, as is shown in Fig. 16.2.

The discharge of the cadmium electrode can be written as an electrochemical reaction

$$Cd + 2(OH)^- = Cd(OH)_2 + 2e^- \qquad (16.4)$$

This shows that there is consumption of water from the electrolyte during discharge, as can also be seen in the neutral species reaction in (16.3). This consumption of water must be considered in the determination of the electrolyte composition.

This is different from the zinc electrode discussed above, where the equivalent discharge reaction is

$$Zn + 2(OH)^- = ZnO + H_2O + 2e^- \qquad (16.5)$$

and there is no net change in the amount of water.

16.3.3 Comments on the Mechanism of Operation of the Cadmium Electrode

There is another matter to be considered in the behavior of the cadmium electrode, since discharge involves the formation of a layer of $Cd(OH)_2$ on top of the Cd. This would require a mechanism to either transport Cd^{2+} or OH^- ions through the growing $Cd(OH)_2$ layer, both of which seem unlikely. This reaction is generally thought to involve the formation of an intermediate species that is soluble in the KOH electrolyte. The most likely intermediate species is evidently $Cd(OH)_3^-$.

The kinetics of the cadmium electrode are sufficiently rapid so that the potential changes relatively little on either charge or discharge. Typical values are a deviation of 60 mV during charge, and 15 mV during discharge at the C/2, or 2 h, rate. In addition, there are small potential overshoots at the beginning in both directions if the full capacity had been employed in the previous step. This is, of course, what would be expected if the microstructure started with only one phase, and the second phase has to be nucleated. This is shown in Fig. 16.3 [2].

One of the questions that had arisen in earlier considerations of the mechanism of this electrode was the possibility of the formation of CdO. X-ray investigations have found no evidence for its presence. Thus, if this phase were present it would have to be either as extremely thin layers or be amorphous.

However, this question can be readily answered by the consideration of the potential of the reaction in the triangle with Cd, CdO, and $Cd(OH)_2$ at its corners. This can be determined simply by the reaction along its edge

$$\frac{1}{2}O_2 + Cd = CdO \qquad (16.6)$$

From the standard Gibbs free energy of formation of CdO, this is found to be -1.188 V relative to the potential of oxygen. This is 38 mV positive of the equilibrium potential of the main reaction. Since it is not expected that the electrode

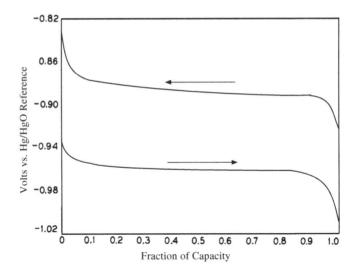

Fig. 16.3 Charge - discharge behavior of cadmium electrode

potential would deviate so far during operation of these electrodes, the formation of CdO is unlikely.

16.4   Metal Hydride Electrodes

16.4.1   Introduction

Metal hydrides are currently used as the negative electrode reactant in large numbers of reversible commercial batteries with aqueous electrolytes, generally in combination with "nickel" positive electrodes.

There are several families of metal hydrides, and the electrochemical properties of some of these materials are comparable to those of cadmium. Developmental efforts have led to the production of small consumer batteries with comparable kinetics, but with up to twice the energy content per unit volume of comparable small "normal" Cd/Ni cells. Typical values for AA size cells are shown in Table 16.1. For this reason, as well as because of the poisonous nature of cadmium, hydride cells are taking a larger and larger portion of this market.

Table 16.1 Typical capacities of AA size cells used in many small electronic devices

Type of cell	mWh/cm^3
"Normal" Cd/Ni	110
"High capacity" Cd/Ni	150
Hydride/Ni	200

16.4.2 Comments on the Development of Commercial Metal Hydride Electrode Batteries

Although there had been research activities earlier in several laboratories, work on the commercialization of small metal hydride electrode cells began in Japan's Government Industry Research Institute (GIRIO) laboratory in Osaka in 1975. By 1991, there were a number of major producers in Japan, and the annual production rate had reached about one million cells. Activities were also underway in other countries. Those early cells had specific capacity values of about 54 Wh/kg and specific powers of about 200 W/kg.

The production rate grew rapidly, reaching an annual rate of about 100 million in 1993, and over one billion cells in 2005. The properties of these small consumer cells also improved greatly. By 2006, the specific capacity had reached 100 Wh/kg, and the specific power 1,200 W/kg. The energy density values also improved, so that they are now up to 420 Wh/L.

They are generally designed with excess negative electrode capacity, that is, $N/P > 1$. This is increased for higher power applications.

The metal hydrides used in small consumer cells are multicomponent metallic alloys, typically containing about 30% rare earths. Prior to this development, the largest commercial use of rare earth materials was for specialty magnets. The major source of these materials is in China, where they are very abundant. Rare earths are also available in large quantities in the United States and South Africa.

In addition to the large current production of small consumer batteries, development efforts have been aimed at the production of larger cells with capacities of 30–100 amp h at 12 V. The primary force that is driving this move toward larger cells is their use in hybrid electric vehicles. In order to meet the high power requirements, the specific capacity of these cells has to be sacrificed somewhat, down to about 45–60 Wh/kg.

16.4.3 Hydride Materials Currently Being Used

There are two major families of hydrides currently being produced that can be roughly identified as AB_5 and AB_2 alloys.

The AB_5 alloys are based upon the pioneering work in the Philips laboratory in the Netherlands that started with the serendipitous discovery of the reaction of gaseous hydrogen with $LaNi_5$. The basic crystal structure is of the layered hexagonal $CaCu_5$ type. Alternate layers contain both lanthanum and nickel, and only nickel. This structure is illustrated in Fig. 16.4.

The reaction of hydrogen with materials of this type can be written as

$$3H_2 + LaNi_5 = LaNi_5H_6 \tag{16.7}$$

Fig. 16.4 Schematic view of
the layered hexagonal lattice
of $LaNi_5$. After [1]

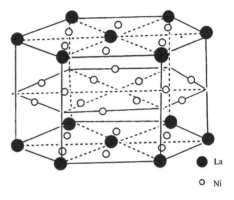

The hydrogen atoms reside in tetrahedral interstitial sites between the host atoms. It has been found that hydrogen can occupy interstitial sites in suitable alloys in which the "holes" have spherical radii of at least 0.4 Å. In larger interstitial positions hydrogen atoms are often "off center", rather than sitting in the middle of the available space.

Developmental work has involved the partial or complete replacement of the lanthanum with other metals, predominantly with Mischmetall (Mm), a mixture of rare earths, or zirconium. A typical composition of the relatively inexpensive Mischmetall is 45–58% Ce, 20–27% La, 13–20% Nd, and 3–8% Pr.

In addition, it has been found advantageous to replace some or all of the nickel with other elements, such as aluminum, manganese, and cobalt. Furthermore, it is possible to change the A/B ratio. One major producer uses a composition that has an different A/B ratio, in the direction of A_2B_7. These materials show relatively flat two-phase discharge voltage plateaus, indicating a reconstitution reaction. Various compositional factors influence the pressure (cell voltage) and the hydrogen (charge) capacity of the electrode, as well as the cycle life. There has also been a lot of developmental work on preparative methods and the influence of microstructure upon the kinetic and cycle life properties of small cells with these materials.

16.4.4 Disproportionation and Activation

Another reaction between hydrogen and these alloys can also take place, particularly at elevated temperatures. It can be written as

$$H_2 + LaNi_5 = LaH_2 + 5Ni \tag{16.8}$$

and is called *disproportionation*. At 298 K, the Gibbs free energies of formation of $LaNi_5$ and LaH_2 are -67 and -171 kJ/mol, respectively, so there is a significant driving force for this to occur, at least on the surface. Experiments have shown that the surface tends to contain regions that are rich in lanthanum, combined with oxygen. In addition, there are clusters of nickel. Because of the presence of these nickel islands, which are permeable to hydrogen, hydrogen can get into the interior of the alloy.

It is often found that a cyclic activation process is necessary in order to get full reaction of hydrogen with the total alloy. As hydrogen works its way into the interior, there is a local volume expansion that often causes cracking and the formation of new fresh surface that is not covered with oxygen. This cracking can cause the bulk material to be converted into a powder, and is called *decrepitation*.

16.4.5 Pressure–Composition Relation

If the particle size is small and there are no surface contamination or activation problems, the LaNi$_5$ alloy reacts readily with hydrogen at a few atmospheres pressure. This is illustrated in Fig. 16.5.

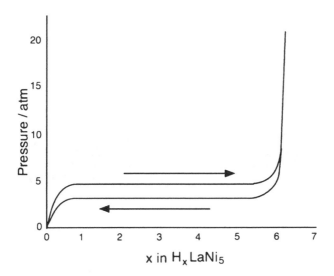

Fig. 16.5 Pressure–composition isotherm for the reaction of LaNi$_5$ with hydrogen

This flat curve is an indication that this is a reconstitution, rather than insertion, reaction.

There is a slight difference in the potential when hydrogen is added from that when hydrogen is removed. This hysteresis is probably related to the mechanical work that must occur due to the volume change in the reaction.

The pressure plateaus for the alloys that are used in batteries are a bit lower, so that the electrochemical potential remains somewhat more positive than that which would cause the evolution of hydrogen gas on the negative electrode.

It has been found that the logarithm of the potential at which this reaction occurs depends linearly upon the lattice parameter of the host material for this family of alloys. This is shown in Fig. 16.6.

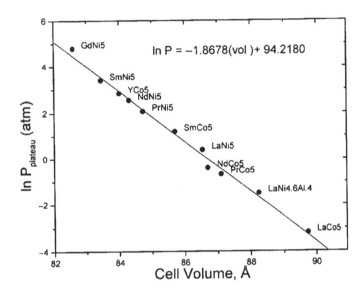

Fig. 16.6 Relation between the logarithm of the plateau pressure and the volume of the crystal structure's unit cell. After [3]

In order to reduce the blocking of the surface by oxygen, as well as to help hold the particles together, thin layers of either copper or nickel are sometimes put on their surfaces by the use of electroless plating methods [3]. PVDF or a similar material is also often used as a binder.

16.4.6   The Influence of Temperature

The equilibrium pressure over all metal hydride materials increases at higher temperatures. This is shown schematically in Fig. 16.7.

The relation between the potential plateau pressure and the temperature is generally expressed in terms of the Van't Hoff equation

$$\ln p(H_2) = (\Delta H/RT) - (\Delta S/R) \qquad (16.9)$$

This can readily be derived from the general relations

$$\Delta G = RT \ln p(H_2) \qquad (16.10)$$

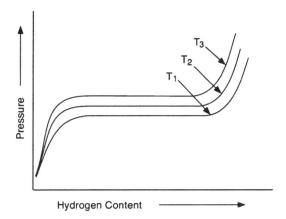

Fig. 16.7 Schematic variation of the equilibrium pressure of a metal hydride system with temperature

and

$$\Delta G = \Delta H - T\Delta S \qquad (16.11)$$

This relationship is shown in Fig. 16.8 for LaNi$_5$ and a commercial mischmetal-containing alloy [3]. It can be seen that the pressure is lower, and thus the electrical potential is higher, in the case of the practical alloy.

This type of representation is often used to compare metal hydride systems that are of interest for the storage of hydrogen from the gas phase. Figure 16.9 is an example of such a plot [4].

It can be seen that the range of temperature and pressure that can be considered for the storage of hydrogen gas is much greater than that which is of interest for the use in aqueous electrolyte battery systems.

Higher pressure in gas systems is equivalent to a lower potential in an electrochemical cell, as can be readily seen from the Nernst equation

$$E = -(RT/zF) \ln p(H_2) \qquad (16.12)$$

The reaction potential must be above that for the evolution of hydrogen, and if it is too high, the cell voltage is reduced. As a result, the range of materials is quite constrained, and a considerable amount of effort has been invested in making minor modifications by changes in the alloy composition.

16.4.7 AB$_2$ Alloys

The other major group of materials that are now being used as battery electrodes are the AB$_2$ alloys. There are two general types of AB$_2$ structures, sometimes called

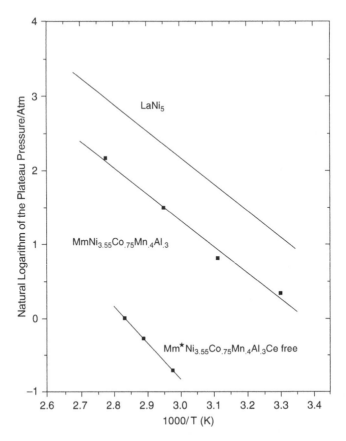

Fig. 16.8 Van't Hoff plot for $LaNi_5H_x$ and two compositions of a $MmNi_{3.55}Co_{0.75}Mn_{0.4}Al_{0.3}H_x$ alloy. After [3]

Friauf–Laves phases; the C14, or $MgZn_2$, type in which the B atoms are in a close packed hexagonal array, and the C15, or $MgCu_2$, type in which the B atoms are arranged in a close packed cubic array. Many materials can be prepared with these, or closely related, structures. The A atoms are generally either Ti or Zr. The B elements can be V, Ni, Cr, Mn, Fe, Co, Mo, Cu, and Zn. Some examples are listed in Table 16.2.

It has generally been found that the C14 type structure is more suitable for hydrogen storage applications. A typical composition can be written as (Ti,Zr) $(V,Ni,Cr)_2$.

16.4.8 General Comparison of These Two Structural Types

Both of these systems can provide a high charge storage density. Present data indicate that this can be slightly (5–10%) higher for some of the AB_2 than the AB_5 materials.

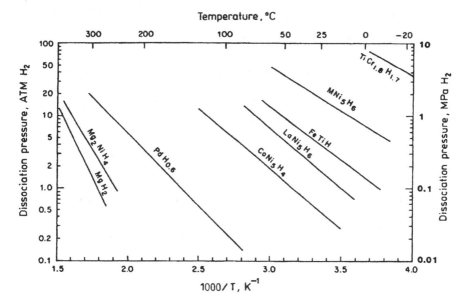

Fig. 16.9 Example of Van't Hoff plot showing data for a wide range of materials. After [4]

Table 16.2 Structures of AB$_2$ phase materials

Material	C14 structure (hexagonal)	C15 structure (cubic)
TiMn$_2$	X	
ZrMn$_2$	X	
ZrV$_2$		X
TiCr$_2$	X	X
ZrCr$_2$	X	X
ZrMo$_2$		X

However, there is a significant difference in the electrochemical characteristics of these two families of alloys. As illustrated in Fig. 16.5, the hydrogen pressure is essentially independent of the composition over a wide range in the AB$_5$ case. Thus, the cell potential is essentially independent of the state of charge, characteristic of a reconstitution reaction. On the other hand, the hydrogen activity generally varies appreciably with the state of charge in the AB$_2$ alloys, giving charge-dependent cell voltages. The fact that the cell voltage decreases substantially during discharge of cells with the AB$_2$ alloy hydrides can be considered to be a significant disadvantage for the use of these materials in batteries.

A serious issue, particularly in the AB$_2$ materials, is the question of oxidation, and subsequent corrosion, particularly of the B metals. This can lead to drastic reductions in the capacity and cycle life, as well as causing a time-dependent increase in gaseous hydrogen pressure in the cell. Because of this, special preetching treatments have been developed to reduce this problem. The vanadium content of the surface is evidently particularly important.

16.4.9   Other Alloys That Have Not Been Used in Commercial Batteries

An alloy of the AB_2 type based upon Ti and Mn, with some V, is currently being used for the storage of gaseous hydrogen, rather than in batteries. One commercial application is in fuel cell-propelled submarines manufactured in Germany. However, its hydrogen activity range is too high to be applicable to use in batteries.

Some years ago, there was a development program in the Battelle laboratory in Switzerland funded by Daimler Benz aimed at the use of titanium–nickel materials of the general compositions A_2B (Ti_2Ni) and AB (TiNi) for use in automobile starter batteries. This early work showed that if electrodes have the right composition and microstructure and are properly prepared, they can perform quite well for many cycles. However, this development was never commercialized.

An interesting side issue is the fact that the TiNi phase, which is very stable in KOH, is one of the materials known to be ferroelastic and to have mechanical memory characteristics. Its mechanical deformation takes place by the formation and translation of twin boundaries, rather than by dislocation motion. As a result, it is highly ductile, yet extremely resistant to fracture. Thus, it would be interesting to consider the use of minor amounts of this phase as a metallic binder in hydride, or other, electrodes. It should be able to accommodate the repeated microscopic mechanical deformation that typically occurs within the electrode structure upon cycling without fracturing.

16.4.10   Microencapsulation of Hydride Particles

A method was developed some years ago in which a metallic coating of either copper or nickel is deposited by electro-less methods upon the hydride particles before the mechanical formation of the hydride electrode [5]. This ductile layer helps the formation of electrodes by pressing, acts as a binder, contributes to the electronic conductivity and thus improves electrode kinetics, and helps against overcharge. It evidently also increases the cycle life. Since both copper and nickel do not corrode in KOH, this layer also acts to prevent oxidation and corrosion.

16.4.11   Other Binders

In addition to the copper or nickel metallic binders, some Japanese cells use PTFE, silicon rubber, or SEBS rubber as a binder. It has been found that this can greatly influence the utilization at high (up to 5C) rates. With the rubber binders (e.g., 3 wt% PTFE), flexible thin sheet electrodes can be made that make the fabrication of small spiral cells easier.

16.4.12   Inclusion of a Solid Electrolyte in the Negative Electrode of Hydride Cells

An interesting development was the work in Japan on the use of a proton-conducting solid electrolyte in the negative electrodes of hydride cells. This material is tetra-methyl ammonium hydroxide pentahydrate, $(CH_3)_4NOH \cdot 5 H_2O$, which has been called TMAH5. It is a clathrate hydrate, and melts (at about 70°C) rather than decomposes when it is heated. Thus it can be melted to impregnate a preformed porous electrode to act as an internal electrolyte. This is typically not true for other solid electrolytes, and can be advantageous in increasing the electrode–electrolyte contact area.

TMAH5 has a conductivity of about 5×10^{-3} S/cm^2 at ambient temperatures. While this value is higher than the conductivity of almost all other known proton-conducting solid electrolytes, it is less than that of the normal KOH aqueous electrolyte. Thus, if this solid electrolyte were to be used, one would have to be concerned with the development of fine scale geometries. This could surely be done, but it would probably involve the use of screen printing or tape casting fabrication methods, rather than conventional electrode fabrication procedures.

Both Hydride/H_xNiO_2 cells and Hydride/MnO_2 cells have been produced using this solid electrolyte. Because of the lower potential of the MnO_2 positive electrode relative to the "nickel" electrode, the latter cells have lower voltages.

16.4.13   Maximum Theoretical Capacities of Various Metal Hydrides

The maximum theoretical specific capacities of various hydride negative electrode materials are listed in Table 16.3. Values are shown for both the hydrogen charged and uncharged weight bases. They include two AB$_5$ type alloys that are being used by major producers, as well as the basic LaNi$_5$ alloy and two AB$_2$ materials.

Table 16.3 Specific capacities of several AB$_5$ and AB$_2$ alloys

Material	Uncharged (mAh/g)	H$_2$ charged (mAh/g)
LaNi$_5$H$_6$	371.90	366.81
MmNi$_{3.5}$Co$_{0.7}$Al$_{0.8}$H$_6$	393.93	388.23
(LaNd)(NiCoSi)$_5$H$_4$	248.80	246.51
TiMn$_2$	509.67	500.16
(Ti,Zr)(V,Ni)$_2$	448.78	441.39

As would be expected, small commercial cells have practical values that are less than the theoretical maxima presented in the last few pages. Hydride electrodes generally have specific capacities of 320–385 mAh/g. For comparison, the H_xNiO_2 "nickel" positive electrodes typically have practical capacities about 240 mAh/g.

References

1. R.F. Scarr, J.C. Hunter and P.J. Slezak, "Alkaline–Manganese Dioxide Batteries", in *Handbook of Batteries*, 3 rd Edition, ed. by D. Linden and T.B. Reddy, McGraw-Hill (2002), p. 10.1.
2. P.C. Milner and U.B. Thomas, "The Nickel-Cadmium Cell", in *Advances in Electrochemistry and Electrochemical Engineering*, 5, 1 (1967).
3. J.J. Reilly, "Metal Hydride Electrodes" in *Handbook of Battery Materials*, ed. by J.O. Besenhard, Wiley-VCH (1999), p. 209.
4. G.D. Sandrock and E.L. Huston, Chemtech *11*, 754 (1981).
5. T. Sakai, H. Yoshinaga, H. Miyamura, N. Kuriyama and H. Ishikawa, J Alloys Compd *180*, 37 (1992).

Chapter 17
Positive Electrodes in Other Aqueous Systems

17.1 Introduction

This chapter will discuss three topics relating to positive electrodes in aqueous electrolyte battery systems, the manganese dioxide electrode, the "nickel" electrode, and the so-called *memory effect* that is found in batteries that have "nickel" positive electrodes.

The first of these deals with a very common material, MnO_2, which is used in the familiar "alkaline" cells that are found in a very large number of small portable electronic devices. This electrode operates by a simple proton insertion reaction.

MnO_2 can have a number of different crystal structures, and it has been known for many years that they exhibit very different electrochemical behavior. It is now recognized that the properties of the most useful version can be explained by the presence of excess protons in the structure, whose charge compensates for that of Mn^{4+} cation vacancies that result from the electrolytic synthesis method.

The so-called "nickel" electrode is discussed in the following section. This electrode is also ubiquitous, as it is used in several types of common batteries. Actually, this electrode is not metallic nickel at all, but a two-phase mixture of nickel hydroxide and nickel oxy-hydroxide. It is reversible and also operates by the insertion and deletion of protons. The mechanism involves proton transport through one of the phases that acts as a solid electrolyte. The result is the translation of a two-phase interface at essentially constant potential.

The third topic in this group is a discussion of what has been a vexing problem for consumers. It occurs in batteries that have "nickel" positive electrodes. The mechanism that results in the appearance of this problem is now understood. In addition, the reason for the success of the commonly used practical solution to it can be understood.

R.A. Huggins, *Energy Storage*,
DOI 10.1007/978-1-4419-1024-0_17, © Springer Science+Business Media, LLC 2010

17.2   Manganese Dioxide Electrodes in Aqueous Systems

17.2.1   Introduction

Manganese dioxide, MnO_2, is the reactant that is used on the positive side of the very common *alkaline* cells that have zinc as the negative electrode material. There are several crystallographic versions of MnO_2, some of which are much better for this purpose than others. Thus, this matter is more complicated than it might seem at first.

MnO_2 is polymorphic, with several different crystal structures. The form found in mineral deposits has the rutile (beta) structure, and is called *pyrolusite*. It is relatively inactive as a positive electrode reactant in KOH electrolytes. It can be given various chemical treatments to make it more reactive, however. One of these produces a modification called *birnessite*, which contains some additional cations. Manganese dioxide can also be produced chemically, and then generally has the delta structure. The material that is currently much more widely used in batteries is produced electrolytically, and is called *EMD*. It has the gamma (*ramsdellite*) structure.

The reason for the differences in the electrochemical behavior of the several morphological forms of manganese dioxide presented a quandary for a number of years. It was known, however, that the electrochemically active materials contain about 4% water in their structures that can be removed by heating to elevated temperatures (100–400°C), but the location and form of that water remained a mystery. This problem was solved by Ruetschi, who introduced a cation vacancy model for MnO_2 [1, 2].

The basic crystal structure of the various forms of MnO_2 contains Mn^{4+} ions in octahedral holes within hexagonally (almost) close packed layers of oxide ions. That means that each Mn^{4+} ion has six oxygen neighbors, and these MnO_6 octahedra are arranged in the structure to share edges and corners. Differences in the edge- and corner-sharing arrangements result in the various polymorphic structures.

If some of the Mn^{4+} ions are missing (cation vacancies), their missing positive charge has to be compensated by something else in the crystal structure. The Ruetschi model proposed that this charge balance is accomplished by the local presence of four protons. These protons would be bound to the neighboring oxide ions, forming a set of four OH^- ions. This local configuration is sometimes called a *Ruetschi defect*. There is very little volume change, as OH^- ions have essentially the same size as O^{2-} ions, and these species play the central role in determining the size of the crystal structure.

On the other hand, reduction of the MnO_2 occurs by the introduction of additional protons during discharge, as first proposed by Coleman [3], and does produce a volume change. The charge of these added mobile protons is balanced by a reduction in the charge of some of the manganese ions present from Mn^{4+} to Mn^{3+}. Mn^{3+} ions are larger than Mn^{4+} ions, and this change in volume during reduction has been observed experimentally.

The presence of protons (or OH^- ions) related to the manganese ion vacancies facilitates the transport of additional protons as the material is discharged. This is why these materials are very electrochemically reactive.

17.2.2   The Open Circuit Potential

The EMD is produced by oxidation of an aqueous solution of manganous sulfate at the positive electrode of an electrolytic cell. This means that the MnO_2 that is produced is in contact with water.

The phase relations, and the related ternary phase stability diagram, for the H–Mn–O system can be determined by use of available thermodynamic information [4, 5], as discussed in previous chapters. From this information, it becomes obvious which neutral species reactions determine the potential ranges of the various phases present and their values.

Following this approach, it is found that the lower end of the stability range of MnO_2 is at a potential that is 1.014 V vs. one atmosphere of H_2. The upper end is well above the potential at which oxygen evolves by the decomposition of water.

Under equilibrium conditions, all oxides exist over a range of chemical composition, being more metal-rich at lower potentials, and more oxygen-rich at higher potentials. In the higher potential case, an increased oxygen content can result from either the presence of cation (Mn) vacancies or oxygen interstitials. In materials with the rutile, and related, structures that have close-packed oxygen lattices, the excess energy involved in the formation of interstitial oxygens is much greater than that for the formation of cation vacancies. As a result, it can be assumed that cation vacancies are present in the EMD MnO_2 that is formed at the positive electrode during electrolysis.

Due to the current that flows during the electrolytic process the potential of the MnO_2 that is formed is actually higher than the equilibrium potential for the decomposition of water. A number of other oxides with potentials above the stability range of water have been shown to oxidize water. Oxygen gas is evolved, and they become reduced by the insertion of protons. Therefore, it is quite reasonable to expect that EMD MnO_2 would have Mn vacancies, and that there would also be protons present, as discussed by Ruetschi [1, 2].

When such positive oxides oxidize water and absorb hydrogen as protons and electrons their potentials decrease to the oxidation limit of water, 1.23 V vs. H_2 at 25°C. This is the value of the open circuit potential of MnO_2 electrodes in Zn/MnO_2 cells.

This water oxidation phenomenon that results in the insertion of protons into MnO_2 is different from the insertion of protons by the absorption of water into the crystal structure of materials that initially contain oxygen vacancies, originally discussed by Stotz and Wagner [6]. It has been shown that both mechanisms can be present in some materials [7, 8].

17.2.3   Variation of the Potential During Discharge

As mentioned above, this electrode operates by the addition of protons into its crystal structure. This is a single-phase insertion reaction, and therefore the potential varies with the composition, as discussed earlier.

If all of the initially present Mn^{4+} ions are converted to Mn^{3+} ions, the overall composition can be expressed as $HMnO_2$, or $MnOOH$.

It is also possible to introduce further protons, so that the composition moves in the direction of $Mn(OH)_2$. In this case, however, there is a significant change in the crystal structure, so that the mechanism involves the translation of a two-phase interface between $MnOOH$ and $Mn(OH)_2$. This is analogous to the main reaction involved in the operation of the "nickel" electrode, as will be discussed later.

The sequence of these two types of reactions during discharge of a MnO_2 electrode is illustrated in Fig. 17.1.

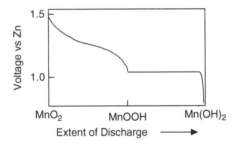

Fig. 17.1 Schematic discharge curve of Zn/MnO_2 cell

The second, two-phase, reaction occurs at such a low cell voltage, that the energy that is available is of low quality, and is generally not used. Such cells are normally considered to only be useful down to about 1.2 V.

17.3   The "Nickel" Electrode

17.3.1   Introduction

The "nickel" electrode is widely used in battery technology, for example, on the positive side of so-called Cd/Ni, Zn/Ni, Fe/Ni, H_2/Ni, and Metal Hydride/Ni cells, in some cases for a very long time. It has relatively rapid kinetics and exhibits unusually good cycling behavior. This is directly related to its mechanism of operation, a solid state insertion reaction involving two ternary phases, $Ni(OH)_2$ and $NiOOH$, with no soluble product. While the attractive properties of this electrode have led to many investigations, there are still a number of aspects of its operation that are not fully understood. This chapter will focus primarily upon the microstructural mechanism of this two-phase insertion reaction and the thermo-dynamic features of the ternary Ni–O–H system that determine the observed potentials and capacities.

17.3.2   Structural Aspects of the Ni(OH)$_2$ and NiOOH Phases

The nanostructure of this electrode can be most simply described as a layer type configuration in which slabs of NiO$_2$ are separated by *galleries* in which various mobile guest species can reside. The structure of the NiO$_2$ layers consists of parallel sheets of hexagonally close-packed O^{2-} ions between which nickel ions occupy essentially all of the octahedral positions.

As will be described below, the mechanism of operation of this electrode involves the transition between Ni(OH)$_2$ and NiOOH upon oxidation, and the reverse upon reduction. Both of these phases are *vario-stoichiometric* (have ranges of stoichiometry). One can thus also describe their compositions in terms of the value of x in the general formula H$_x$NiO$_2$.

In the case of stoichiometric β-Ni(OH)$_2$, the equilibrium crystal structure, which is isomorphous with *brucite*, Mg(OH)$_2$, has galleries that contain a proton concen-tration such that one can consider it as consisting of nickel-bonded layers of OH$^-$ ions instead of O^{2-} ions. The nominal stoichiometry could thus be written as H$_2$NiO$_2$. Stoichiometric NiOOH has half as many protons in the galleries, and thus can be thought of as having an ordered mixture of O^{2-} and OH$^-$ ions. Its nominal composition would then be HNiO$_2$. This terminology will be used later to emphasize that the structural changes in this system occur by insertion reactions.

When it is initially prepared, Ni(OH)$_2$ is often in the α modification, with a substantial amount of hydrogen-bonded water in the galleries. This structure is, however, not stable, and it gradually loses this water and converts to the equilibrium β-Ni(OH)$_2$ structure, in which the galleries are free of water and contain only protons.

The equilibrium form of NiOOH, likewise called the β form, also has only protons in the galleries. However, there is also a γ modification of the NiOOH phase that contains water, as well as other species from the electrolyte, in the galleries. This γ modification forms at high charge rates or during prolonged overcharge in the alkaline electrolyte. In both cases, the potential is quite positive. It can also be formed by electrochemical oxidation of the α-Ni(OH)$_2$ phase.

One can understand the transition of the β-NiOOH to the γ modification at high potentials under overcharge conditions qualitatively in terms of the structural instability of the H$_x$NiO$_2$ type phase when the proton concentration is reduced substantially. Under those conditions, the bonding between adjacent slabs will be primarily of the relatively weak van der Waals type. This allows the entry of species from the electrolyte into the gallery space. This type of behavior is commonly found in other insertion reaction materials if the interslab forces are weak and the electrolyte species are compatible.

The general relations between these various phases are generally described in terms of the scheme presented by Bode and co-workers [9].

A number of very good papers were published by the Delmas group in Bordeaux [10–14] that were aimed at the stabilization of the α-Ni(OH)$_2$ phase by the presence of cobalt so that one might be able to cycle between the α-Ni(OH)$_2$ and γ-NiOOH phases. Since both of these phases have water, as well as other species, in the galleries, they have faster kinetics than the proton-conducting γ phases, although the potential is less positive. An important feature of their work has been the synthesis of sodium analogs by solid state preparation methods and the use of solid state ion exchange techniques (*chimie douce*, or *soft chemistry*) to replace the sodium with other species [15].

The available information concerning the interslab spacing, the critical feature of the crystallographic structure of these phases in the "nickel" electrode, is presented in Table 17.1. It is readily seen that the crystallographic changes involved in the β-Ni(OH)$_2$–β-NiOOH reaction are very small, as they have almost the same value of interslab spacing. This is surely an important consideration in connection with the very good cycle life that is generally experienced with these electrodes. It can also be seen that the structural change involved in the α-Ni(OH)$_2$ – γ-NiOOH transformation is somewhat larger. There are also differences in the slab stacking sequence in these various phases, but that factor will not be considered here.

Table 17.1 Interslab distances for a number of phases related to the "nickel" electrode

Phase	Spacing (Å)
β-Ni(OH)$_2$	4.6
β-NiOOH	4.7
NaNiO$_2$	5.2
Na$_y$(H$_2$O)$_z$NiO$_2$	5.5
Na$_y$(H$_2$O)$_z$CoO$_2$	5.5
γ-H$_x$Na$_y$(H$_2$O)$_z$NiO$_2$	7.0
γ-H$_x$K$_y$(H$_2$O)$_z$NiO$_2$	7.0
γ'-H$_x$Na$_y$(H$_2$O)$_{2z}$NiO$_2$	9.9

Both the α and β versions of the Ni(OH)$_2$ phase are predominantly ionic, rather than electronic, conductors, and have a pale green color. The NiOOH phase, on the other hand, is a good electronic conductor, and both the β and γ versions are black.

17.3.3 Mechanism of Operation

The normal cycling reaction of commercial cells containing this electrode involves back and forth conversion between the β-Ni(OH)$_2$ structure and the β-NiOOH structure. It has been well established that these are separate, although variostoichiometric, phases, rather than end members of a continuous solid solution. The experimental evidence for this conclusion involves both X-ray measurements that show no gradual variation in lattice parameters with the extent of reaction [16], as well as similar IR observations [17] that indicate only changes in the amounts of the two separate phases as the electrode is charged and discharged.

Although the electrode potential when this two-phase structure is present is appreciably above the potential at which water should be oxidized to form oxygen gas, as recognized long ago by Conway [18], gaseous oxygen evolution cannot happen if the solid electrolyte $Ni(OH)_2$ separates the water from the electronic conductor NiOOH. Oxygen evolution can only occur when the electronically conducting NiOOH phase is present on the surface in contact with the aqueous electrolyte.

Therefore, as a first approximation, one can describe the microstructural changes occurring in the electrode in terms of the translation of the $Ni(OH)_2$/NiOOH interface. When the electrode is fully reduced, its structure consists of only $Ni(OH)_2$, whereas upon full oxidation, only NiOOH is present. This is shown schematically in Fig. 17.2. The crystallographic transition between the $Ni(OH)_2$ and NiOOH structures, with their different proton concentrations in the galleries, is shown schematically in Fig. 17.3.

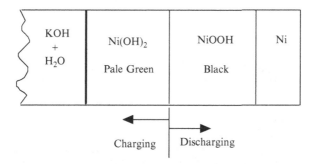

Fig. 17.2 Schematic drawing of the microstructure of the "nickel" electrode. The major phases present are $Ni(OH)_2$, which is a proton-conducting solid electrolyte, and NiOOH, a proton-conducting mixed conductor. The electrochemical reaction takes place by the translation of the $Ni(OH)_2$/NiOOH interface and the transport of protons through the $Ni(OH)_2$ phase

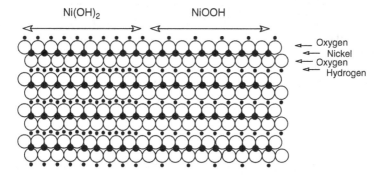

Fig. 17.3 Schematic drawing of the crystallographic transition between the $Ni(OH)_2$ and NiOOH structures, showing the step in the proton concentration in the galleries

It has long been known [19] that the NiOOH forms first at the interface between the $Ni(OH)_2$ and the underlying electronic conductor, rather than at the electrolyte/$Ni(OH)_2$ interface. Other authors (e.g., [20, 21]) have observed the motion of the color boundary during charge and discharge of such electrodes.

17.3.4   Relations Between Electrochemical and Structural Features

It is useful to consider the operation of this electrode in terms of the net reaction in which hydrogen is either added to, or deleted from, the layer structure. In the case of oxidation, this can be written as a neutral chemical reaction:

$$Ni(OH)_2 - \frac{1}{2}H_2 = NiOOH \qquad (17.1)$$

However, in electrochemical cells this oxidation reaction takes place electrochemically, and since this normally involves an alkaline electrolyte, it is generally written in the electrochemical literature as

$$Ni(OH)_2 + OH^- = NiOOH + H_2O + e^- \qquad (17.2)$$

However, the general rule is that electrochemical reactions take place at the boundary where there is a transition between ionic conduction and electronic conduction. Since $Ni(OH)_2$ is predominantly an ionic conductor (a solid electrolyte), the electrochemical reaction occurs at the $Ni(OH)_2$/NiOOH interface, where neither H_2O nor OH^- are present. The electrochemical reaction should therefore more properly be written as

$$Ni(OH)_2 - H^+ = NiOOH + e^- \qquad (17.3)$$

In order for the reaction to proceed, the protons must be transported away from the interface through the galleries in the $Ni(OH)_2$ phase and into the electrolyte. However, in the alkaline aqueous electrolyte environment, hydrogen is not present as either H^+ or H_2. Instead, hydrogen is transferred between the electrolyte and the $Ni(OH)_2$ phase by the interaction of neutral H_2O molecules and OH^- ions in the electrolyte with the H^+ ions at the electrolyte/$Ni(OH)_2$ interface. Thus, the reaction at the electrolyte/$Ni(OH)_2$ interface must be electrically neutral and can be written as:

$$OH^-_{(electrolyte)} + H^+_{(Ni(OH)_2)} = H_2O \qquad (17.4)$$

The variation of the electrical potential of the "nickel" electrode with the amount of hydrogen in its structure is shown in Fig. 17.4. It shows that under low potential, highly reducing conditions only pale green $Ni(OH)_2$ phase is present, and it has a relatively steep potential-composition relation. However, this phase can have a range of composition concentration. It was shown some time ago that up to about 0.25 electrons (and thus 0.25 protons) per mol can be deleted from this phase before the nucleation of NiOOH, and the onset of the two-phase $Ni(OH)_2/NiOOH$ equilibrium [21]. Translated to the crystallographic picture, this means that the proton concentration in the phase nominally called $Ni(OH)_2$ can deviate significantly from the stoichiometric value, up to a proton vacancy fraction of some 12.5%. The proton-deficient composition limit for the $Ni(OH)_2$ phase can thus be expressed as $H_{(1.75}NiO_2)$. The presence of these proton vacancies makes proton transport in this phase very rapid.

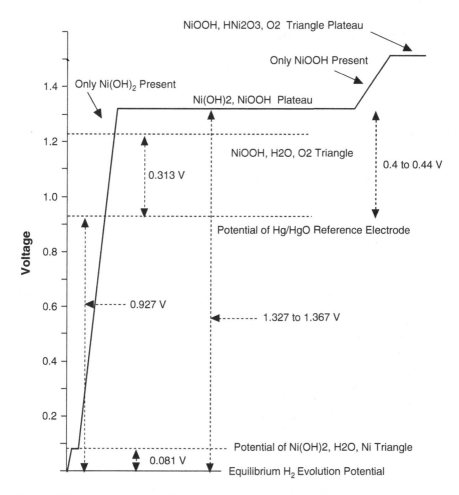

Fig. 17.4 Variation of the potential as hydrogen is removed from the nickel electrode during oxidation when the cell is being charged. These are near-equilibrium data

When both phases are present, there is a relatively long constant-potential plateau during the $Ni(OH)_2$ - $NiOOH$ reaction. This extends from the proton-deficient concentration limit in the $Ni(OH)_2$ phase ($H_{(1.75}NiO_2)$) to the maximum proton concentration in the $NiOOH$ phase. According to Barnard et al. [21], this is when about 0.75 electrons (or protons) per mol are deleted from the electrode. This is equivalent to an end composition of $H_{1.25}NiO_2$ (or $H_{1.25}NiO_2$). Under more oxidizing conditions, when further protons are deleted, only one phase, $NiOOH$, is present, and its potential becomes more positive.

The apparent length of the constant potential two-phase plateau that is observed experimentally depends upon when the $NiOOH$ phase reaches the electrolyte/electrode interface, and thus upon the thickness of the $Ni(OH)_2$ phase and the geometrical shape of the $Ni(OH)_2/NiOOH$ interface. The morphology of this interface, which is often not flat [22], is dependent upon several factors. As will be discussed subsequently, a flat interface is inherently unstable during this oxidation reaction. On the other hand, the interface will tend toward a smooth shape when it translates in the reduction direction. In both cases, the current density is a critical parameter.

Under more oxidizing conditions, when only the $NiOOH$ phase is present, the electrode is black and electronically conducting. This phase has wide ranges of both composition and potential. As mentioned above, the upper limit of proton concentration has been found to be approximately $H_{1.25}NiO_2$ for the β modification. Upon further oxidation in the $NiOOH$ single-phase regime the gallery proton concentration is reduced. It is generally found that the proton concentration can be substantially lower for the γ modification than in the β case. These can thus be far from the nominal composition of $NiOOH$.

17.3.5 Self-Discharge

Since the $NiOOH$ phase is a good mixed-conductor, with a high mobility of both ionic and electronic species, equilibrium with the adjacent electrolyte is readily attained. In the absence of current through the external circuit, there will be a chemical reaction at the $NiOOH$ surface with water in the electrolyte that results in the addition of hydrogen to the electrode. This causes a shift in the direction of a less positive potential. This increase in the hydrogen content and decrease of the potential thus results in a gradual self-discharge of the electrode.

The electrochemical literature generally assumes that this self-discharge reaction involves the generation of oxygen, since the potential of the electrode is more positive than that necessary for the evolution of oxygen by the decomposition of water, as mentioned above.

There are two possible oxygen evolution reactions involving species in the electrolyte:

$$2NiOOH + H_2O = 2Ni(OH)_2 + \frac{1}{2}O_2 \qquad (17.5)$$

which can also be written as

$$H_2O = \frac{1}{2}O_2(\text{into electrolyte or gas}) + H_2(\text{into electrode}) \qquad (17.6)$$

and

$$2OH^- = H_2O + \frac{1}{2}O_2 + 2e^- \qquad (17.7)$$

However, the latter reaction does not provide any hydrogen to the electrode, and thus cannot contribute to self-discharge. Instead, it is the electrochemical oxygen evolution reaction, involving passage of current through the outer circuit, as mentioned later.

The rate of the self-discharge reaction can be simply measured for any value of electrode potential in the single-phase NiOOH regime, where the potential is state-of-charge dependent by using a potentiostat to hold the potential at a constant value, and measuring the anodic current through the external circuit that is required to maintain that value of the potential (and thus also the corresponding proton concentration in the electrode). This is the opposite of the self-discharge process, and can be written as

$$2Ni(OH)_2 + OH^- = 2NiOOH + H_2O + e^- \qquad (17.8)$$

Measurements of the self-discharge current as a function of potential in the NiOOH regime for the case of electrodes produced by two different commercial manufacturers are shown in Fig. 17.5. The differences between the two curves are not important, as they are related to differences between the microstructures of the two electrodes.

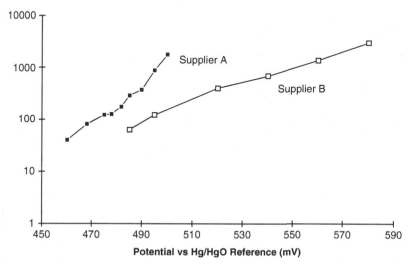

Fig. 17.5 Self discharge current as a function of potential in the NiOOH regime measured on electrodes produced by two different commercial manufacturers

If anodic current is passed through the NiOOH electrode, part will be used to counteract the self-discharge mentioned above. If the magnitude of the current is greater than the self-discharge current, additional protons will be removed from the electrode's crystal structure, making the potential more positive. This results in an increased rate of self-discharge. Thus a steady state will evolve in which the applied current will be just balanced by the rate of self-discharge and the proton concentration in the galleries will reach a new steady (lower) value.

17.3.6 Overcharge

If the applied current density exceeds that which can be accommodated by the kinetics of the compositional change and the self-discharge process, another mechanism must come into play. This is the direct generation of oxygen gas at the electrolyte/NiOOH interface by the decomposition of water in the electrolyte. This can be described by the reaction

$$2OH^- = H_2O + \frac{1}{2}O_2 + 2e^- \tag{17.9}$$

in which the electrons go into the current collector.

17.3.7 Relation to Thermodynamic Information

The available thermodynamic data relating to the various phases in the Ni–O–H system can be used to produce a ternary phase stability diagram. From this information, one can also readily calculate the potentials of the various possible stable phase combinations. This general methodology [23–26] has been used with great success to understand the stability windows of a number of electrolytes, as well as the potential-composition behavior of many electrode materials in lithium, sodium, oxide, and other systems.

With this information, the microstructural model discussed above, and the available information about the stoichiometric ranges of the important phases, one should be able to explain the observed electrochemical behavior of the nickel electrode.

Unfortunately, reliable thermodynamic information for this system is rather scarce. The data that have been used are included in Table 17.2, taken mostly from the compilation in [27]. No recognition was given to the question of stoichiometry or the ordered/disordered state of crystal perfection. Or even to the differences between the α and β structures of $Ni(OH)_2$ and the β and γ structures of NiOOH. Therefore, the calculated potentials can only be considered semiquantitative at present.

Table 17.2 Thermodynamic data

Phase	ΔG_f^0 (25°C)
NiO	−211.5
NiOOH	−329.4
Ni(OH)$_2$	−458.6
H$_2$O	−237.14

The results of these calculations are shown in the partial ternary phase stability diagram of Fig. 17.6, in which all phases are assumed to have their nominal compositions, and are represented as points. The two-phase tie lines and three-phase triangles indicate the phases that are stable in the presence of each other at ambient temperature. Also shown are the potentials of the various relevant three-phase equilibria versus the hydrogen evolution potential at one atmosphere. The composition of the electrode during operation on the main plateau follows along the heavy line that points toward the hydrogen corner of the diagram and lies on the edge of the triangle in which all compositions have a potential of 1.339 V vs. hydrogen.

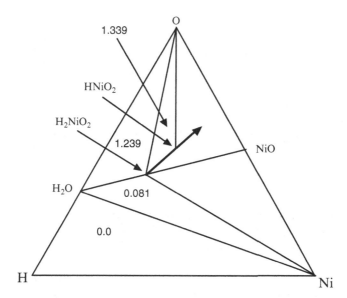

Fig. 17.6 Partial Gibbs triangle. The main charge–discharge reaction takes place along the *thick line* at 1.339 V vs. H$_2$. The overall composition moves along that line upon further charging

Parts of this figure are incomplete, for the data available did not indicate what happens during overcharge when additional hydrogen is removed from HNiO$_2$ and the potential exceeds 1.339 V at the time that it was first written [28]. It was obvious, of course, that the overall composition follows further along the arrow, but the species at the corners of the sub-triangle that has the observed higher potential could not be identified. The compositions of the ternary phases are written in H_xNiO_2 format to emphasize the importance of the hydrogen concentration in the galleries of these layer structure materials. Subsequent information, which led to an explanation of the so-called *memory effect*, is discussed later in this chapter.

One important question is whether there is a stable tie line between NiOOH and H_2O. The alternative is a tie line between $Ni(OH)_2$ and oxygen. Only one of these can be stable, as tie lines cannot cross. The Gibbs free energy change involved in the determining reaction can be calculated

$$2NiOOH + H_2O = 2Ni(OH)_2 + \frac{1}{2}O_2 \tag{17.10}$$

This is found to be -21.26 kJ at 25°C, which means that the NiOOH–H_2O tie line is not stable, and that there is a stable three-phase equilibrium involving Ni $(OH)_2$, NiOOH, and oxygen. A situation in which both $Ni(OH)_2$ and NiOOH are in contact with water can only be metastable.

It is possible to calculate the potential of the $Ni(OH)_2$, NiOOH, O_2 triangle relative to the one atmosphere hydrogen evolution potential from the reaction

$$\frac{1}{2}H_2 + NiOOH = Ni(OH)_2 \tag{17.11}$$

that is the primary reaction of the "nickel" electrode, as discussed above, since this reaction occurs along one of the sides of this three-phase equilibrium triangle. From the data in Table 17.2, it can be determined that the Gibbs free energy change ΔG_r^0 accompanying this reaction is -129.2 kJ per mol. From the relation

$$\Delta G_r^0 = -zFE \tag{17.12}$$

since z is unity for this reaction, the equilibrium potential is 1.34 V positive of the hydrogen evolution potential, as is found from experiments.

Since the potential of a Hg/HgO reference electrode is 0.93 V positive of the reversible hydrogen evolution potential (RHE), this calculation predicts that the equilibrium value of the two-phase constant potential plateau for the main reaction of the nickel electrode should occur at about 0.41 V positive of the Hg/HgO reference. This is also about 0.11 V more positive than the equilibrium potential of oxygen evolution from water.

This result can be compared with the experimental information from Barnard et al. [21] on the potentials of both the *activated* (highly disordered) and *deactivated* (more highly ordered) β-$Ni(OH)_2$–β-NiOOH reaction. Their data fell in the range 0.44–0.47 V positive of the Hg/HgO electrode potential. They found the comparable values for the α-$Ni(OH)_2$–γ-NiOOH reaction to be in the range 0.39–0.44 V relative to the Hg/HgO reference. Despite the lack of definition of the structures to which the thermodynamic data relate, this should be considered to be a quite good correlation.

Further oxidation causes the electrode composition to move along the arrow further away from the hydrogen corner of the ternary diagram, and leads to an electrode structure in which the $Ni(OH)_2$ phase is no longer stable, as is found experimentally. The potential moves to more positive values as the stoichiometry of the NiOOH phase changes, and if no other reaction interferes, should eventually arrive at another, higher, plateau in which the lower proton concentration limit of NiOOH is in equilibrium with some other phase or phases.

Another complicating fact is that electrolyte enters the β-NiOOH at high potentials, converting it to the γ modification. As mentioned earlier, the water-containing α-$Ni(OH)_2$ and γ-NiOOH phases are not stable, and during normal cycling are gradually converted to the corresponding β phases that have only protons in their galleries. When these metastable phases are present, the electrode potential of the reaction plateau is less positive, as is characteristic of insertion structures with larger interslab spacings. Correspondingly, the apparent capacity of the electrode prior to rapid oxygen evolution is greater. These several factors are discussed further in the next section.

17.4   Cause of the Memory Effect in "Nickel" Electrodes

17.4.1   Introduction

It is often found that batteries with "nickel" positive electrodes, for example, Cd/Ni, Hydride/Ni, Zn/Ni, Fe/Ni, and H_2/Ni cells, have a so-called *memory effect*, in which the available capacity apparently decreases if they are used under conditions in which they are repeatedly only partially discharged before recharging. In many cases, these batteries are kept connected to their chargers for long periods of time. It is also widely known that this problem can be "cured" by subjecting them to a slow, deep discharge.

The phenomena that take place in such electrodes have been studied by a number of investigators over many years, but no rational and consistent explanation of the *memory effect* related to "nickel" electrodes emerged until recently. Although it has important implications for the practical use of such cells, some of the major reviews in this area donot even mention this problem, and others give it little attention and/ or no explanation.

In studying this apparent loss of capacity, a number of investigators have shown that a second plateau appears at a lower potential during discharge of "nickel" electrodes [29–45]. Importantly, it is found that under low current conditions, the total length of the two plateaus remains constant. As the capacity on the lower one, sometimes called *residual capacity*, becomes greater, the capacity of the higher one shrinks. The relative lengths of the two plateaus vary with the conditions of prior charging. This is shown schematically in Fig. 17.7.

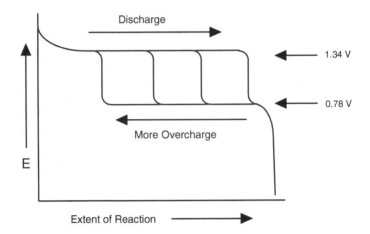

Fig. 17.7 Schematic representation of the two discharge plateaus. With increasing overcharge, the length of the upper one decreases and the lower one increases

Since the capacity of the lower plateau is only about 0.78 V positive of the reversible hydrogen electrode potential, it is generally not useful for most of the applications for which "nickel" electrode batteries are employed. The user does not see this capacity, but instead, sees only the dwindling capacity on the upper plateau upon discharge. Thus, it is quite obvious that the appearance of this lower plateau and reduction in the length of the upper plateau are important components of the memory effect.

It is also found that this lower plateau and the memory effect both disappear if the cell is deeply discharged. Thus the existence of the lower plateau and its disappearance are both obviously related to the *curing* of the memory effect.

These phenomena can now be explained on the basis of available thermo-dynamic and structural information by using the ternary Gibbs phase stability diagram for the H–Ni–O system as a thinking tool [46, 47].

17.4.2 Mechanistic Features of the Operation of the "Nickel" Electrode

The microscopic mechanism of the basic operation of these electrodes was discussed earlier in this chapter. However, it is important for this discussion, and will be briefly reviewed here. It involves an insertion reaction that results in the translation of a two-phase interface between H_2NiO_2 and $HNiO_2$, both of which are vario-stoichiometric (have ranges of stoichiometry). The H_2NiO_2 is in contact with the alkaline electrolyte, and the $HNiO_2$ is in contact with the metallic current collector. The outer layer of the H_2NiO_2 phase is pale green and is predominantly an ionic conductor, allowing the transport of protons to and from the two-phase $H_2NiO_2/HNiO_2$ boundary. $HNiO_2$ is a good electronic conductor, and is black. The electrochemical reaction takes place at that ionic/electronic two-phase interface. This boundary is displaced as the reaction proceeds, and the motion of the color boundary has been experimentally observed. When the electrode is fully reduced, its structure consists of only H_2NiO_2, whereas oxidation causes the interface to translate in the opposite direction until only $HNiO_2$ is present. Although these are both ternary phases, the only compositional change involves the amount of hydrogen present, and the structure of the host "NiO_2" does not change. Thus, this is a pseudo-binary reaction, although it takes place in a ternary system, and the potential is independent of the overall composition; that is, the state of charge.

Once the H_2NiO_2 has been completely consumed and the $HNiO_2$ phase comes into contact with the aqueous electrolyte, it is possible to obtain further oxidation. This involves a change in the hydrogen content of the $HNiO_2$ phase. The variation of the composition of this single phase results in an increase in the potential from this two-phase plateau to higher values, as is expected from the Gibbs phase rule.

After the low-hydrogen limit of the composition of the $HNiO_2$ phase is reached, further oxidation can still take place. Another potential plateau is observed, and oxygen evolution occurs. This is often called "overcharging", and obviously involves another process.

A number of authors have shown that the length of the lower plateau observed upon discharging is a function of the amount of the γ-NiOOH phase formed during overcharging [38]. However, other authors [12] have shown that it is possible to prevent the formation of the γ phase during overcharging by using a dilute electrolyte. Yet the lower discharge potential plateau still appears. There is also evidence that the γ phase can disappear upon extensive overcharging, but the lower discharge plateau is still observed [38].

Neutron diffraction studies [43], which see only crystalline structures, showed a gradual transition between the γ- and β-NiOOH structures upon discharge, with no discontinuity at the transition between the upper and lower discharge plateaus. There was no evidence of a change in the compositions of either of the two phases, just a variation in their amounts, which changes continuously along both discharge plateaus. These authors attributed the presence of the lower plateau to undefined "technical parameters".

Several other authors have explained the presence of the lower discharge plateau in terms of the formation of some type of "barrier layer" [30, 36], and there is evidence for the formation of β-H_2NiO_2, which is not electronically conducting, on the lower plateau [41]. This can, of course, be interpreted as a barrier.

These studies all seem to assume that the oxygen that is formed during operation upon the upper plateau during charging comes only from the decomposition of the aqueous electrolyte. However, something else is obviously happening that leads to the formation of the lower plateau that is observed upon discharge. It must also relate to a change in the amounts, compositions, or structure of the solid phase, or phases, present.

Although the electrochemical behavior of the "nickel" electrode upon the lower potential plateau can be understood in terms of a pseudo-binary insertion/extraction hydrogen reaction, the evolution of oxygen and the formation of the second discharge plateau indicate that the assumption that the oxygen comes from (only) the electrolysis of the aqueous electrolyte during overcharge cannot be correct. In order to understand this behavior, recognition must be given to the fact that the evolution of oxygen indicates that at this potential, this electrode should be thought of in terms of the ternary H–Ni–O system, rather than as a simple binary phase reaction.

Use of the Gibbs triangle as a *thinking tool* to understand the basic reactions in the H–Ni–O system has been discussed in several places [48–50]. The major features of the lower-potential portion of this system can be readily determined from available thermodynamic information. A major part of the Gibbs phase stability triangle for this system is again shown in Fig. 17.8, copied from the discussion earlier in this chapter.

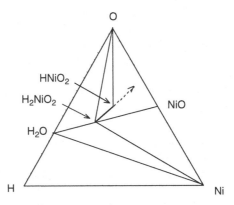

Fig. 17.8 Partial Gibbs triangle. The main charge–discharge reaction takes place along the *thick line*. The overall composition moves along the *dashed line* upon further charging

Since the two phases H_2NiO_2 and $HNiO_2$ are on a tie line that points to the hydrogen corner, neither hydrogen insertion nor deletion involve any change in the Ni/O_2 ratio, and this can be considered to be a pseudo-binary reaction. The tie line between those two phases is one side of a triangle that has pure oxygen at its other corner. This means that both of these phases are stable in oxygen, as is well known.

As the result of the Gibbs phase rule, movement of the overall composition along this tie line occurs at a constant potential plateau. It was shown earlier that the potential of this plateau is 1.34 V versus pure hydrogen at 25°C.

Thus the equilibrium electrode potential of the basic H_2NiO_2–$HNiO_2$ reaction is not only composition-independent, but also more positive than the potential of the decomposition of water, as is experimentally observed. Also, because the H_2NiO_2 that is between the $HNiO_2$ and the water is a solid electrolyte, there is little or no oxygen evolution.

As additional hydrogen is removed, the potential moves up the curve where only $HNiO_2$ is present. When the overall composition exceeds the stability range of that phase it moves further from the hydrogen corner and enters another region in the phase diagram, as indicated by the dashed line in Fig. 17.8.

17.4.3   Overcharging Phenomena

The potential then moves along the upper charging (or overcharging) plateau. Since all of the area within a Gibbs triangle must be divided into sub-triangles, the overall composition must be moving into a new sub-triangle. One corner of this new triangle must be $HNiO_2$, and another must be oxygen. This is consistent with the observation that oxygen is evolved at this higher charging potential. The question is then, what is the composition of the phase that is at the third corner?

If gaseous oxygen is evolved from the electrode, not just from decomposition of the water, the third-corner composition must be below (i.e., have less oxygen) than all compositions along the dashed line.

One possibility might be the phase Ni_3O_4, another could be NiO. However, neither of these phases, which readily crystallize, has been observed. There must be another phase with a reduced ratio of oxygen to nickel.

Although evidently not generally recognized by workers interested in the "nickel" electrode, it has been found [51, 52] that a phase with a composition close to HNi_2O_3 can be formed under conditions comparable to those during charging of the electrode on the upper voltage plateau. This phase can form as an amorphous product during the oxidation of $HNiO_2$. Its crystal structure and composition were determined after hydrothermal crystallization. In addition, the mean nickel oxidation state was found by active oxygen analysis to be only 2.65.

The composition HNi_2O_3 lies on a line connecting $HNiO_2$ and NiO. This would then lead to a sub-triangle as shown in Fig. 17.9, which meets the requirement that there be another phase in equilibrium with both $HNiO_2$ and oxygen that has a reduced ratio of oxygen to nickel.

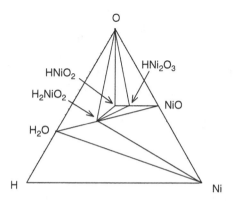

Fig. 17.9 Gibbs triangle showing the presence of the HNi_2O_3 phase

The gradual formation of amorphous HNi_2O_3 during oxygen evolution upon the upper overcharging plateau, and its influence upon behavior during discharge, is the key element in the memory effect puzzle.

As overcharge continues, oxygen is evolved, and more and more of the HNi_2O_3 phase forms. Thus, the overall composition of the solid gradually shifts along the line connecting $HNiO_2$ and HNi_2O_3.

Upon discharge, the overall composition moves in the direction of the hydrogen corner of the Gibbs triangle. This is indicated by the dashed line in Fig. 17.10.

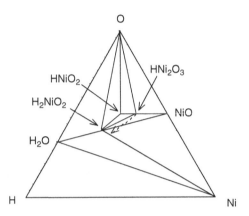

Fig. 17.10 Composition path during the discharge of the HNi_2O_3 formed during overcharge

It is seen that the HNi_2O_3 portion of the total solid moves into a different sub-triangle that has H_2NiO_2, HNi_2O_3, and NiO at its corners. From the available thermodynamic data, one can calculate that the potential in this sub-triangle is 0.78 V versus hydrogen. That is essentially the same as experimentally found for the lower discharge plateau mentioned earlier. The larger the amount of HNi_2O_3 that has been formed during overcharging, the longer the corresponding lower discharge plateau will be. The upper discharge plateau becomes correspondingly shorter.

After traversing this triangle, the overall composition of what had been HNi_2O_3 moves into another sub-triangle that has H_2NiO_2, NiO and Ni at its corners. The HNi_2O_3 disappears, and the major product is H_2NiO_2. The potential in this sub-triangle can be calculated to be 0.19 V versus hydrogen.

If the electrode is now recharged, its potential does not go back up to the 0.8 V plateau, since HNi_2O_3 is no longer present, but goes to the potential for the oxidation of its major component, H_2NiO_2, The overall composition again moves away from the hydrogen corner, and the H_2NiO_2 loses hydrogen and gets converted to $HNiO_2$. This is the standard charging cycle low potential plateau. This also means that the lower reduction plateau is no longer active, for the HNi_2O_3 has disappeared, and the *memory effect* has been *cured*.

The calculated potentials in the various sub-triangles of the H–Ni–O system are shown in Fig. 17.11.

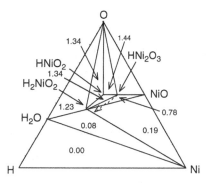

Fig. 17.11 Calculated values of the potential in the various sub-triangles in the H–Ni–O ternary system versus hydrogen. It is seen how the composition path during discharge of HNi_2O_3 leads to the observation of the lower discharge plateau at about 0.78 V, and the disappearance of that phase when the potential moves to a much lower value

The reactions in the H–Ni–O system obviously have very rapid kinetics, for this electrode can be both charged and discharged at high rates. Therefore, it is quite reasonable to expect the phases present to be at or near their equilibrium amounts and compositions. This is indicated by the very good correlation between experimental results and the information obtained by the use of ternary phase stability diagrams based upon the available thermodynamic data.

17.4.4   Conclusions

The basic mechanisms that are involved in causing the memory effect have been identified. The key element is the formation of an amorphous HNi_2O_3 phase upon overcharging into the potential range where oxygen is evolved. Upon subsequent reduction, the presence of this phase produces the potential plateau at about 0.8 V versus hydrogen, reducing the available capacity at the normal higher reduction potential. The more the overcharge, the more HNi_2O_3 that is formed, and the longer the lower plateau. If the electrode undergoes further reduction, this phase disappears, and the potential drops to a much lower value. Subsequent charging of the electrode brings the composition back to the initial state, and the *memory effect is cured.*

This model provides an understanding of the main features of the memory effect, and also explains the several confusing and apparently contradictory observations in the literature. It is expected that further experimental work will address this matter. Additional confirmation of the presence of HNi_2O_3 in the microstructures of overcharged electrodes would be especially useful.

The implication from this mechanism is that the major reason for the memory effect, a decrease in the capacity at the normal discharge potential, is related to extensive overcharging, rather than to the use of shallow discharge cycles.

References

1. P. Ruetschi, J. Electrochem. Soc. *131*, 2737 (1984)
2. P. Ruetschi and R. Giovanoli, J. Electrochem. Soc. *135*, 2663 (1988)
3. J.J. Coleman, Trans. Electrochem. Soc. *90*, 545 (1946)
4. I. Barin, *Thermochemical Data of Pure Substances*, VCH (1995)
5. M. Pourbaix, *Atlas of Electrochemical Equilibria*, Pergamon Press (1966)
6. S. Stotz and C. Wagner, Ber. Bunsenges. Physik. Chem. *70*, 781 (1966)
7. A. Netz, W.F. Chu, V. Thangadurai, R.A. Huggins and W. Weppner, Ionics *5*, 426 (1999)
8. R.A. Huggins, J. Power Sources *153*, 365 (2006)
9. H. Bode, K. Dehmelt and J. Witte, Electrochim. Acta *11*, 1079 (1966)
10. P. Oliva, et al., J. Power Sources *8*, 229 (1982)
11. C. Faure, et al., J. Power Sources *35*, 249 (1991)
12. C. Faure, C. Delmas and P. Willmann, J. Power Sources *35*, 263 (1991)
13. C. Faure, C. Delmas and M. Fouassier, J. Power Sources *35*, 279 (1991)
14. C. Delmas, in *Solid State Ionics II*, G.-A. Nazri, D.F. Shriver, R.A. Huggins and M. Balkanski, Eds., Materials Research Society (1991), p. 335.
15. C. Delmas, et al., Solid State Ionics *28–30*, 1132 (1988)
16. G.W.D. Briggs, E. Jones and W.F.K. Wynne-Jones, Trans. Faraday Soc. *51*, 394 (1955)
17. F.P. Kober, J. Electrochem. Soc. *112*, 1064 (1965)
18. B.E. Conway and P.L. Bourgault, Can. J. Chem. *37*, 292 (1959)
19. E.M. Kuchinskii and B.V. Erschler, J. Phys. Chem. (USSR) *14*, 985 (1940)
20. G.W.D. Briggs and M. Fleischmann, Trans. Faraday Soc. *67*, 2397 (1971)
21. R. Barnard, C.F. Randell and F.L. Tye, J. Appl. Electrochem. *10*, 109 (1980)
22. R.W. Crocker and R.H. Muller, Presented at the Meeting of the Electrochemical Society, Toronto (1992)
23. R.A. Huggins, in *Fast Ion Transport in Solids*, P.P. Vashishta, J.N. Mundy and G.K. Shenoy, Eds., North-Holland (1979), p. 53
24. W. Weppner and R.A. Huggins, Solid State Ionics *1*, 3 (1980)
25. N.A. Godshall, I.D. Raistrick and R.A. Huggins, Mat. Res. Bull. *15*, 561 (1980)
26. N.A. Godshall, I.D. Raistrick and R.A. Huggins, J. Electrochem. Soc. *131*, 543 (1984)
27. J. Balej and J. Divisek, Presented at the Meeting of the Bunsengesellschaft, Wien (1992)
28. R.A. Huggins, M. Wohlfahrt-Mehrens and L. Jörissen, Presented at *Symposium on Intercalation Chemistry and Intercalation Electrodes*, Meeting of the Electrochemical Society in Hawaii, Spring 1993
29. P.C. Milner and U.B. Thomas, in *Advances in Electrochemistry and Electrochemical Engineering*, C.W. Tobias, Ed. (1967), p. 1
30. R. Barnard, G.T. Crickmore, J.A. Lee and F.L. Tye, J. Appl. Electrochem. *10*, 61 (1980)
31. B. Klapste, K. Mickja, J. Mrha and J. Vondrak, J. Power Sources *8*, 351 (1982)
32. A.H. Zimmerman and P.K. Effa, J. Electrochem. Soc. *131*, 709 (1984)
33. H.S. Lim and S.A. Verzwyvelt, J. Power Sources *22*, 213 (1988)
34. H. Vaidyanathan, J. Power Sources *22*, 221 (1988)
35. J. McBreen, Modern Aspects of Electrochemistry *21*, 29 (1990)
36. A.H. Zimmerman, in *Nickel Hydroxide Electrode*, D.A. Corrigan and A.H. Zimmerman, Eds. Electrochem. Soc. Proc. *90–94*, 311 (1990)
37. A.H. Zimmerman, Proc. IECEC *4*, 63 (1994)
38. P. Wilde, PhD Thesis, University of Ulm, Germany (1996)
39. N. Sac-Epee, M.R. Palacín, B. Beaudoin, A. Delahaye-Vidal, T. Jamin, Y. Chabre, and J.-M. Tarascon, J. Electrochem. Soc. *144*, 3896 (1997)
40. N. Sac-Epee, M.R. Palacìn, A. Delahaye-Vidal, Y. Chabre, and J.-M. Tarâscon, J. Electrochem. Soc. *145*, 1434 (1998)

41. C. Leger, C. Tessier, M. Ménétrier, C. Denage, and C. Delmas, J. Electrochem. Soc. *146*, 924 (1999)
42. F. Fourgeot, S. Deabate, F. Henn and M. Costa, Ionics *6*, 364 (2000)
43. S. Deabate, F. Fourgeot and F. Henn, Ionics *6*, 415 (2000)
44. F. Barde, M.R. Palacin, Y. Chabre, O. Isnard, and J.-M. Tarascon, Chem. Mater. *16*, 3936 (2004)
45. R.A. Huggins, Solid State Ionics *177*, 2643 (2006)
46. R.A. Huggins, J. Power Sources *165*, 640 (2007)
47. R.A. Huggins, M. Wohlfahrt-Mehrens and L. Jörissen, Presented at Meeting of the Electro-chemical Society, Hawaii (1992)
48. R.A. Huggins, M. Wohlfahrt-Mehrens, and L. Jörissen, in *Solid State Ionics III*, G-A. Nazri, J-M. Tarascon and M. Armand, Eds., Materials Research Society Proc. 293 (1993), p. 57
49. R.A. Huggins, H. Prinz, M. Wohlfahrt-Mehrens, L. Jörissen and W. Witschel, Solid State Ionics *70/71*, 417 (1994)
50. C. Greaves, M.A. Thomas and M. Turner, Power Sources *9*, 163 (1983)
51. C. Greaves, A.M. Malsbury and M.A. Thomas, Solid State Ionics *18/19*, 763 (1986)
52. A.M. Malsbury and C. Greaves, J. Solid State Chem. *71*, 418 (1987)

Chapter 18
Negative Electrodes in Lithium Systems

18.1 Introduction

A great deal of attention is currently being given to the development and use of batteries in which lithium plays an important role. Looked at very simply, there are two major reasons for this. One is that lithium is a very electropositive element, and its employment in electrochemical cells can lead to larger voltages than are possible with the other alkali metals. The second positive aspect of lithium systems is the possibility of major reductions in weight, at least partly due to the light weight of elemental lithium and many of its compounds.

Although there are now many lithium-based batteries available commercially, there is a large amount of research and development effort underway. There are two general targets, the achievement of significant improvements in performance and safety, and a great reduction in costs. Since this technology has not matured and stabilized, the discussion here will focus upon phenomena and components, rather than complete systems. This chapter deals with negative electrodes in lithium systems. Positive electrode phenomena and materials are treated in the next chapter.

Early work on the commercial development of rechargeable lithium batteries to operate at or near ambient temperatures involved the use of elemental lithium as the negative electrode reactant. As discussed below, this leads to significant problems. Electrodes currently employed on the negative side of lithium cells involve a solid solution of lithium in one of the forms of carbon.

Lithium cells that operate at temperatures above the melting point of lithium must necessarily use alloys instead of elemental lithium. These are generally binary or ternary metallic phases.

There is also increasing current interest in the possibility of the use of metallic alloys instead of carbons at ambient temperatures, with the goal of reducing the electrode volume, as well as achieving significantly increased capacity.

There are differences in principle between the behavior of elemental and binary phase materials as electrodes. It is the purpose of this chapter to elucidate these principles, as well as to present some examples.

R.A. Huggins, *Energy Storage*,
DOI 10.1007/978-1-4419-1024-0_18, © Springer Science+Business Media, LLC 2010

18.2   Elemental Lithium Electrodes

It is obvious that elemental lithium has the lowest potential, as well as the lowest weight per unit charge, of any possible lithium reservoir material in an electrochemical cell. Materials with lower lithium activities have higher potentials, leading to lower cell voltages, and they also carry along extra elements as dead weight.

There are problems with the use of elemental lithium, however. These are due to phenomena that occur during the recharging of all electrodes composed of simple metallic elements. In the particular case of lithium, however, this is not just a matter of increasing electrode impedance and reduced capacity, as are typically found with other electrode materials. In addition, severe safety problems can ensue. Some of these phenomena will be discussed in the following sections.

In the case of an electrochemical cell in which an elemental metal serves as the negative electrode, the process of recharging may seem to be very simple, for it merely involves the electrodeposition of the metal from the electrolyte onto the surface of the electrode. This is not the case, however.

In order to achieve good rechargability, a consistent geometry must be maintained on both the macroscopic and microscopic scales. Both electrical disconnection of the electroactive species and electronic short circuits must also be avoided. In addition, thermal runaway must not occur.

Phenomena related to the inherent microstructural and macrostructural instability of a growth interface and related thermal problems will now be briefly reviewed.

18.2.1   Deposition at Unwanted Locations

In the absence of a significant nucleation barrier, deposition will tend to occur anywhere at which the electric potential is such that the element's chemical potential is at, or above, that corresponding to unit activity. This means that electrodeposition may take place upon current collectors and other parts of an electrochemical cell that are at the same electrical potential as the negative electrode, as well as upon the electrode structure where it is actually desired. This was a significant problem during the period in which attempts were being made to use pure (molten) lithium as the negative electrode in high-temperature molten halide salt electrolyte cells. Another problem with these high temperature cells was the fact that alkali metals dissolve in their halides at elevated temperatures. This leads to electronic conduction and self discharge.

18.2.2   Shape Change

Another difficulty is the *shape change* phenomenon, in which the location of the electrodeposit is not the same as that where the discharge (deplating) process took place. Thus, upon cycling the electrode metal gets preferentially transferred to new locations. For the most part, this is a problem of current distribution and hydrodynamics, rather than being a materials issue. Therefore, it will not be discussed further here.

18.2.3 Dendrites

An additional type of problem relates to the inherent instability of a flat interface on a microscopic scale during electrodeposition, even in the case of a chemically clean surface. It has been shown that there can be an electrochemical analog of the constitutional supercooling that occurs ahead of a growth interface during thermally driven solidification [1].

This will be the case if the current density is such that solute depletion in the electrolyte near the electrode surface causes the local gradient of the element's chemical potential in the electrolyte immediately adjacent to the solid surface to be positive. Under such a condition, there will be a tendency for any protuberance upon the surface to grow at a faster rate than the rest of the interface. This leads to exaggerated surface roughness, and eventually to the formation of either dendrites or filaments. In more extreme cases, it leads to the nucleation of solid particles in the liquid electrolyte ahead of the growing solid interface.

This is also related to the inverse phenomenon, the formation of a flat interface during electropolishing, as well as the problem of morphology development during the growth of an oxide layer upon a solid solution alloy [2, 3]. Another analogous situation is present during the crystallization of the solute phase from liquid metal solutions.

The protuberances upon a clean growing interface can grow far ahead of the general interface, often developing into dendrites. A general characteristic of dendrites is a tree-and-branches type of morphology, which has very distinct geometric and crystallographic characteristics, due to the orientation dependence of either the surface energy or the growth velocity.

18.2.4 Filamentary Growth

A different phenomenon that is often mistakenly confused with dendrite formation is the result of the presence of a reaction product layer upon the growth interface if the electrode and electrolyte are not stable in the presence of each other. The properties of these layers can have an important effect upon the behavior of the electrode. In some cases, they may be useful solid electrolytes, and allow electrodeposition by ionic transport through them. Such layers upon negative electrodes in lithium systems have been given the name "SEI", and will be discussed in a later chapter. But in other cases, reaction product layers may be ionically blocking, and thus significantly increase the interfacial impedance.

Interfacial layers often have defects in their structure that can lead to local variations in their properties. Regions of reduced impedance can cause the formation of deleterious filamentary growths upon recharge of the electrode. This is an endemic problem with the use of organic solvent electrolytes in contact with lithium electrodes at ambient temperatures.

When a protrusion grows ahead of the main interface, the protective reaction product layer will typically be locally less thick. This means that the local impedance to the passage of ionic current is reduced, resulting in a higher current density and more rapid growth in that location. This behavior can be exaggerated if the blocking layer is somewhat soluble in the electrolyte, with a greater solubility at elevated temperatures. When this is the case, the higher local current leads to a higher local temperature and a greater solubility. The result is then a locally thinner blocking layer, and an even higher local current.

Furthermore, the current distribution near the tip of a protrusion that is well ahead of the main interface develops a 3-dimensional character, leading to even faster growth than the main electrode surface, where the mass transport is essentially one-dimensional. Especially in relatively low concentration solutions, this leads to a runaway type of process, so that the protrusions consume most of the solute, and grow farther and farther ahead of the main, or bulk, interface.

This phenomenon can result in the metal deposit having a hairy or spongy character. During a subsequent discharge step, the protrusions often get disconnected from the underlying metal, so that they cannot participate in the electrochemical reaction, and the rechargeable capacity of the electrode is reduced.

This unstable growth is a major problem with the rechargeability of elementary negative electrodes in a number of electrochemical systems and constitutes an important limitation upon the development of rechargeable lithium batteries using elemental lithium as the negative electrode reactant.

18.2.5 Thermal Runaway

The organic solvent electrolytes that are typically used in lithium batteries are not stable in the presence of high lithium activities. This is a common problem when using elemental lithium negative electrodes in contact with electrolytes containing organic cationic groups, regardless of whether the electrolyte is an organic liquid or a polymer [4].

They react with lithium and form either crystalline or amorphous product layers upon the surface of the electrode structure. These reactions are exothermic and cause local heating. Experiments using an *accelerating rate calorimeter* have shown that this problem increases dramatically as cells are cycled, presumably due to an increase in the surface area of the lithium due to morphological instability during repetitive recharging [5]. This is a fundamental difficulty with elemental lithium electrodes, and has led to serious safety problems.

The exothermic formation of reaction product films also occurs when carbon or alloy electrodes are used that operate at potentials at which the electrolyte reacts with lithium. However, if their morphology is constant the surface area does not change substantially, so that it can lead to heating, but typically does not lead to thermal runaway at the negative electrode.

18.3 Alternatives to the Use of Elemental Lithium

Because of these safety and cycle life problems with the use of elemental lithium, essentially all commercial rechargeable lithium batteries now use lithium–carbon alloys as negative electrode reactants today.

A considerable amount of research attention is now also being given to the possibility of the use of metallic lithium alloys instead of the carbons, because of the expectation that this may lead to significant increases in capacity. The large volume changes that accompany increased capacity present a significant problem, however. These matters, as well as the possibility of the use of novel micro- or nano-structures to alleviate this difficulty, will be briefly discussed later in this chapter.

18.4 Lithium–Carbon Alloys

18.4.1 Introduction

Lithium–carbons are currently used as the negative electrode reactant in the very common small rechargeable lithium batteries used in consumer electronic devices. As will be seen in this chapter, a wide range of structures, and therefore of properties, is possible in this family, depending upon how the carbon is produced. The choices made by the different manufacturers are not all the same. Several good reviews of the materials science aspects of this topic can be found in the literature [6, 7].

The crystal structure of pure graphite is shown schematically in Fig. 18.1. It consists of parallel sheets containing interconnected hexagons of carbon, called

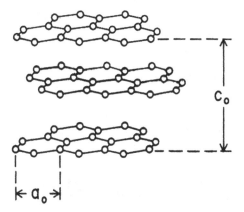

Fig. 18.1 Model of a portion of the crystal structure of graphite

graphene layers or sheets. They are stacked with alternate layers on top of one another. This is described as A-B-A-B-A stacking.

Graphite is amphoteric, and either cations or anions can be inserted into it between the graphene layers. When cations are inserted, the host graphite structure takes on a negative charge. Cation examples are: Li^+, K^+, Rb^+, Cs^+. When anions are inserted, the host graphite structure takes on a positive charge, and anion examples are: Br^-, SO_4^{2-}, or SbF_6^-.

The insertion of alkali metals into carbon was first demonstrated in 1926 [8], and the chemical synthesis of lithium–carbons was demonstrated in 1955 [9]. X-ray photoemission spectroscopy experiments showed that the inserted lithium gives up its electron to the carbon, and thus the structure can be viewed as Li^+ ions contained between the carbon layers of the graphite structure [10]. A general review of the early work on the insertion of species into graphite can be found in [11].

Insertion often is found to occur in "stages", with nonrandom filling of positions between the layers of the host crystal structure. This ordering can occur in individual layers, and also in the filling of the stack of layers.

The possibility of the use of graphite as a reactant in the negative electrode of electro chemical cells containing lithium was first investigated some 30 years ago [12]. The experiments were, however, unsuccessful. Swelling and defoliation occurred due to cointercalation of species from the organic solvent electrolytes that were used at that time.

This problem has been subsequently solved by the use of other liquid electrolytes.

Attention was again brought to this possibility by a conference paper that was presented in 1983 [13] that showed that lithium can be reversibly inserted into graphite at room temperatures when using a polymeric electrolyte. Although not publicly known at that time, two patents relating to the use of the insertion of lithium into graphite as a reversible negative electrode in lithium systems, at both elevated [14] and ambient [15] temperatures, had already been submitted by Bell Laboratories. Royalties paid for the use of these patents have become very large.

This situation changed abruptly as the result of the successful development by SONY in 1990 of commercial rechargeable batteries containing negative electrodes based upon materials of this family and their commercial introduction as the power source in camcorders [16, 17].

There has been a large amount of work on the understanding and development of graphites and related carbon-containing materials for use as negative electrode materials in lithium batteries since that time.

Lithium–carbon materials are, in principle, no different from other lithium-containing metallic alloys. However, since this topic is treated in more detail later, only a few points that specifically relate to carbonaceous materials will be discussed here.

One is that the behavior of these materials is very dependent upon the details of both the nanostructure and the microstructure. Therefore, the composition and the thermal and mechanical treatment of the electrode materials all play important roles in determining the resulting thermodynamic and kinetic properties. Materials with a more graphitic structure have properties that are much different from those with less well-organized structures. The materials that are used by the various commercial producers are not all the same, as they reflect the different choices that they have made for their specific products. However, the major producers of small consumer lithium batteries generally now use relatively graphitic carbons.

An important consideration in the use of carbonaceous materials as negative electrodes in lithium cells is the common observation of a considerable loss of capacity during the first charge–discharge cycle due to irreversible lithium absorption into the structure, as will be seen later. This has the distinct disadvantage that it requires that an additional amount of lithium be initially present in the cell. If this irreversible lithium is supplied from the positive electrode, an extra amount of the positive electrode reactant material must be put into the cell during its fabrication. As the positive electrode reactant materials often have relatively low specific capacities, for example, around 140 mAh/g, this irreversible capacity in the negative electrode leads to a requirement for an appreciable amount of extra reactant material weight and volume in the total cell.

18.4.2 Ideal Structure of Graphite Saturated With Lithium

Lithium can be inserted into the graphite structure up to a maximum concentration of 1 Li per 6 carbons, or LiC_6. One of the major influences of the presence of lithium in the graphite crystal structure is that the stacking of graphene layers is changed by the insertion of lithium. It changes from A-B-A-B-A stacking to A-A-A-A-A stacking. This is illustrated schematically in Fig. 18.2.

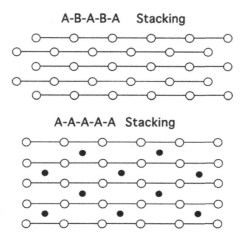

Fig. 18.2 Difference between the A-B-A-B-A and A-A-A-A-A stacking of the graphene layers when lithium is inserted. The *black circles* are the lithium ions

The distribution of lithium ions within the gallery space between the graphene layers is illustrated schematically in Fig. 18.3.

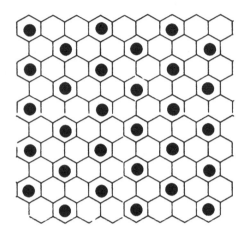

Fig. 18.3 Schematic representation of the lithium distribution in the gallery space in relation to the carbon hexagonal network in the adjacent graphene layers

18.4.3 Variations in the Structure of Graphite

There is actually a wide range of lithium–carbon structures, and most such materials do not actually have the ideal graphite structure. The ones that are closest are made synthetically by vapor transport, and are called highly ordered pyrolytic graphite (HOPG). This is a slow and very expensive process. The graphites that are used commercially range from natural graphite to materials formed by the pyrolysis of various polymer or hydrocarbon precursors. They are often divided into two general types, designated as *soft or graphitizing carbons*, and *hard carbons* [18].

At modest temperatures and pressures, there is a strong tendency for carbon atoms to be arranged in a planar graphene-type configuration, rather than a three-dimensional structure such as that in diamond.

Soft carbons are generally produced by the pyrolysis of liquid materials such as petroleum pitch, which is the residue from the distillation of petroleum fractions.

The carbon atoms in their structure are initially arranged in small graphene-type groups, but there is generally a significant amount of imperfection in their two-dimensional honeycomb networks, as well as randomness in the way that the layers are vertically stacked upon each other. In addition, there is little coordination in the rotational orientation of nearby graphene layers. The term "turbostratic" is generally used to describe this general type of three-dimensional disorder in carbons [18].

The three types of initial disorder, in-plane defects, interplane stacking defects, and rotational misorientation, gradually become healed as the temperature is raised. The first two earlier than the rotational disorder between adjacent layers, for that requires more thermal energy.

The microstructure of such materials that have been heated to intermediate temperatures is shown schematically in Fig. 18.4.

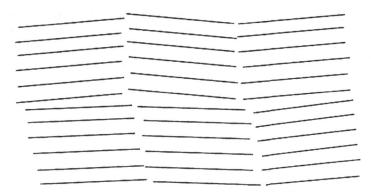

Fig. 18.4 Schematic drawing of the microstructure of graphite after heating to intermediate temperatures

At this intermediate stage, the structure contains many small three-dimensional subgrains. In addition to containing some internal imperfections, they differ from their neighbors in both vertical and horizontal orientations. They are separated by subgrain walls (boundaries) that have surface energy. This subgrain wall surface energy gradually gets reduced as the individual subgrains grow in size and the overall graphitic structure becomes more perfect.

The *hard carbons*, which are typically produced by the pyrolysis of solid materials, such as chars or glassy carbon, initially have a significant amount of cross-linking, related to the structure of their precursors. In addition, they can have a substantial amount of nanoporosity. As a result, it is more difficult to make these structural rearrangements, and turbostratic disorder is more persistent. The result is the requirement for more thermal energy, that is, higher temperatures.

The structure that results from the pyrolysis of carbonaceous precursors depends greatly upon the maximum temperature that is reached. Heating initially amorphous, or *soft,* carbons to the range of 1,000–2,000°C produces microstructures in which graphene sheets form and begin to grow, with diameters up to about 15 nm, and they become assembled into small stacks of 50–100 sheets. These subgrains initially have a turbostratic arrangement, but their alignment into larger ordered, that is, graphitic regions gradually takes place as the temperature is increased from 2,000 to 3,000°C.

18.4.4 Structural Aspects of Lithium Insertion into Graphitic Carbons

One of the important features in the interaction of lithium with graphitic materials is the phenomenon of *staging*. Lithium that enters the graphite structure is not distributed uniformly between all the graphene layers at ambient temperatures. Instead, it resides in certain interlayer *galleries*, but not others, depending upon the total amount of lithium present.

The distribution is described by the number of graphene layers between those that have the lithium guest ions present. A stage 1 structure has lithium between all of the graphene layers, a stage 2 structure has an empty gallery between each occupied gallery, and a stage 4 structure has four graphene layers between each gallery containing lithium. This will be discussed further later in this chapter. This is obviously a simplification, for in any real material there will be regions with predominately one structure and other regions with another.

The phenomenon of nonrandom gallery occupation is found in a number of other materials, and can be attributed to a catalytic effect, in which the ions that initially enter a gallery pry open the van der Waals-bonded interlayer space, making it easier for following ions to enter.

However, the situation is a bit more complicated, for there must be communication between nearby galleries in order for the structure to adopt the ordered stage structure. This is related to the intertunnel communication in the *hollandite* structure described in Chap. 13, but will not be further discussed here.

18.4.5   Electrochemical Behavior of Lithium in Graphite

The electrochemical behavior of lithium in carbon materials is highly variable, depending upon the details of the graphitic structure. Materials with a more perfect graphitic structure react with lithium at more negative potentials, whereas those with less well organized structures typically operate over much wider potential ranges, resulting in cell voltages that are both lower and more state-of-charge dependent.

In a number of cases, the carbons that are used in commercial batteries have been heated to temperatures over about 2,400°C, where they become quite well graphitized. Capacities typically range from 300 to 350 mAh/g, whereas the maximum theoretical value (for LiC_6) is 372 mAh/g.

A typical discharge curve under operating conditions, with currents as large as 2–4 mA/cm^2, is shown in Fig. 18.5.

This behavior is not far from what is found under near equilibrium conditions, as shown in Fig. 18.6. It can be seen that there is a difference between the data during charge, when lithium is being added, and discharge, when lithium is being deleted. This displacement (hysteresis) between the charge and discharge curves is at least partly due to the mechanical energy involved in the structural changes.

It can be seen that these data show plateaus, indicating the presence of three ranges of composition within which reconstitution reactions take place. As the composition changes along these plateaus, the relative amounts of material with the two end compositions vary. This means that there will be regions, or "domains", where the graphene layer stacking is of one type, and regions in which it has the other. The relative volumes of these two domains vary as the overall composition

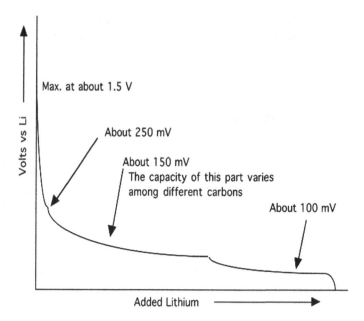

Fig. 18.5 Typical discharge curve of a lithium battery negative electrode

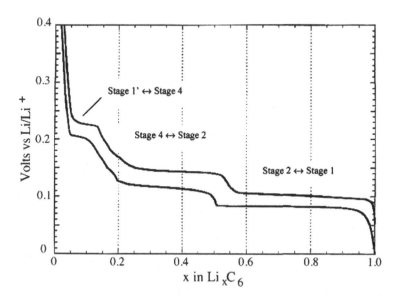

Fig. 18.6 Potential versus composition during lithiation and delithiation of a graphite electrode at the C/50 rate at ambient temperature. After [19]

traverses these *two phase regions*. The differences in stacking result in differences in inter-layer spacing, and therefore a considerable amount of distortion of the structure. Such a model was presented some time ago by Daumas and Herold [20].

18.4.6   Electrochemical Behavior of Lithium in Amorphous Carbons

The electrochemical behavior is quite different when the carbon has not been heated so high, and the structure is not so well ordered. There is a wide range of possible sites in which the lithium can reside, with different local structures, and therefore different energies. The result is that the potential varies gradually, rather than showing the steps characteristic of more ordered structures. This is shown in Fig. 18.7. It can be seen that, in addition to varying with the state of charge, the potential is significantly greater than is found in the graphitic materials. This means that the cell voltages are correspondingly lower.

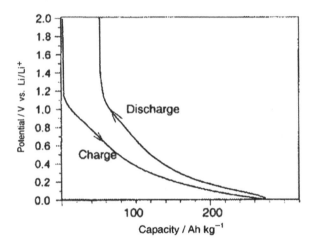

Fig. 18.7 Typical data for the reaction of lithium with an amorphous carbon

It can be seen that there was some capacity loss on the first cycle. The capacity upon the first charging (that is not useful) was greater than the capacity in the subsequent discharge cycle. The source of this phenomenon is not yet understood, but there must be some lithium that is "trapped" in the structure and does not come out during discharge. Because of this extra (useless) capacity during the initial charging of this negative electrode it is necessary to put extra capacity in the positive electrode. This is unfortunate, for the specific capacity of the positive electrodes in such systems is less than that in the negative electrodes. As a result, a significant amount of extra weight and volume is necessary.

18.4.7 Lithium in Hydrogen-Containing Carbons

It is often found that there is a considerable amount of hydrogen initially present in various carbons, depending upon the nature of the precursor. This gradually disappears as the temperature is raised.

If the precursor is heated to 500–700°C, there is still a lot of hydrogen present in the structure. It has been found experimentally that this can lead to a very large capacity for lithium, which is proportional to the amount of hydrogen present [21–23]. There is a loss in this capacity upon cycling, perhaps due to the gradual loss of hydrogen in the structure.

The large capacity may be due to lithium binding to hydrogen-terminated edges of small graphene fragments. The local configuration would then be analogous to that in the organolithium molecule $C_2H_2Li_2$. This is consistent with the experimental observation of the dependence of the lithium capacity upon the amount of hydrogen present. This would also result in a change in the local bonding of the host carbon atoms from sp^2 to sp^3.

In addition to a large capacity, experiments have shown a very large hysteresis with these materials [23]. Hysteresis is generally considered to be a disadvantage, as the discharge potential is raised, reducing the useful cell voltage.

Hysteresis is characteristic of reactions that involve mechanical energy loss as the result of shape and volume changes. However, in this case it is more likely due to the energy involved in the change of the bonding of the nearby carbon atoms [23].The result of experiments performed on one example of a hydrogen-containing material is shown in Fig. 18.8. It can be seen that there was a very large capacity loss on the first cycle. The capacity upon the first charging (that is not useful) was much greater than the capacities in subsequent cycles. As mentioned

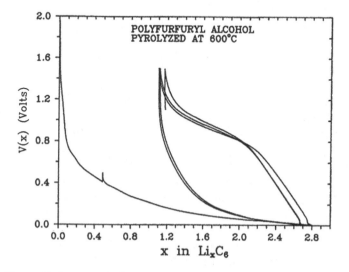

Fig. 18.8 Charge–discharge curves for a material containing hydrogen. After [7]

above, this extra lithium must be supplied by the positive electrode. The source of this phenomenon is not yet understood, but there must be a lot of lithium that is *trapped* in the structure and does not come out during the first and subsequent discharges.

18.5 Metallic Lithium Alloys

18.5.1 Introduction

Attention has been given to the use of lithium alloys as an alternative to elemental lithium for some time. Groups working on batteries with molten salt electrolytes that operate at temperatures of 400–450°C, well above the melting point of lithium, were especially interested in this possibility. Two major directions evolved. One involved the use of lithium–aluminum alloys [24, 25], and another was concerned with lithium–silicon alloys [26–28].

Whereas this approach can avoid the problems related to lithium melting, as well as the others mentioned above, there are always at least two disadvantages related to the use of alloys. Because they reduce the activity of the lithium, they necessarily reduce the cell voltage. In addition, the presence of additional species that are not directly involved in the electrochemical reaction always brings additional weight, and often, volume. Thus, the maximum theoretical values of the specific energy are always reduced compared to what might be attained with pure lithium. The energy density is also often reduced. But metallic lithium has a large specific volume, so that this is not always the case.

In practical situations, however, the excess weight and volume due to the use of alloys may not be very far from those required with pure lithium electrodes, for it is generally necessary to have a large amount of excess lithium in rechargeable cells in order to make up for the capacity loss related to the dendrite or filament growth problem upon cycling.

Lithium alloys have been used for a number of years in the high temperature "thermal batteries" that are produced commercially for military purposes. These devices are designed to be stored for long times at ambient temperatures before use, where their self-discharge kinetic behavior is very slow. They must be heated to elevated temperatures when their energy output is desired. An example is the Li alloy/FeS_2 battery system that employs a chloride molten salt electrolyte. In order to operate, the temperature must be raised above the melting point of the electrolyte. This type of cell typically uses either Li–Si or Li–Al alloys in the negative electrode.

The first use of lithium alloys as negative electrodes in commercial batteries to operate at ambient temperatures was the employment of Wood's metal alloys in lithium-conducting button type cells by Matsushita in Japan. Development work on the use of these alloys started in 1983 [29], and they became commercially available somewhat later.

18.5.2 Equilibrium Thermodynamic Properties of Binary Lithium Alloys

Useful starting points when considering lithium alloys as electrode reactants are their phase diagrams and equilibrium thermodynamic properties. In some cases, this information is available, so that predictions can be made of their potentials and capacities. In other cases, experimental measurements are required. Relevant principles were discussed in earlier chapters, and will not be repeated here.

Elevated temperature data for a number of phases in the Li–Al, Li–Bi, Li–Cd, Li–Ga, Li–In, Li–Pb, Li–Sb, Li–Si, and Li–Sn binary lithium alloy systems, made using a LiCl–KCl molten salt electrolyte are listed in Table 18.1.

Table 18.1 Plateau potentials and composition ranges of a number of binary lithium alloys at 400°C

Voltage versus Li/Li^+	System	Range of y
0.910	Li_ySb	0–2.0
0.875	Li_ySb	2.0–3.0
0.760	Li_yBi	0.6–1.0
0.750	Li_yBi	1.0–2.82
0.570	Li_ySn	0.57–1.0
0.455	Li_ySn	1.0–2.33
0.430	Li_ySn	2.33–2.5
0.387	Li_ySn	2.5–2.6
0.283	Li_ySn	2.6–3.5
0.170	Li_ySn	3.5–4.4
0.565	Li_yGa	0.15–0.82
0.122	Li_yGa	1.28–1.48
0.09	Li_yGa	1.53–1.93
0.558	Li_yCd	0.12–0.21
0.373	Li_yCd	0.33–0.45
0.058	Li_yCd	1.65–2.33
0.507	Li_yPb	0–1.0
0.375	Li_yPb	1.1–2.67
0.271	Li_yPb	2.67–3.0
0.237	Li_yPb	3.0–3.5
0.089	Li_yPb	3.8–4.4
0.495	Li_yIn	0.22–0.86
0.145	Li_yIn	1.74–1.92
0.080	Li_yIn	2.08–2.67
0.332	Li_ySi	0–2.0
0.283	Li_ySi	2.0–2.67
0.156	Li_ySi	2.67–3.25
0.047	Li_ySi	3.25–4.4
0.300	Li_yAl	0.08–0.9

18.5.3 Experiments at Ambient Temperature

Experiments have also been performed to determine the equilibrium values of the electrochemical potentials and capacities in a smaller number of binary lithium systems at ambient temperatures [30, 31]. Because of slower kinetics at lower temperatures, these experiments took longer to perform. Data are presented in Table 18.2.

Table 18.2 Plateau potentials and composition ranges of lithium alloys at ambient temperatures under equilibrium conditions

Voltage versus Li/Li$^+$	System	Range of y
0.956	Li$_y$Sb	1.0–2.0
0.948	Li$_y$Sb	2.0–3.0
0.828	Li$_y$Bi	0–1.0
0.810	Li$_y$Bi	1–3.0
0.680	Li$_y$Cd	0–0.3
0.352	Li$_y$Cd	0.3–0.6
0.055	Li$_y$Cd	1.5–2.9
0.660	Li$_y$Sn	0.4–0.7
0.530	Li$_y$Sn	0.7–2.33
0.485	Li$_y$Sn	2.33–2.63
0.420	Li$_y$Sn	2.6–3.5
0.380	Li$_y$Sn	3.5–4.4
0.601	Li$_y$Pb	0–1.0
0.449	Li$_y$Pb	1.0–3.0
0.374	Li$_y$Pb	3.0–3.2
0.292	Li$_y$Pb	3.2–4.5
0.256	Li$_y$Zn	0.4–0.5
0.219	Li$_y$Zn	0.5–0.67
0.157	Li$_y$Zn	0.67–1.0
0.005	Li$_y$Zn	1.0–1.5

18.5.4 Liquid Binary Alloys

Although the discussion here has involved solid lithium alloys, similar considerations apply to those based on sodium or other species. In addition, it is not necessary that the active material be solid. The same principles hold for liquids.

An example was discussed in Chap. 11 relating to the so-called sodium–sulfur battery that operates at about 300°C. In this case, both of the electrodes are liquids, and the electrolyte is a solid sodium ion conductor. This configuration can thus be described as an L/S/L system. It is the inverse of conventional systems with solid electrodes and a liquid electrolyte, S/L/S systems.

18.5.5 Mixed-Conductor Matrix Electrodes

In order to be able to achieve appreciable macroscopic current densities while maintaining low local microscopic charge and particle flux densities, many battery electrodes that are used in conjunction with liquid electrolytes are produced with porous microstructures containing very fine particles of the solid reactant materials. The high reactant surface area porous structure is permeated with the electrolyte.

This porous fine-particle approach has several characteristic disadvantages, however. Among these are difficulties in producing uniform and reproducible microstructures, and limited mechanical strength when the structure is highly porous. In addition, they often suffer Ostwald ripening, sintering, or other time-dependent changes in both microstructure, and properties during cyclic operation.

Furthermore, it is often necessary to have an additional material present in order to improve the electronic transport within an electrode. Various highly dispersed carbons are often used for this purpose.

A quite different approach was introduced some years ago [32–34], in which it was demonstrated that a rather dense solid electrode can be fabricated that has a composite microstructure in which particles of the reactant phase, or phases, are finely dispersed within a solid electronically-conducting matrix in which the electroactive species is also mobile; that is, within a mixed conductor. There is thus a large internal reactant/mixed-conducting matrix interfacial area. The electroactive species is transported through the solid matrix to this interfacial region, where it undergoes the chemical part of the electrode reaction. Since the matrix material is also an electronic conductor, it can also act as the electrode's current collector. The electrochemical part of the reaction takes place on the outer surface of the composite electrode.

When such an electrode is discharged by the deletion of the electroactive species, the residual particles of the reactant phase remain as relics in the microstructure. This provides fixed permanent locations for the reaction to take place during following cycles, when the electroactive species again enters the structure. Thus, this type of configuration has the additional advantage that it can provide a mechanism for the achievement of true microstructural reversibility.

In order for this concept to be applicable, the matrix and the reactant phases must be thermodynamically stable in contact with each other. One can evaluate this possibility if information about the relevant phase diagrams as well as the titration curves of the component binary systems, is available. The stability window of the matrix phase must span the reaction potential of the reactant material. It has been shown that one can evaluate the possibility that these conditions are met from knowledge of the binary titration curves.

Since there is generally a common component, these two binaries can also be treated as a ternary system. Although ternary systems are not explicitly discussed here, it can be simply stated that the two materials must lie at corners of the same constant-potential tie triangle in the relevant isothermal ternary phase diagram in order to not interact. The potential of the tie triangle determines the electrode reaction potential, of course. An additional requirement is that the reactant material must have two phases present in the tie triangle, but the matrix phase only one.

The kinetic requirements for a successful application of this concept are readily understandable. The primary issue is the rate at which the electroactive species can reach the matrix/reactant interfaces. The critical parameter is the chemical diffusion coefficient of the electroactive species in the matrix phase. This can be determined by various techniques, as discussed in later chapters.

The first example that was demonstrated was the use of the phase with the nominal composition $Li_{13}Sn_5$ as the matrix, in conjunction with reactant phases in the lithium– silicon system at temperatures near 400°C. This is an especially favorable case, due to the very high chemical diffusion coefficient of lithium in the $Li_{13}Sn_5$ phase.

The relation between the potential-composition data for these two systems under equilibrium conditions is shown in Fig. 18.9 [32]. It is seen that the phase $Li_{2.6}Sn$ ($Li_{13}Sn_5$) is stable over a potential range that includes the upper two-phase reconstitution reaction plateau in the lithium–silicon system. Therefore, lithium can react with Si to form the phase $Li_{1.71}Si$ ($Li_{12}Si_7$) inside an all-solid composite electrode containing the $Li_{2.6}Sn$ phase, which acts as a lithium-transporting, but electrochemically inert matrix.

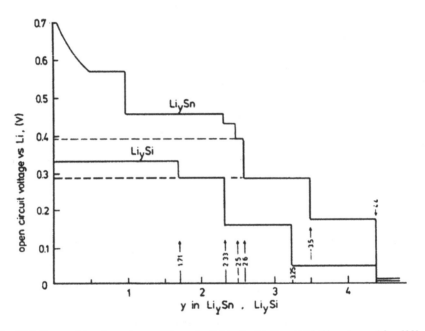

Fig. 18.9 Composition dependence of the potential in the Li–Sn and Li–Si systems. After [32]

Figure 18.10 shows the relatively small polarization that is observed during the charge and discharge of this electrode, even at relatively high current densities [32]. It is seen that there is a potential overshoot due to the free energy involved in the nucleation of a new second phase if the reaction goes to completion in each direction. On the other hand, if the composition is not driven quite so far, so that there is some of the reactant phase remaining, this nucleation-related potential overshoot does not appear, as seen in Fig. 18.11 [32].

This concept has also been demonstrated at ambient temperature in the case of the Li–Sn–Cd system [35, 36]. The composition-dependence of the potentials in the

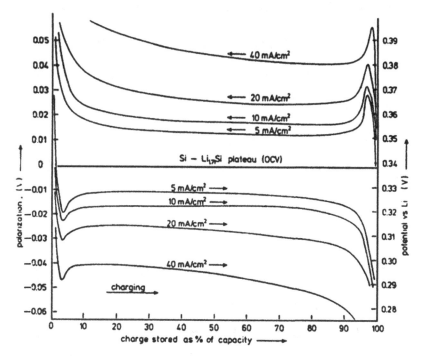

Fig. 18.10 Charge and discharge curves of the Li–Si alloy in the matrix of the electrochemically inert mixed-conducting Li–Sn alloy at different current densities. After [32]

two binary systems at ambient temperatures are shown in Fig. 18.12, and the calculated phase stability diagram for this ternary system is shown in Fig. 18.13. It was shown that the phase $Li_{4.4}Sn$, which has fast chemical diffusion for lithium [37], is stable at the potentials of two of the Li–Cd reconstitution reaction plateaus, and therefore can be used as a matrix phase. The behavior of this composite electrode, in which Li reacts with the Cd phases inside of the Li–Sn phase, is shown in Fig. 18.14.

In order to achieve good reversibility, the composite electrode microstructure must have the ability to accommodate any volume changes that might result from the reaction that takes place internally. This can be taken care of by clever microstructural design and alloy fabrication techniques.

18.5.6 Decrepitation

A phenomenon called "decrepitation", that is also sometimes called "crumbling", can occur in materials that undergo significant volume changes upon the insertion of guest species. These dimensional changes cause mechanical strain in the

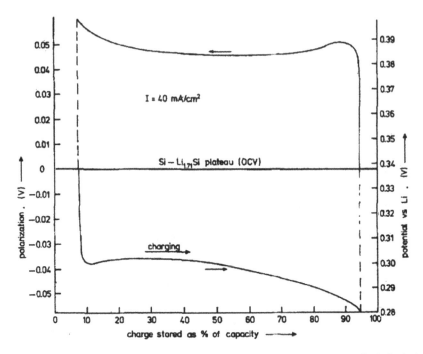

Fig. 18.11 Charge and discharge curves of the Li–Si, Li–Sn composite if the capacity is limited so that the reaction does not go to completion in either direction. There is no large nucleation overshoot in this case. After [32]

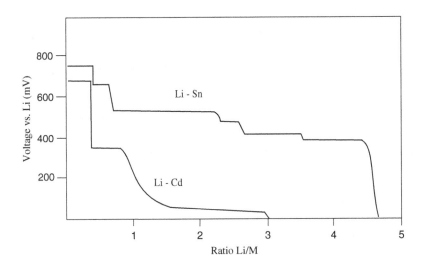

Fig. 18.12 Potential versus composition for Li–Sn and Li–Cd systems at ambient temperature. After [36]

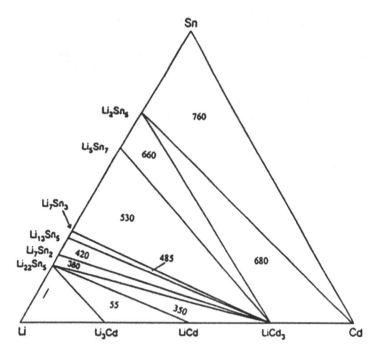

Fig. 18.13 Calculated phase stability diagram for the Li–Cd–Sn system at ambient temperature. Numbers are voltages versus Li. After [36]

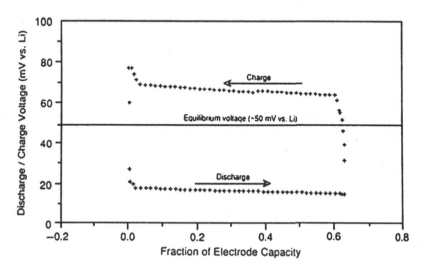

Fig. 18.14 Charge–discharge curve of the Li–Cd system with a fast mixed-conducting phase in the lithium–tin system at ambient temperature. After [37]

microstructure, often resulting in the fracture of particles in an electrode into smaller pieces.

This can be a striking, and sometimes disastrous, phenomenon, for it is not specifically related to fine particles, or even to electrochemical systems. As an example, it has been shown that some bulk solid metals can be caused to fracture, and can even be converted into powders, by repeated exposure to hydrogen gas if they form metal hydrides under the particular thermodynamic conditions present. This is, of course, different from the hydrogen embrittlement problem in metals with body-centered cubic crystal structures, which involves the segregation of hydrogen to dislocations within the microstructure, influencing their mobility.

Decrepitation is often particularly evident during cycling of electrochemical systems. It can readily result in the loss of electronic contact between reactive constituents in the microstructure and the current collector. As a consequence, the reversible capacity decreases.

This phenomenon has long been recognized in some electrochemical systems in which metal hydrides are employed as negative electrode reactants.

Similar phenomena also occur in lithium systems employing alloy electrodes, some of which undergo very large changes in specific volume if the composition is varied over a wide range in order to achieve a large charge capacity.

Because of its potentially large capacity, a considerable amount of attention has been given recently to the Li–Sn system, which is a fine example of this phenomenon. The phase diagram of the Li–Sn system shows that there are six intermediate phases. The thermodynamic and kinetic properties of the different phases in this system were investigated some time ago at elevated temperatures [37, 38] and also at ambient temperatures [30, 31, 35, 36]. The volume changes that occur in connection with phase changes in this alloy system are large. The phase that forms at the highest lithium concentration, $Li_{4.4}Sn$, has a specific volume that is 283% of that of pure tin. Thus, the Li–Sn electrodes swell and shrink, or "breathe", a lot as lithium is added or deleted.

Observations on metal hydrides that undergo larger volume changes have shown that this process does not continue indefinitely. Instead, it is found that there is a terminal particle size that is characteristic of a particular material. Particles with smaller sizes do not fracture further.

Experiments on lithium alloy electrodes have also shown that the electrochemical cycling behavior is significantly improved if the initial particle size is already very small [39], and it is reasonable to conclude that this is related to the terminal particle size phenomenon.

A theoretical study of the mechanism and the influence of the important parameters related to decrepitation utilized a simple one-dimensional model to calculate the conditions under which fracture will be caused to occur in a two-phase structure due to a specific volume mismatch [40]. This model predicts that there will be a terminal particle size below which further fracture will not occur. The value of this characteristic dimension is material-specific, depending upon two parameters, the magnitude of a strain parameter related to the volume mismatch and the fracture toughness of the lower-specific-volume phase. For the same value of volume mismatch, the tendency to fracture will be reduced. Therefore, the larger the terminal particle size the greater the toughness of the material. The results of this model calculation are shown in Fig. 18.15 [40].

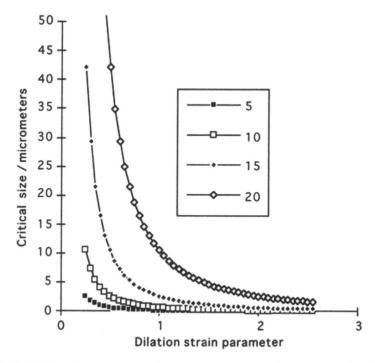

Fig. 18.15 Variation of the critical particle size as a function of the dilation strain for several values of the fracture toughness of the phase in tension. After [40]

The magnitude of the volume change depends upon the amount of lithium that has entered the alloy crystal structure, and is essentially the same for all lithium alloys. This is shown in Fig. 18.16 [41].

18.5.7 Modification of the Micro- and Nanostructure of the Electrode

Some innovative approaches have been employed to ameliorate the decrepitation problem due to the large volume changes inherent in the use of metal alloy and silicon negative electrodes in lithium systems. If that can be done, there is the possibility of a substantial improvement in the electrode capacity.

The general objective is to give the reactant particles room to "breathe", so that they do not impinge upon each other. However, this has to be done so that they are maintained in electrical contact with the current collector system. Thus, they cannot be physically isolated.

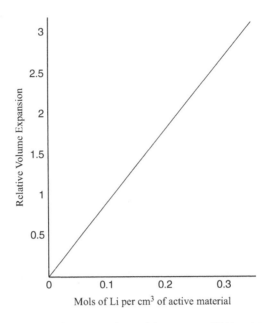

Fig. 18.16 Relation between volume expansion and the amount of lithium introduced into lithium alloys. After [41]

One interesting direction involves the modification of the shape of the surface upon which thin films of active material are deposited [42]. When the reactant film is dense, the volume changes and related stresses parallel to the surface cause a separation from the substrate and loss of electronic contact. But if the surface is rough, there are high spots and low spots that have different local values of current density when the active material is electrodeposited. The deposition rate is greater at the higher locations, and less elsewhere. The result is that the active material is mostly deposited at the high spot locations, and grows in a generally columnar shape away from the substrate. This leaves some space between the columnar growths to allow for their volume changes during the operation of the electrode. This is illustrated schematically in Fig. 18.17.

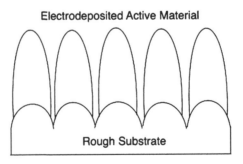

Fig. 18.17 Schematic drawing of the preferential deposition of reactant material upon protrusions on the substrate surface

Another alternative would be to make separated conductive spots on the surface, perhaps by the use of photolithography, that become the preferred locations for the deposition of reactant. By control of the spot arrangement, the electrodeposition can result in the formation of reactant material with limited impingement, thus allowing more "breathing room" when it undergoes charge and discharge.

It has been recently shown that a very attractive potential solution to this cycling problem is the use of reactant material in the form of nanowires. This is illustrated schematically in Fig. 18.18.

Electrolyte

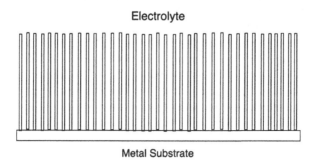

Metal Substrate

Fig. 18.18 Schematic drawing of electrode with a large number of nanowires

The particular example has been silicon [43]. Such wires can be grown directly upon a metallic substrate, so that they are all in good electronic contact. Because there is some space between the individual wires, they can expand and contract as lithium is added or deleted without the constraints present in either thin film or powdered electrode structures. Experiments showed that such fine wires can attain essentially the theoretical capacity of the Li–Si system.

18.5.8 Formation of Amorphous Products at Ambient Temperatures

This chapter has been primarily concerned with understanding the behavior of negative electrode materials under equilibrium or near-equilibrium conditions, from which the potential and capacity limits can be determined. Actual behavior in real applications always deviates from these limiting values, of course.

It was mentioned earlier that repeated cycling can cause crystalline materials to become amorphous. The spectrum of materials in which amorphous phases have been formed under these conditions is now quite broad, and includes some materials of potential interest as positive electrode reactants, such as some vanadium-based materials with the general formula RVO_4, which R is Al, Cr, Fe, In, or Y [44].

There have been a number of observations that the operation of negative electrode materials at very high lithium activities can result in the formation of amorphous, rather than crystalline, products. The properties of these amorphous materials are different from those of the corresponding crystalline materials. This is very different from the amorphization of positive electrode materials under cycling conditions.

One example is a group of nitride alloys with structures related to that of Li_3N, which is known to be a fast ionic conductor for lithium, but in which some of the lithium is replaced by a transition metal such as Co, that has been found to become amorphous upon the first insertion of lithium [45–48].

Experimental evidence for the electrochemical amorphization of alloys in the Li–Si, Li–Sn or Li–Ag systems was presented by Limthongkul [49]. In the latter two cases, this was only a transient phenomenon.

Especially interesting, however, have been experiments that gave evidence for the formation of amorphous silicon during the initial lithiation of a number of silicon-containing precursors, including SiB_3, SiO, $CaSi_2$, and $NiSi_2$ [50–52]. The electrochemical behavior of these materials after the initial lithiation cycle was essentially the same as that found in Si powder that was initially amorphous. There was, however, an appreciable amount of irreversible capacity in the first cycles of these precursors, about 1 mol of Li in the case of SiB_3 and the disilicides, which was evidently due to an irreversible displacement reaction with Li to form one mol of amorphous silicon. In the case of SiO, the irreversible capacity amounted to about two mols of Li, which was surely related to the irreversible formation of Li_2O as well as the amorphous silicon.

Some of these materials with amorphous Si are of considerable potential interest as negative electrode reactants in lithium systems, as their charge/discharge curves are in an attractive potential range, they have reasonable kinetics, and their reversible capacities are quite high. The materials with silicon nanowire structure appear to be particularly attractive.

References

1. R.A. Huggins and D. Elwell, J. Crystal Growth *37*, 159 (1977)
2. C. Wagner, J. Electrochem. Soc. *101*, 225 (1954)
3. C. Wagner, J. Electrochem. Soc. *103*, 571 (1956)
4. G. Deublein and R.A. Huggins, Solid State Ionics *18/19*, 1110 (1986)
5. U. von Sacken, E. Nodwell and J.R. Dahn, Solid State Ionics *69*, 284 (1994)
6. M. Winter, K.-C. Moeller and J.O. Besenhard, "Carbonaceous and Graphitic Anodes", in *Lithium Batteries, Science and Technology*, ed. by G-A Nazri and G. Pistoia, Kluwer Academic Publishers (2004), p. 144
7. J.R. Dahn, A.K. Sleigh, H. Shi, B.M. Way, W.J. Weydanz, J.N. Reimers, Q. Zhong and U. von Sacken, "Carbons and Graphites as Substitutes for the Lithium Anode", in *Lithium Batteries*, ed. by G. Pistoia, Elsevier (1994), p. 1
8. K. Fredenhagen and G. Cadenbach, Z. Anorg. Allg. Chem. *158*, 249 (1926)
9. D. Guerard and A. Herold, Carbon *13*, 337 (1975)
10. G.K. Wertheim, P.M.Th.M. Van Attekum and S. Basu, Solid State Commun. *33*, 1127 (1980)
11. L.B. Ebert, "Intercalation Compounds of Graphite", in Annual Review of Materials Science, Vol. 6, ed. by R.A. Huggins, Annual Reviews, Inc. (1976), p. 181

12. J. O. Besenhard and H. P. Fritz, J. Electroanal. Chem. *53*, 329 (1974)
13. R. Yazami and P. Touzain, J. Power Sources *9*, 365 (1983)
14. S. Basu, U. S. Patent No 4,304,825 (Dec. 8, 1981)
15. S. Basu, U. S. Patent No 4,423,125 (Dec. 27, 1983)
16. T. Nagaura and K. Tozawa, in *Progress in Batteries and Solar Cells*, JEC Press, Inc. *9*, 209 (1990)
17. T. Nagaura, in *Progress in Batteries and Solar Cells*, JEC Press, Inc. *10*, 218 (1991)
18. R.E. Franklin, Proc. Roy Soc (London) *A209*, 196 (1951)
19. R. Yazami, personal communication
20. N. Daumas and A. Herold, C. R. Acad. Sci. C *286*, 373 (1969)
21. T. Zheng, Y. Liu, E.W. Fuller, S. Tseng, U. von Sacken and J.R. Dahn, J. Electrochem. Soc. *142*, 2581 (1995)
22. T. Zheng, J.S. Xue and J.R. Dahn, Chem. Mat. *8*, 389 (1996)
23. T.Zheng, W.R. McKinnon and J.R. Dahn, J. Electrochem. Soc. *143*, 2137 (1996)
24. N. P Yao, L.A. Heredy and R.C. Saunders, J. Electrochem. Soc. *118*, 1039 (1971)
25. E.C. Gay, et al., J. Electrochem. Soc. *123*, 1591 (1976)
26. S.C. Lai, J. Electrochem. Soc. *123*, 1196 (1976)
27. R.A. Sharma and R.N. Seefurth, J. Electrochem Soc. *123*, 1763 (1976)
28. R.N. Seefurth and R.A. Sharma, J. Electrochem. Soc. *124*, 1207 (1977)
29. H. Ogawa, Proceedings of 2nd International Meeting on Lithium Batteries, (Elsevier Sequoia) (1984), p. 259
30. J. Wang, P. King and R.A. Huggins, Solid State Ionics *20*, 185 (1986)
31. J. Wang, I.D. Raistrick and R.A. Huggins, J. Electrochem. Soc. *133*, 457 (1986)
32. B.A. Boukamp, G.C. Lesh and R.A. Huggins, J. Electrochem. Soc. *128*, 725 (1981)
33. B.A. Boukamp, G.C. Lesh and R.A. Huggins, in Proc. Symp. on Lithium Batteries, ed. by H.V. Venkatasetty, Electrochem. Soc. (1981), p. 467.
34. R.A. Huggins and B.A. Boukamp, US Patent 4,436,796
35. A. Anani, S. Crouch-Baker and R.A. Huggins, in Proc. Symp. on Lithium Batteries, ed. by A.N. Dey, Electrochem. Soc. (1987), p. 382
36. A. Anani, S. Crouch-Baker and R.A. Huggins, J. Electrochem. Soc. *135*, 2103 (1988)
37. C. J. Wen and R. A. Huggins, J. Solid State Chem. *35*, 376 (1980)
38. C.J. Wen and R.A. Huggins, J. Electrochem. Soc. *128*, 1181 (1981)
39. J. Yang, M. Winter, and J. O. Besenhard, Solid State Ionics *90*, 281 (1996)
40. R. A. Huggins and W. D. Nix, Ionics *6*, 57 (2000)
41. A. Timmons, PhD Dissertation, Dalhousie University (2007)
42. M. Fujimoto, S. Fujitani, M. Shima, et al., US Patent 7,195,842 (March 27, 2007)
43. C.K. Chan, H. Peng, G. Liu, K. McIlwrath, X. Feng Zhang, R.A. Huggins and Y. Cui, Nat. Nanotechnol. *3*, 31 (2008)
44. Y. Piffard, F. Leroux, D. Guyomard, J.-L. Mansot and M. Tournoux, J. Power Sources *68*, 698 (1997)
45. M. Nishijima, T. Kagohashi, N. Imanishi, Y. Takeda, O. Yamamoto and S. Kondo, Solid State Ionics *83*, 107 (1996)
46. T. Shodai, S. Okada, S-i. Tobishima, and J-i. Yamaki, Solid State Ionics *86–88*, 785 (1996)
47. M. Nishijima, T. Kagohashi, Y. Takeda, N. Imanishi and O. Yamamoto, in 8th International Meeting on Lithium Batteries, (1996), p. 402
48. T. Shodai, S. Okada, S. Tobishima and J. Yamaki, in 8th International Meeting on Lithium Batteries, (1996), p. 404
49. P. Limthongkul, PhD Thesis, Mass. Inst. of Tech. (2002)
50. B. Klausnitzer, PhD Thesis, University of Ulm (2000)
51. A. Netz, PhD Thesis, University of Kiel (2001)
52. A. Netz, R.A. Huggins and W. Weppner, Presented at 11th International Meeting on Lithium Batteries, (2002). Abstract No. 47

Chapter 19
Positive Electrodes in Lithium Systems

19.1 Introduction

Several types of lithium batteries are used in a variety of commercial products, and are produced in very large numbers. According to various reports, the sales volume in 2008 was approximately 10 billion dollars per year, and it was growing rapidly. Most of these products are now used in relatively small electronic devices, but there is also an extremely large potential market if lithium systems can be developed sufficiently to meet the requirements for hybrid, or even plug-in hybrid vehicles.

As might be expected, there is currently a great deal of interest in the possibility of the development of improved lithium batteries in both the scientific and technological communities. An important part of this activity is aimed at the improvement of the positive electrode component of lithium cells, where improvements can have large impacts upon the overall cell performance.

However, before giving attention to some of the details of positive electrodes for use in lithium systems, some comments will be made about the evolution of lithium battery systems in recent years.

Modern advanced battery technology actually began with the discovery of the high ionic conductivity of the solid phase $NaAl_{11}O_{17}$, called sodium beta alumina, by Kummer and co-workers at the Ford Motor Co. laboratory [1]. This led to the realization that ionic transport in solids can actually be very fast, and that it might lead to a variety of new technologies. Shortly thereafter, workers at Ford also showed that one can use this highly conducting solid electrolyte to produce an entirely new type of battery, using molten sodium at the negative electrode and a molten solution of sodium in sulfur as the positive electrode, with the sodium-conducting solid electrolyte in between [2].

This attracted a lot of attention, and scientists and engineers from a variety of other fields began to get interested in this area, which was very different from conventional aqueous electrochemistry, of the late 1960s. This concept of a liquid electrode, solid electrolyte (L/S/L) system was quite different from conventional S/L/S batteries. The development of the $Na/NiCl_2$ "Zebra" battery system, which

R.A. Huggins, *Energy Storage*,
DOI 10.1007/978-1-4419-1024-0_19, © Springer Science+Business Media, LLC 2010

has since turned out to be more attractive than the Na/Na_xS version, came along somewhat later [3–5]. This is discussed elsewhere in this text.

As might be expected, consideration was soon given to the possibility of analogous lithium systems, for it was recognized that an otherwise equivalent lithium cell should produce higher voltages than a sodium cell. In addition, lithium has a lower weight than sodium, another potential plus. There was a difficulty, however, for no lithium-conducting solid electrolyte was known that had a sufficiently high ionic conductivity, and sufficient stability, to be used for this purpose.

Instead, a concept employing a lithium-conducting molten salt electrolyte, a eutectic solution of LiCl and KCl that has a melting point of 356°C, seemed to be an attractive alternative. However, because a molten salt electrolyte is a liquid, the electrode materials had to be solids. That is, the lithium system had to be of the S/L/S type.

Elemental lithium could not be used, because of its low melting point. Instead, solid lithium alloys, primarily the Li/Si and Li/Al systems, were investigated [6], as discussed elsewhere in this text.

A number of materials were pursued as positive electrode reactants at that time, with most attention given to the use of either FeS or FeS_2. Upon reaction with lithium, these materials undergo *reconstitution reactions*, with the disappearance of the initial phases and the formation of new ones [7].

19.2   Insertion Reaction, Instead of Reconstitution Reaction, Electrodes

An important next step was the introduction of the concept that one can reversibly insert lithium into solids to produce electrodes with useful potentials and capacities. This was first demonstrated by Whittingham in 1976, who investigated the addition of lithium to the layer-structured TiS_2 to form Li_xTiS_2, where x went from 0 to 1 [8, 9].

Evidence that this insertion-driven solid solution redox process is quite reversible, even over many cycles, is shown in Fig. 19.1, where the charge and discharge behavior of a Li/TiS_2 cell is shown after 76 cycles [10].

Subsequently, the insertion of lithium into a significant number of other materials, including V_2O_5, LiV_3O_8, and V_6O_{13}, was investigated in many laboratories. In all of these cases, this involved the assumption that one should assemble a battery with pure lithium negative electrodes and with small amounts of, or no, lithium initially in the positive electrode. That is, the electrochemical cell is assembled in the charged state.

The fabrication method generally involved the use of glove boxes and a molten salt or organic liquid electrolyte. This precluded operation at high potentials, and the related oxidizing conditions, as discussed elsewhere.

That work involved the study of potential positive electrode materials by the addition of lithium, and thus scanned their behavior at potentials lower than about 3 V vs. Li, for this is the starting potential for most electrode materials that are

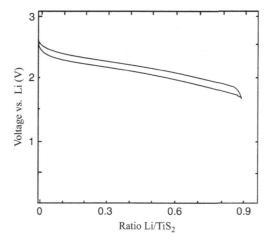

Fig. 19.1 Charge/discharge behavior of a Li/TiS$_2$ cell after 76 cycles. After [10]

synthesized in air. As lithium is added and the cell is discharged, the potential of the positive electrode goes down toward that of pure lithium.

19.2.1   More Than One Type of Interstitial Site or More Than One Type of Redox Species

The variation of the potential depends upon the distribution of available interstitial places that can be occupied by the Li guest ions. If all sites are not the same in a given crystal structure, the result can be the presence of more than one plateau in the voltage-composition curve. An example of this is the equilibrium titration curve for the insertion of lithium into the V$_2$O$_5$ structure shown in Fig. 19.2 [11].

It will be seen later that similar voltage/composition behavior can result from the presence of more than one species that can undergo a redox reaction as the amount of inserted lithium is varied.

19.3   Cells Assembled in the Discharged State

On the other hand, if a positive electrode material initially contains lithium, and some or all of the lithium is deleted, the potential goes up, rather than down, as it does upon the insertion of lithium. Therefore, it is possible to have positive electrode materials that react with lithium at potentials above 3 V, if they already contain lithium, and this lithium can be electrochemically extracted.

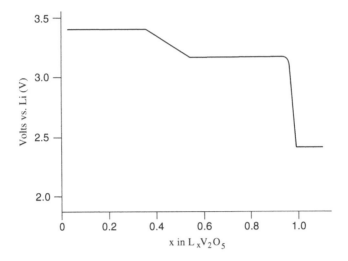

Fig. 19.2 Variation of the potential with the concentration of lithium guest species in the V_2O_5 host structure. After [11]

This concept is shown schematically in Fig. 19.3 for a hypothetical material that is *amphoteric*, and can react at both high and low potentials.

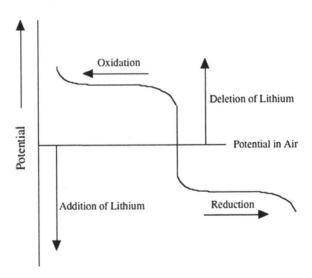

Fig. 19.3 Schematic representation of the behavior of a material that is amphoteric, i.e. that can be both electrochemically oxidized at high potentials by the deletion of lithium, and electrochemically reduced at lower potentials by the addition of lithium

This approach, involving the use of materials in which lithium is already present, was first demonstrated in Prof. Goodenough's laboratory in Oxford. The first examples of materials initially containing lithium, and electrochemically deleting lithium from them, was the work on $Li_{1-x}CoO_2$ [12] and $Li_{1-x}NiO_2$ [13] in 1980. They showed that it is possible in this way to obtain high reaction potentials, up to over 4 V.

It was not attractive to use such materials in cells with metallic Li negative electrodes, however, and this approach did not attract any substantial interest at that time. This abruptly changed as the result of the surprise development by SONY Energytek [14, 15] of a lithium battery containing a carbon negative electrode and a $LiCoO_2$ positive electrode, an organic solvent electrolyte, that became commercially available in 1990. These cells were initially assembled in the discharged state. They were activated by charging, whereby lithium left the positive electrode material, raising its potential, and moved to the carbon negative electrode, whose potential was concurrently reduced.

This cell can be represented as

$$Li_xC/organic\ solvent\ electrolyte/Li_{1-x}CoO_2$$

and the cell reaction can be written as

$$C + LiCoO_2 = Li_xC + Li_{1-x}CoO_2 \tag{19.1}$$

19.4 Solid Positive Electrodes in Lithium Systems

19.4.1 Introduction

In almost every case, the materials that are now used as positive electrode reactants in reversible lithium batteries operate by the use of insertion reactions.

It is interesting that the most commonly used positive electrode in small consumer electronics batteries is now still $LiCoO_2$, although a considerable amount of research is underway in the quest for a more desirable material.

A charge/discharge curve showing the reversible extraction of lithium from $LiCoO_2$ is shown in Fig. 19.4. It is seen that approximately 0.5 Li per mol of $LiCoO_2$ can be reversibly deleted and reinserted. The charge involved in the transfer of lithium ions is balanced by the Co^{3+}/Co^{4+} redox reaction. This process cannot go further, because the layered crystal structure becomes unstable, and there is a transformation into another structure, with a loss of oxygen.

Quite a number of materials are now known from which it is possible to delete lithium at high potentials. Some of these will be described briefly below. This is a very active research area at the present time, and no such discussion can be expected to be complete.

There are a number of interesting materials that have a *face-centered cubic packing* of oxide ions, including both those with the *spinel* structure, such as,

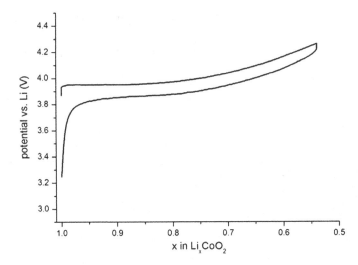

Fig. 19.4 Charge/discharge behavior of Li_xCoO_2

$LiMn_2O_4$. There are also variants containing more than one redox ion, and those with *ordered cation distributions*, which are often described as having *layered structures* (e.g. $LiCoO_2$ and $LiNiO_2$). Another group of materials have *hexagonal close packed oxide ion packing*, including some with ordered *olivine-related structures* (e.g. $LiFePO_4$).

In addition, there are a number of interesting materials that have more open crystal structures, sometimes called *framework*, or *skeleton* structures. These are sometimes described as containing *polyanions*. Examples are some sulfates, molybdates, tungstates, and phosphates, as well as Nasicon, and Nasicon-related materials (e.g. $Li_3V_2(PO_4)_3$ and $LiFe_2(SO_4)_3$). In these materials, lithium ions can occupy more than one type of interstitial position. Especially interesting are materials with more than one type of polyanion. In some cases the reaction potentials are related to the potentials of the redox reactions of ions in octahedral sites, which are influenced by the charge and crystallographic location of other highly charged ions on tetrahedral sites in their vicinity.

Since the reaction potentials of these positive electrode materials are related to the redox reactions that take place within them, consideration should be given to this matter.

The common values of the formal valence of a number of redox species in solids are given in Table 19.1. In some cases, the capacity of a material can be enhanced by the use of more than one redox reaction. In such cases, the issue is whether this can be done without a major change in the crystal structure.

Table 19.1 Common valences of redox ions in solids

Element	Valences	Valence range	Comments
Ti	2,3,4	2	
V	2,3,4,5	3	
Cr	2,3,6	1	6 is poisonous
Mn	2,3,4,6,7	2	6, 7 usable?
Fe	2,3	1	
Co	2,3	1	
Ni	2,3,4	2	
Cu	1,2	1	

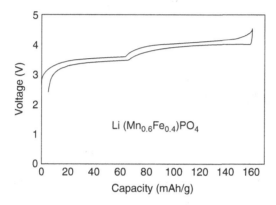

Fig. 19.5 Charge–discharge curve of the reaction of lithium with an example of a double-cation olivine structure material. After [16]

An example of the reaction of lithium with an electrode material containing two redox ions, a Li–Mn–Fe phosphate with the olivine structure, is shown in Fig. 19.5 [16].

Table 19.2 Potentials of redox reactions in a number of host materials/volts vs. lithium

Redox system	Nasicon framework phosphates	Layered close-packed oxides	Cubic close-packed spinels	Hexagonal close-packed olivines
V^{2+}/V^{3+}	1.70–1.75			
Nb^{3+}/Nb^{4+}	1.7–1.8			
Nb^{4+}/Nb^{5+}	2.2–2.5			
Ti^{3+}/Ti^{4+}	2.5–2.7		1.6	
Fe/Fe^{2+}	2.65			
Fe^{2+}/Fe^{3+}	2.7–3.0			3.4
V^{3+}/V^{4+}	3.7–3.8			
Mn^{2+}/Mn^{3+}		4.0	1.7	>4.3
Co^{2+}/Co^{3+}		4.2	1.85	>4.3
Ni^{2+}/Ni^{3+}		4.8		>4.3
Mn^{3+}/Mn^{4+}			4.0	
Fe^{3+}/Fe^{4+}	4.4			
Co^{3+}/Co^{4+}			5.0	

Not all redox reactions are of practical value in electrode materials, and in some cases, their potentials depend upon their environments within the crystal structure. Some experimental data are presented in Table 19.2.

When lithium or other charged mobile, guest ions are inserted into the crystal structure, their electrostatic charge is balanced by a change in the oxidation state of one or more of the redox ions contained in the structure of the host material. The reaction potential of the material is determined by the potential at which this oxidation or reduction of these ions occurs in the host material. In some cases, this redox potential is rather narrowly defined, whereas in others redox occurs over a range of potential, due to the variation of the configurational entropy with the guest species concentration, as well as the site distribution.

19.4.2 Influence of the Crystallographic Environment on the Potential

It has been shown that the crystallographic environment in which a given redox reaction takes place can affect the value of its potential. This matter has been investigated by comparing the potentials of the same redox reactions in a number of oxides with different polyanions, but with the same type of crystal structure. Some of the early references to this topic are [17, 18].

These materials all have crystal structures in which the redox ion is octahedrally surrounded by oxide ions, and these oxide ions also have cations with a different charge in tetrahedral environments on their other side. The electron clouds around the oxide ions are displaced by the presence of adjacent cations with different

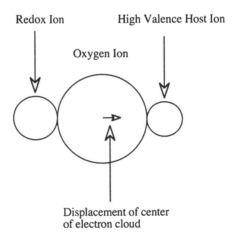

Fig. 19.6 Schematic representation of the displacement of the electron cloud around an oxide ion by the charge upon nearby cations

charges. This is called an "induction effect". It will be further discussed later. It shown schematically in Fig. 19.6.

One of the first cathode materials with a polyanion structure to be investigated was $Fe_2(SO_4)_3$. It can apparently reversibly incorporate up to 2 Li per formula unit, has a very flat discharge curve, indicating a reconstitution reaction, at 3.6 V vs. Li/Li^+ [19, 20].

19.4.3 Oxides with Structures in Which the Oxygen Anions Are in a Face-Centered Cubic Array

19.4.3.1 Materials with Layered Structures

As mentioned above, the positive electrode reactant in the SONY cells was $Li_{x-}CoO_2$, whose properties were first investigated at Oxford [8]. It can be synthesized so that it is stable in air, with $x = 1$. Its crystal structure can be described in terms of a close-packed face-centered cubic arrangement of oxide ions, with the Li^+ and Co^{3+} cations occupying octahedrally coordinated positions in between layers of oxide ions. The cation positions are ordered such that the lithium ions and the transition metal ions occupy alternate layers between close-packed (111) planes of oxide ions. As a result, these materials are described as having layered, rather than simple cubic, structures. This is shown schematically in Fig. 19.7. However, there is a

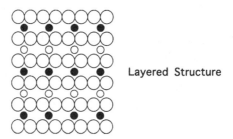

Layered Structure

Fig. 19.7 Simplified schematic drawing of a layered structure in which there is alternate occupation of the cation layers between the close-packed oxide ion layers. The solid and open small circles represent two different types of cations. The larger circles are oxide ions

slight distortion of the cubic oxide stacking because of the difference between the bonding of the monovalent and trivalent cations.

When lithium ions move between octahedral sites within the layers of this structure, they must go through nearby tetrahedral sites that lie along the jump path.

Li_xCoO_2 can be cycled many times over the range $1 > x > 0.5$, but there is a change in the structure and a loss of some oxygen from the structure if more Li^+ ions are deleted. This causes a gradual reduction in the capacity of this electrode material.

Because it has an inherently lower cost and is somewhat less poisonous, it would be preferable to use $LiNiO_2$ instead of $LiCoO_2$. However, it has been found that Li_xNiO_2 is difficult to prepare with the right stoichiometry, as there is a tendency for nickel ions to reside in the lithium layers. This results in a loss of capacity. It was also found that $LiNiO_2$ readily loses oxygen at high potentials, destroying its layer structure, and tending to lead to safety problems because of an exothermic reaction with the organic solvent electrolyte.

There have been a number of investigations of the modification of Li_xNiO_2 by the substitution of other cations for some of the Ni^{3+} ions. It has been found that the replacement of 20–30% of the Ni^{3+} by Co^{3+} ions will impart sufficient stability [21]. Other aliovalent alternatives have also been explored, including the introduction of Mg^{2+} or Ti^{4+} ions.

In the case of $LiMnO_2$, that also has the alpha $NaFeO_2$ structure, it has been found that if more than 50% of the lithium ions are removed during charging, conversion to the spinel structure tends to occur. About 25% of the Mn ions move from octahedral sites in their normal layers into the alkali metal layers, and lithium is displaced into tetrahedral sites [22]. But this conversion to the spinel structure can be avoided by the replacement of half of the Mn ions by chromium [23]. In this case, the capacity (190 mAh/g) is greater than can be accounted for by a single redox reaction, such as Mn^{3+} to Mn^{4+}. This implies that the chromium ions are involved, whose oxidation state can go from Cr^{3+} to Cr^{6+}. Unfortunately, the use of chromium is not considered desirable because of the toxicity of Cr^{6+}.

The replacement of some of the manganese ions in $LiMnO_2$ by several other ions in order to prevent the conversion to the spinel structure has been investigated [24].

A number of other layer-structure materials have also been explored. Some of them contain two or more transition metal ions at fixed ratios, often including Ni, Mn, Co, and Al. In some cases, there is evidence of ordered structures at specific compositions and well-defined reaction plateaus, at least under equilibrium or near-equilibrium conditions. This indicates reconstitution reactions between adjacent phases.

There have been several investigations of layer structure phases with manganese and other transition metals present. A number of these, including $LiMn_{1-y}Co_yO$, have been found to not be interesting, as they convert to the spinel structure rather readily.

However, the manganese–nickel materials, $Li_xMn_{0.5}Ni_{0.5}O_2$ and related compositions, have been found to have very good electrochemical properties, with indications of a solid solution insertion reaction in the potential range 3.5–4.5 V vs. Li [25–28]. It appears that the redox reaction involves a change from Ni^{2+} to Ni^{4+}, whereas the Mn remains as Mn^{4+}. This means that there is no problem with Jahn-Teller distortions, which are related to the presence of Mn^{3+}. The stability of the manganese ions is apparently useful in stabilizing this structure.

At higher manganese concentrations, these materials adopt the spinel structure and apparently react by reconstitution reactions, as will be discussed later in this chapter.

Success with this cation combination apparently led to considerations of compositions containing three cations, such as Mn, Ni, and Co. One of these is $LiMn_{1/3}Ni_{1/3}Co_{1/3}O_2$ [29, 30]. The presence of the cobalt ions evidently stabilizes the layer structure against conversion to the spinel structure. These materials have good electrochemical behavior, and have been studied in many laboratories, but one concern is that they evidently have limited electronic conductivity.

When they are fully lithiated, the nickel is evidently predominantly divalent, the cobalt trivalent, and the manganese tetravalent in these materials. Thus, the major electrochemically active species is nickel, with the cobalt playing an active role only at high potentials. The manganese evidently does not play an active role. It does reduce the overall cost, however.

An extensive discussion of the various approaches to the optimization of the layer structure materials can be found in [31].

19.4.3.2 Materials with the Spinel Structure

The spinel class of materials, with the nominal formula AB_2O_4, has a related structure that also has a close-packed face-centered cubic arrangement of oxide ions. Although this structure is generally pictured in cubic coordinates, it also has parallel layers of oxide ions on (111) planes, and there are both octahedrally coordinated sites and tetrahedrally coordinated sites between the oxide ion planes. The number of octahedral sites is equal to the number of oxide ions, but there are twice as many tetrahedral sites. The octahedral sites reside in a plane intermediate between every two oxide ion planes. The tetrahedral sites are in parallel planes slightly above and below the octahedral site planes between the oxide ion planes.

In *normal spinels*, the A (typically monovalent or divalent) cations occupy 1/8 of the available tetrahedral sites, and the B (typically trivalent or quadrivalent) cations 1/2 of the B sites. In *inverse spinels*, the distribution is reversed.

The spinel structure is quite common in nature, indicating a large degree of stability. As mentioned above, there is a tendency for the materials with the layer

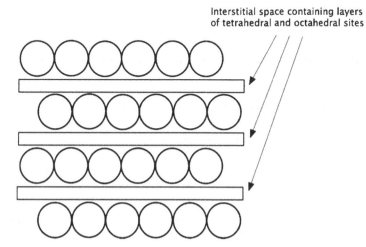

Fig. 19.8 Schematic drawing of the spinel structure in which the cations between the close-packed (111) planes of oxide ions are distributed among both tetrahedral and octahedral sites

structures to convert to the closely related spinel structure. This structure is shown schematically in Fig. 19.8.

A wide range of materials with different A and B ions can have this structure, and some of them are quite interesting for use in lithium systems. An especially important example is $Li_xMn_2O_4$. There can be both lithium insertion and deletion from the nominal composition in which $x = 1$. This material has about 10% less capacity than Li_xCoO_2, but it has somewhat better kinetics and does not have as great a tendency to evolve oxygen.

$Li_xMn_2O_4$ can be readily synthesized with x equal to unity, and this composition

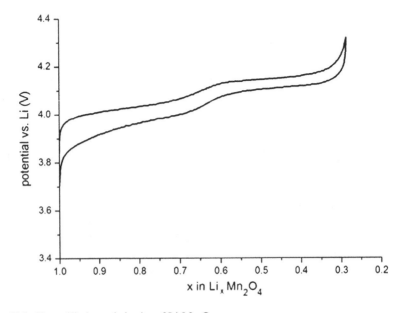

Fig. 19.9 Charge/discharge behavior of $Li_xMn_2O_4$

can be used as a positive electrode reactant in lithium batteries. A typical charge/discharge curve is shown in Fig. 19.9.

It is seen that there are two plateaus. This is related to an ordering reaction of the lithium ions on the tetrahedral sites when x is about 0.5.

Although the $Li_xMn_2O_4$ system, first investigated by Thackeray et al. [32, 33], has the inherent advantages of low cost, good kinetics, and being non-poisonous, it has been found to have some problems that can result in a gradual loss of capacity [34]. Thorough discussions of early work to optimize this material can be found in [35, 36].

One of the problems with this material is the loss of Mn^{2+} into the organic solvent electrolyte as the result of a disproportionation reaction when the potential is low near the end of discharge.

$$2Mn^{3+} = Mn^{4+} + Mn^{2+} \tag{19.2}$$

These ions travel to the carbon negative electrode, with the result that a layer of manganese metal is deposited on its surface that acts to block lithium ion transport.

Another problem that can occur at low potentials is the local onset of Jahn-Teller distortion. This can cause mechanical damage to the crystal structure. On the other hand, if the electrode potential becomes too high as the result of the extraction of too much lithium, oxygen can escape and react with the organic solvent electrolyte.

These problems are reduced by the modification of the composition of the electrode by the presence of additional lithium and a reduction of the manganese [37]. This increase in stability comes at the expense of the capacity. Although the theoretical capacity of $LiMn_2O_4$ is 148 mAh/g, this modification results in a capacity of only 128 mAh/g.

There have also been a number of investigations in which various other cations have been substituted for part of the manganese ions. But in order to avoid the loss of a substantial amount of the normal capacity, it was generally thought that the extent of this substitution must be limited to relatively small concentrations.

At that time, the tendency was to perform experiments only up to a voltage about 4.2 V above the Li/Li^+ potential, as had been done for safety reasons when using Li_xCoO_2. But it was soon shown that it is possible to reach potentials up to 5.4 V vs. Li/Li^+ using some organic solvent electrolytes [38, 39].

Experiments on the substitution of some of the Mn^{2+} ions in Li_xMnO_2 by Cr^{3+} ions [40] showed that the capacity upon the 3.8 V plateau was decreased in proportion to the concentration of the replaced Mn ions. But when the potential was raised to higher values, it was found that this missing capacity at about 4 V reappeared at potentials about 4.9 V that was obviously related to the oxidation of the Cr ions that had replaced the manganese ions in the structure. This particular option, replacing inexpensive and non-toxic manganese with more expensive and toxic chromium is, of course, not favorable.

In both the cases of chromium substitution and nickel substitution, the sum of the capacities of the higher potential plateau and the lower plateau are constant. This implies that there is a one-to-one substitution, and thus that the oxidation that occurs in connection with the chromium and nickel ions is a one-electron process. This is in contradiction to the normal expectation that these ions undergo a three-electron (Cr^{3+} to Cr^{6+}) or a two-electron (Ni^{2+} to Ni^{4+}) oxidation step.

Another example is work on lithium manganese spinels in which some of the manganese ions have been replaced by copper ions. One of these is $LiCu_xMn_{2-x}O_4$ [41–43]. Investigations of materials in which up to a quarter of the manganese ions are replaced by Cu ions have shown that a second plateau appears at 4.8–5.0 V vs. Li/Li^+ that is due to a Cu^{2+}/Cu^{3+} reaction, in addition to the normal behavior of the Li–Mn spinel in the range 3.9–4.3 V vs. Li/Li^+ that is related to the Mn^{3+}/Mn^{4+} reaction. Data for this case are shown in Fig. 19.10 [41]. Unfortunately, the overall

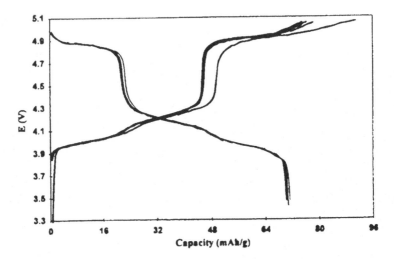

Fig. 19.10 Potential-composition curves for $LiCu_{0.5}Mn_{1.5}O_4$. After [41]

capacity seems to be reduced when there is a substantial amount of copper present in this material [42]. When x is 0.5, the total capacity is about 70 mAh/g, with only

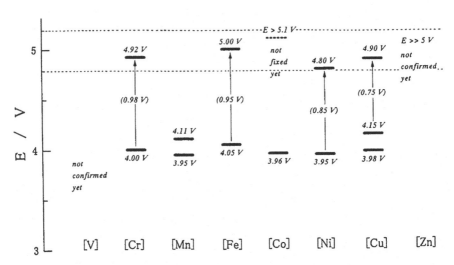

Fig. 19.11 Potential ranges, vs. Li, of redox potentials found as the result of the introduction of a number of cations into lithium manganese spinels. The operating potential range of lithium manganese spinel itself is also shown. After [44]

about 25 mAh/g obtainable in the higher potential region.

The redox potentials that are observed when a number of elements are substituted into lithium manganese spinel structure materials are shown in Fig. 19.11 [44].

An especially interesting example is the spinel structure material with a composition $Li_xNi_{0.5}Mn_{1.5}O_4$. Its electrochemical behavior is different from the others, showing evidence of two reconstitution reactions, rather than solid solution behavior [45].

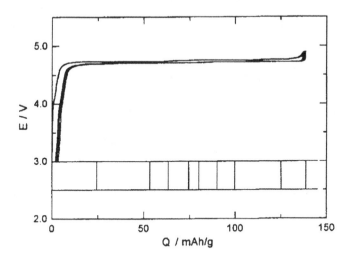

Fig. 19.12 Charge/discharge curves for $Li_xNi_{0.5}Mn_{1.5}O_4$. After [45]

The constant potential charge/discharge curve for this material in the high potential range is shown in Fig. 19.12 [45]. Careful coulometric titration experiments showed that this apparent plateau is actually composed of two reactions with a potential separation of only 20 mV.

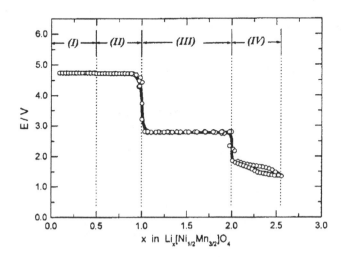

Fig. 19.13 Coulometric titration curve for the reaction of lithium with $Li_xNi_{0.5}Mn_{1.5}O_4$. After [45]

In addition to this high potential reaction, this material also has a reconstitution reaction with a capacity of 1 Li per mol at 2.8 V vs. Li, as well as further lithium uptake via a single phase reaction below 1.9 V. These features are shown in Fig. 19.13 [45]. It is not fully known what redox reactions are involved in this behavior, but it is believed that those at the higher potentials relate to nickel, and the lower ones to manganese.

19.4.3.3   Lower Potential Spinel Materials with Reconstitution Reactions

Whereas this discussion has centered about lithium-containing materials that exhibit high potential reactions, and thus are useful as reactants in the positive electrode, attention should also be given to another related spinel structure material that has a reconstitution reaction at 1.55 V vs. Li [46, 47]. This is $Li_{1.33}Ti_{1.67}O_4$, that can also be written as $Li_x[Li_{0.33}Ti_{1.67}O_4]$ for some of the lithium ions share the octahedral sites in an ordered arrangement with the titanium ions. It also sometimes appears in the literature as $Li_4Ti_5O_{12}$.

This spinel structure material is unusual in that there is essentially no change in the lattice dimensions with the variation of the amount of lithium in the crystal structure, and it has been described as undergoing a *zero-strain insertion reaction* [48]. This is an advantage in that there is almost no volume change-related hysteresis, resulting in very good reversibility upon cycling.

As was mentioned in Chap. 19, this material can also be used on the negative electrode side of a battery. Although there is a substantial voltage loss compared to the use of carbons, the good kinetic behavior can make this option attractive for high power applications, where the lithium-carbons can be dangerous because their reaction potential is rather close to that of elemental lithium.

A charge/discharge curve for this interesting material is shown in Fig. 19.14.

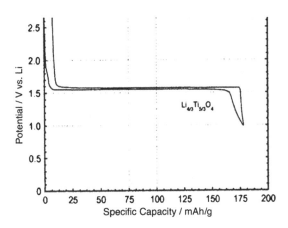

Fig. 19.14 Charge/discharge curve for $Li_4Ti_5O_{12}$. After [47]

19.4.4 Materials in Which the Oxide Ions Are in a Close-Packed Hexagonal Array

Whereas in the spinel and the related layered materials such as Li_xCoO_2, Li_xNiO_2, and Li_xMnO_2, the oxide ions are in a cubic close-packed array, there are also many materials in which the oxide ions are in a hexagonal close-packed configuration. Some of these are currently of great interest for use as positive electrode reactants in lithium batteries, but are generally described as having *framework structures*. They are sometimes also called "scaffold," "skeleton," "network", or "polyanion" structures.

19.4.4.1 The Nasicon Structure

The *Nasicon structure* first attracted attention within the solid state ionics community because some materials with this structure were found to be very good solid electrolytes for sodium ions. One such composition was $Na_3Zr_2Si_2PO_{12}$.

This structure has monoclinic symmetry, and can be considered as consisting of MO_6 octahedra sharing corner oxide ions with adjacent XO_4 tetrahedra. Each octahedron is surrounded by six tetrahedral, and each tetrahedron by four octahedra. These are assembled as a three-dimensional network of M_2X_3 groups. Between these units is three-dimensional interconnected interstitial space, through which small cations can readily move. This structure is shown schematically in Fig. 19.15.

Fig. 19.15 Schematic representation of the Nasicon structure

Unfortunately, Nasicon was found to not be thermodynamically stable vs. elemental sodium, so that it did not find use as an electrolyte in the Na/Na_xS and $Na/NiCl_2$ cells, which are discussed elsewhere in this text, at that time.

However, by using M cations whose ionic charge can be varied, it is possible to make materials with this same structure that undergo redox reactions upon the insertion or deletion of lithium within the interstitial space. The result is that although Nasicon materials may not be useful for the function for which they were first investigated, they may be found to be useful for a different type of application.

As mentioned earlier, it has been found that the identity, size and charge of the X ions in the tetrahedral parts of the structure influences the redox potential of the M ions in the adjacent octahedra [49, 50]. This has been called an *induction effect*.

A number of compositions with this structure have been investigated for their potential use as positive electrode reactants in lithium cells [49–52]. An example is $Li_3V_2(PO_4)_3$, whose potential vs. composition data are shown in Fig. 19.16 [52]. The related differential capacity plot is shown in Fig. 19.17.

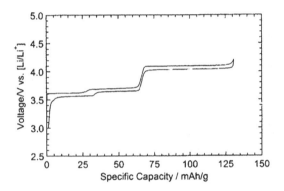

Fig. 19.16 Charge/discharge curve for $Li_3V_2(PO_4)_3$, that has the Nasicon structure. After [52]

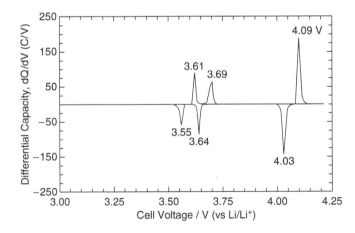

Fig. 19.17 Differential capacity plot corresponding to the charge–discharge data for $Li_3V_2(PO_4)_3$ shown in Fig. 19.10. After [52]

It is seen that the titration curve shows three two-phase plateaus, corresponding to the extraction of two of the lithium ions in the initial structure. The first two plateaus indicate that there are two slightly different configurations for one of the two lithium ions. The potential must be increased substantially, to over 4 V, for the deletion of the second. Experiments showed that it is possible to extract the third lithium from this material by going up to about 5 V, but that this process is not readily reversible, whereas the insertion/extraction of the first two lithium ions is highly reversible.

These phosphate materials all show significantly more thermal stability than is found in some of the other positive electrode reactants, such as those with layer or spinel structures. This is becoming ever more important as concerns about the safety aspects of high-energy batteries mount.

19.4.4.2 Materials with the Olivine Structure

Another group of materials that have a hexagonal stacking of oxide ions are those with the Olivine structure. These materials have caused a great deal of excitement, as well as controversy, in the research community since it was first shown that they can reversibly react with lithium at ambient temperature [53]. The most interesting of these materials is $LiFePO_4$, which has the obvious advantage of being composed of safe and inexpensive materials.

The olivine structure can be described as M_2XO_4, in which the M ions are in half of the available sites of the close-packed hexagonal oxygen array. The more highly charged X ions occupy one-eighth of the tetrahedral sites. Thus, it is a hexagonal analog of the cubic spinel structure discussed earlier. However, unlike spinel, the two octahedral sites in olivine are crystallographically distinct, and have different sizes. This results in a preferential ordering if there are two M ions of different sizes and/or charges. Thus, $LiFePO_4$ and related materials containing lithium and transition metal cations have an ordered cation distribution. The M_1 sites containing lithium are in linear chains of edge-shared octahedral that are parallel to the c-axis in the hexagonal structure in alternate a–c planes. The other (M_2) sites are in a zig–zag arrangement of corner-shared octahedral parallel to the c-axis in the other a–c planes. The result is that lithium transport is highly directional in this structure.

Experiments showed that the extraction of lithium did not readily occur with olivines containing the Mn, Co, or Ni, but proceeded readily in the case of $LiFePO_4$. The deletion of lithium from $LiFePO_4$ occurs by a reconstitution reaction with a moving two-phase interface in which $FePO_4$ is formed at a potential of 3.43 V vs. Li. Although the initial experiments only showed the electrochemical removal of about 0.6 Li ions per mol, subsequent work has shown that greater values can be attained. A reaction with one lithium ion per mol would give a theoretical specific capacity of 170 mAh/g, which is higher than that obtained with $LiCoO_2$. It has been found that the extraction/insertion of lithium in this material can be quite reversible over many cycles.

These phases have the mineralogical names triphylite and heterosite, although the latter was given to a mineral that also contains manganese. Although this reaction potential is significantly lower than those of many of the materials discussed earlier in this chapter, other properties of this class of materials make them attractive for application in lithium-ion cells. There is active commercialization activity, as well as a measure of conflict over various related patent matters.

These materials do not tend to lose oxygen and react with the organic solvent electrolyte nearly so much as the layer structure materials, and they are evidently much safer at elevated temperatures. As a result, they are being considered for larger format applications, such as in vehicles or load leveling, where there are safety questions with some of the other positive electrode reactant materials.

It appeared that the low electronic conductivity of these materials might limit their application, so work was undertaken in a number of laboratories aiming at the development of two-phase microstructures in which electronic conduction within the electrode structure could be enhanced by the presence of an electronic conduction, such as carbon [54]. Various versions of this process quickly became competitive and proprietary.

A different approach is to dope the material with highly charged (supervalent) metal ions, such as niobium, that could replace some of the lithium ions on the small M_1 sites in the structure, increasing the n-type electronic conductivity [55]. On the other hand, experimental evidence seems to indicate that the electronic conduction in the doped Li_xFePO_4 is p-type, not n-type [55, 56]. This could be possible if the cation doping is accompanied by a deficiency of lithium.

This interpretation has been challenged, however, based upon observations of the presence of a highly conductive iron phosphide phase, Fe_2P under certain conditions [57]. Subsequent studies of phase equilibria in the Li–Fe–P–O quaternary system [58] seem to contradict that interpretation.

Regardless of the interpretation, it has been found that the apparent electronic conductivity in these Li_xFePO_4 materials can be increased by a factor of 10^8, reaching values above 10^{-2} S/cm in this manner. These are higher than those found in some of the other positive electrode reactants, such as $LiCoO_2$ (10^{-3} S/cm) and $LiMn_2O_4$ ($2–5 \times 10^{-5}$ S/cm).

An interesting observation is that very fine scale cation-doped Li_xFePO4 has a restricted range of composition at which the two phases "LiFePO4" and "FePO4" are in equilibrium, compared to undoped and larger particle-size material [59]. Thus, there is more solid solubility in each of the two end phases. This may play an important role in their increased kinetics, for in order for the moving interface reconstitution phase transformation involved in the operation of the electrode to proceed, there must be diffusion of lithium through the outer phase to the interface. The rate of diffusional transport is proportional to the concentration gradient. A wider compositional range allows a greater concentration gradient and, thus, faster kinetics.

These materials have been found to be able to react with lithium at very high power levels, greater than those that are typical of common hydride/H_xNiO_2 cells, and commercial applications of this material are being vigorously pursued.

19.4.5 Materials Containing Fluoride Ions

Another interesting variant has also been explored somewhat. This involves the replacement of some of the oxide ions in lithium transition metal oxides by fluoride ions. An example of this is the lithium vanadium fluorophosphate $LiVPO_4F$, which was found to have a triclinic structure analagous to the mineral tavorite, $LiFePO_4 \cdot OH$ [60]. As in the case of the Nasicon materials mentioned earlier, the relevant redox reaction in this material involves the V^{3+}/V^{4+} couple. The charge/discharge behavior of this material is shown in Fig. 19.18 [48], and the related differential capacity results are presented in Fig. 19.19.

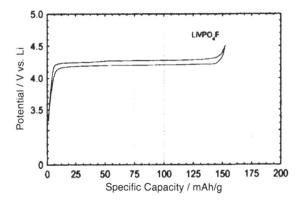

Fig. 19.18 Charge/discharge behavior of $LiVPO_4F$. After [48]

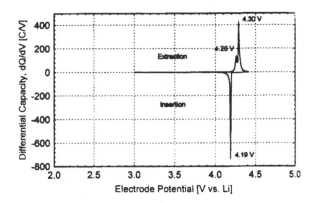

Fig. 19.19 Differential capacity plot corresponding to the charge/discharge data for $LiVPO_4F$ shown in Fig. 19.18. After [48]

19.4.6   Hybrid Ion Cells

An additional variant involves the use of positive electrode reactants that contain other mobile cations. An example of this were the reports of the use of $Na_3V_2(PO_4)F_3$ as the positive electrode reactant and either graphite [61] or $Li_{4/3}Ti_{5/3}O_4$ [62, 63] as the negative reactant in lithium-conducting electrolyte cells. It appears as though the mobile insertion species in the positive electrodes gradually shifts from Na^+ to Li^+. The consideration of this type of mixed-ion materials may lead to a number of interesting new materials.

19.4.7   Amorphization

It was pointed out in Chap. 13 that crystal structures can become amorphous as the result of multiple insertion/extraction reactions. A simple explanation for this phenomenon can be based upon the dimensional changes that accompany the variation in the composition. These dimensional changes are typically not uniform throughout the material, so quite significant local shear stresses can result that disturb the regularity of the atomic arrangements in the crystal structure, resulting in regions with amorphous structures. The degree of amorphization should increase with cycling, as is found experimentally.

There is also another possible cause of this effect that has to do with the particle size. As particles become very small, a significant fraction of their atoms actually reside on the surface. Thus, the surface energy present becomes a more significant fraction of the total Gibbs free energy. Amorphous structures tend to have lower values of surface energy than their crystalline counterparts. As a result, it is easy to understand that there will be an increasing tendency for amorphization as particles become smaller.

19.4.8   The Oxygen Evolution Problem

It is generally considered that a high cell voltage is desirable, and the more the better, since the energy stored is proportional to the voltage, and the output power is proportional to the square of the voltage. However, there are other matters to consider as well. One of these is the evolution of oxygen from a number of the higher potential positive electrode materials.

There is a direct relationship between the electrical potential and the chemical potential of oxygen in materials containing lithium. In this connection, it is useful to remember that the chemical potential was called the *escaping tendency* in the well-known book on thermodynamics by Pitzer and Brewer.

Experiments have shown that a number of the high potential positive electrode reactant materials lose oxygen into the electrolyte. It is also generally thought that the presence of oxygen in the organic solvent electrolytes is related to thermal runaway and the safety problems that are sometimes encountered in lithium cells. An example of experimental measurements that clearly show oxygen evolution is shown in Fig. 19.20.

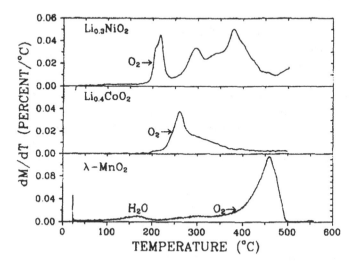

Fig. 19.20 The influence of temperature upon the derivative of the sample weight vs. temperature for three different layer structure materials. After [64]

The relationship between the electrical potential and the chemical potential of oxygen in electrode materials was investigated a number of years ago, but under conditions that are somewhat different from those in current ambient temperature lithium cells. Nevertheless, the principles are the same, and thus, it is useful to review what was found about the thermodynamics of such systems at that time [65].

As discussed in this chapter, many of the positive electrode materials in lithium batteries are ternary lithium transition metal oxides. Since there are three kinds of atoms, that is, three components present, compositions in these systems can be represented on an Isothermal Gibbs Triangle. As discussed in Chaps. 10 and 12, the Gibbs Phase Rule can be written as

$$F = C - P + 2 \tag{19.3}$$

where F is the number of degrees of freedom, C the number of components, and P the number of phases present. At constant temperature and overall pressure, $F = 0$ when $C = P = 3$. This means that all of the intensive variables have fixed values when three phases are present in such three-component systems. Since the electrical potential is an intensive property, this means that the potential has the same value, independent of how much of each of the three phases is present.

It has already been pointed out that the *isothermal phase stability diagram*, an approximation of the Gibbs Triangle in which the phases are treated as though they have fixed, and very narrow, compositions, is a very useful thinking tool to use when considering ternary materials.

The compositions of all of the relevant phases are located on the triangular coordinates, and the possible two-phase tie lines identified. Tie lines cannot cross, and the stable ones can readily be determined from the energy balance of the appropriate reactions. The stable tie lines divide the total triangle into sub-triangles that have two phases at the ends of the tie lines along their boundaries. There are different amounts of the three corner phases at different locations inside the sub-triangles. All of these compositions have the same values of the intensive properties, including the electrical potential.

The potentials within the sub-triangles can be calculated from thermodynamic data on the electrically neutral phases at their corners. From this information, it is possible to calculate the voltages vs. any of the components. This means that one can also calculate the equilibrium oxygen activities and pressures for the phases in equilibrium with each other in each of the sub-triangles. As was shown in Chap. 12, one can also do the reverse, and measure the equilibrium potential at selected compositions in order to determine the thermodynamic data, including the oxygen pressure. The relation between the potential and the oxygen pressure is of special interest because of its practical implications for high voltage battery systems.

The experimental data that are available for ternary lithium-transition metal oxide systems are, however, limited to only a few systems and one temperature. The Li–Mn–O, Li–Fe–O and Li–Co–O systems were studied quantitatively using molten salt electrolytes at 400°C [65]. Because of the sensitivity of lithium to both oxygen and water, they were conducted in a helium-filled glove box. The maximum oxygen pressure that could be tolerated was limited by the formation of Li_2O in the molten salt electrolyte, which was determined to occur at an oxygen partial pressure of 10^{-25} atmospheres at 400°C. This is equivalent to 1.82 V vs. lithium at that temperature. Thus, it was not possible to study materials with potentials above 1.82 V vs. lithium at that temperature.

As an example, the results obtained for the Li–Co–O system under those conditions are shown in Fig. 19.21 [65], which was also included in Chap. 12.

The general equilibrium equation for a ternary sub-triangle that has two binary transition metal oxides (MO_y and MO_{y-x}) and lithium oxide (Li_2O) at its corners can be written as

$$2x\mathrm{Li} + \mathrm{MO}_y = x\mathrm{Li}_2\mathrm{O} + \mathrm{MO}_{y-x} \qquad (19.4)$$

According to Hess's law, this can be divided into two binary reactions, and the Gibbs free energy change ΔG_r is the sum of the two

$$\Delta G_r = \Delta G_r{}^1 + \Delta G_r{}^2 \qquad (19.5)$$

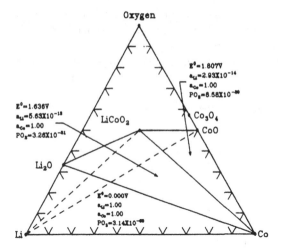

Fig. 19.21 Equilibrium data for the Li–Co–O ternary system at 400°C. After [65]

One is the reaction

$$MO_y = (x/2)O_2 + MO_{y-x} \tag{19.6}$$

The related Gibbs free energy change is given by

$$\Delta G_r{}^1 = - RT \ \ln K \tag{19.7}$$

where K is the equilibrium constant.

The other is the formation of Li_2O that can be written as

$$2xLi + (x/2)O_2 = xLi_2O \tag{19.8}$$

for which the Gibbs free energy change is the standard Gibbs free energy of formation of Li_2O.

$$\Delta G_r{}^2 = x \ \Delta G_f{}^0(Li_2O) \tag{19.9}$$

The potential is related to ΔG_r by

$$E = - \ \Delta G_r/zF \tag{19.10}$$

that can also be written as

$$E = RT/(4F) \ \ln(pO_2) - \ \Delta G_f{}^0(Li_2O)/2F \tag{19.11}$$

This can be simplified to become a linear relation between the potential E and $\ln p$ (O_2), with a slope of $RT/(4F)$ and an intercept related to the Gibbs free energy of formation of Li_2O at the temperature of interest.

Experimental data were obtained on the polyphase equilibria within the sub-triangles in the Li–Mn–O, Li–Fe–O and Li–Co–O systems by electrochemical titration of lithium into various Li_xMO_y materials to determine the equilibrium potentials and compositional ranges. The results are plotted in Fig. 19.22 [65].

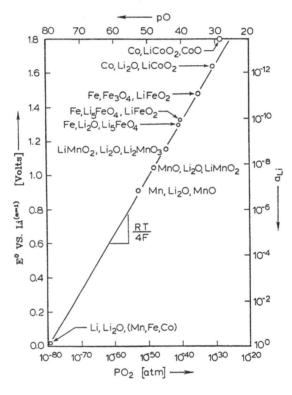

Fig. 19.22 Experimental data on the relation between the potential and the oxygen pressure in phase combinations in the Li–Mn–O, Li–Fe–O and Li–Co–O systems at 400°C. After [65]

It is seen that there is a clear correlation between the potentials and the oxygen pressure in all cases. The equation for the line through the data is

$$E = 3.34 \times 10^{-2} \log p(O_2) + 2.65 \text{ V} \qquad (19.12)$$

The data fit this line very well, even though the materials involved had a variety of compositions and crystal structures. Thus, the relation between the potential and the oxygen pressure is obviously independent of the identity and structures of the materials involved.

Extrapolation of the data in Fig. 19.22 shows that the oxygen pressure would be 1 atmosphere at a potential of 2.65 V vs. Li/Li$^+$ at 400°C.

At 25°C, the Gibbs free energy of formation is -562.1 kJ/mol, so the potential at 1 atmosphere oxygen is 2.91 V vs. Li/Li$^+$. This is about what is observed as the initial open circuit potential in measurements on many transition metal oxide materials when they are fabricated in air.

Evaluating (19.11) for a temperature of 298 K, it becomes

$$E = 1.476 \times 10^{-2} \log p(O_2) + 2.91 \text{ V} \tag{19.13}$$

At this temperature, the slope of the potential vs. oxygen pressure curve is somewhat less than at the higher temperature. But considering it the other way around, the pressure increases more rapidly as the potential is raised.

This result shows that the equilibrium oxygen pressures in the Li–M–O oxide phases increase greatly as the potential is raised. Values of the equilibrium oxygen pressure as a function of the potential are shown in Table 19.3. These data are plotted in Fig. 19.23.

Table 19.3 Values of the equilibrium oxygen pressure over oxide phases in Li–M–O systems at 298 K

E vs. Li/Li$^+$/V	Logarithm of equilibrium oxygen pressure/atm
1	-129
1.5	-95
2	-62
2.5	-28
3	6
3.5	40
4	73.7
4.5	107.6
5	141.4

It can be seen that these values become very large at high electrode potentials, and from the experimental data taken under less extreme conditions, it is obvious that the critical issue is the potential, not the identity of the electrode reactant material or the crystal structure.

One can understand the tendency toward the evolution of oxygen from oxides at high potentials from a different standpoint. Considerations of the influence of the potential on the point defect structure of oxide solid electrolytes has shown that electronic holes tend to be formed at higher potentials, and excess electrons at lower potentials. The presence of holes means that some of the oxide ions have a charge of 1-, rather than 2-. That is, they become peroxide ions, O$^-$. This is an intermediate state on the way to neutral oxygens, as in the neutral oxygen gas molecule O$_2$.

Such ions have been found experimentally on the oxygen electrode surface where the transition between neutral oxygen molecules and oxide ions takes place at the positive electrodes of fuel cells.

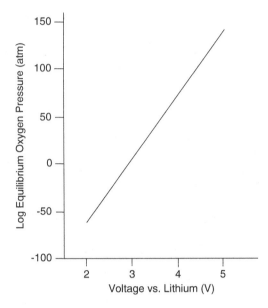

Fig. 19.23 Dependence of the logarithm of the equilibrium oxygen pressure upon the potential in lithium–transition metal–oxide systems

19.4.9   Final Comments on This Topic

It is evident that this is a very active research area, with a number of different avenues being explored in the pursuit of higher potentials, greater capacity, longer cycle life, greater safety, and lower cost. It will be interesting to see which of these new materials, if any, actually come into commercial application.

19.5   Hydrogen and Water in Positive Electrode Materials

19.5.1   Introduction

The electrochemical insertion and deletion of hydrogen is a major feature in some important types of aqueous batteries. The use of metal hydrides as negative electrode reactants in aqueous systems is discussed in Chap. 17, and the hydrogen-driven $H_2NiO_2/HNiO_2$ phase transformation is the major reaction in the positive electrode of a number of "nickel" cells, as described in Chap. 18.

It is generally known that alkali metals react vigorously with water, with the evolution of hydrogen. In addition, a number of materials containing lithium are sensitive to air and/or water, and thus have to be handled in dry rooms or glove boxes. Yet most of the lithium-containing oxides now used as positive electrode reactants in lithium battery systems are synthesized in air, often with little heed given to this problem.

It has long been known that hydrogen (protons) can be present in oxides, including some that contain lithium, and that water (a combination of protons and extra oxide ions) can be absorbed into some selected cases. There are several different mechanisms whereby these can happen.

19.5.2   Ion Exchange

It is possible to simply exchange one type of cationic species for another of equal charge without changing the ratio of cations to anions or introducing other defects in oxides. For example, the replacement of some or all of the sodium cations present in oxides by lithium cations is discussed in several places in this text.

Especially interesting is the exchange of lithium ions by protons. One method is chemically-driven ion exchange, in which there is inter-diffusion in the solid state between native ionic species and ionic species from an adjacent liquid phase. An example of this is the replacement of lithium ions in an oxide solid electrolyte or mixed-conductor by protons as the result of immersion in an acidic aqueous solution. Protons from the solution diffuse into the oxide, replacing lithium ions, which move back into the solution. The presence of anions in the solution that react with lithium ions to form stable products, such as LiCl, can provide a strong driving force. An example could be a lithium transition metal oxide, $LiMO_2$, placed in an aqueous solution of HCl. In this case, the ion exchange process can be written as a simple chemical reaction

$$HCl + LiMO_2 = HMO_2 + LiCl \qquad (19.14)$$

The LiCl product can either remain in solution or precipitate as a solid product.

One can also use electrochemical methods to induce ion exchange. That is, one species inside a solid electrode can be replaced in the crystalline lattice by a different species from the electrolyte electrochemically. The species that is displaced leaves the solid and moves into the electrolyte or into another phase. This electrochemically-driven displacement process is now sometimes called "extrusion" by some investigators.

19.5.3   Simple Addition Methods

Instead of exchanging with lithium, hydrogen can be simply added to a solid in the form of interstitial protons. The charge balance requirement can be accomplished by the co-addition of either electronic or ionic species, that is, either by the introduction of extra electrons or the introduction of negatively charged ionic species, such as O^{2-} ions. If electrons are introduced, the electrical potential of the material will become more negative, with a tendency toward n-type conductivity.

Similarly, oxygen, as oxide ions, can be introduced into solids, either directly from an adjacent gas phase or by reaction with water, with the concurrent formation of gaseous hydrogen molecules. Oxide ions can generally not reside upon interstitial sites in dense oxides because of their size, and thus their introduction requires the presence of oxygen vacancies in the crystal lattice. If only negatively-charged oxide ions are introduced, electroneutrality requires the simultaneous introduction of electron holes. Thus, the electrical potential of the solid becomes more positive, with a tendency toward p-type conductivity.

There is another possibility, first discussed by Stotz and Wagner [66, 67]. This is the simultaneous introduction of species related to both the hydrogen component and the oxygen component of water, that is, both protons and oxide ions. This requires, of course, mechanisms for the transport of both hydrogen and oxygen species within the crystal structure. As mentioned already, hydrogen can enter the crystal structure of many oxides as mobile interstitial protons. The transport of oxide ions, which move by vacancy motion, requires the pre-existence of oxide ion vacancies. This typically involves cation doping. In this dual mechanism, the electrical charge is balanced. Neither electrons nor holes are involved, so the electrical potential of the solid is not changed. The concurrent introduction of both protons and oxide ions is, of course, compositionally equivalent to the addition of water to the solid, although the species H_2O does not actually exist in the crystal structure.

19.5.4 Thermodynamics of the Lithium–Hydrogen–Oxygen System

A number of the features of the interaction between lithium, hydrogen, and oxygen in solids can be understood in terms of the thermodynamics of the ternary Li–H–O system. A useful thinking tool that can be used for this purpose is the *ternary phase stability diagram* with these three elements at the corners. This was discussed in some detail in Chap. 12.

The ternary phase stability diagram for the Li–H–O system at ambient temperature was determined [68] by using chemical thermodynamic data from Barin [69], and assuming that all relevant phases are in their standard states. An updated version is shown in Fig. 19.24.

Using the methods discussed in Chap. 12, the calculated voltages for the potentials of all compositions in the sub-triangles are shown relative to pure lithium.

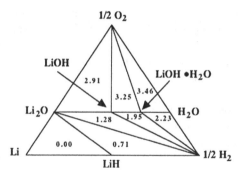

Fig. 19.24 Calculated phase stability diagram for the Li–H–O system at 298 K, assuming unit activities of all phases. The numbers within the triangles are their respective potentials vs. pure lithium. After [68]

If one considers an electrochemical cell with pure lithium at the negative electrode, the potential of water that is saturated with LiOH . H_2O will be 2.23 V when hydrogen is present at one atmosphere. On the other hand, water saturated with LiOH . H_2O will have a potential of 3.46 V vs. Li if one atmosphere of oxygen is present. It can be seen that under these conditions, water has a stability window of 1.23 V, as is the case in the binary hydrogen–oxygen system.

These results may seem to be in conflict with the general conclusion in the literature that the potential of lithium is -3.05 V relative to that of the standard hydrogen electrode (SHE) potential in aqueous electrochemical systems. This can be reconciled by recognizing that the values calculated here are for the case that the water is in equilibrium with LiOH . H_2O, which is very basic, with a pH of 14. The potentials of both the RHE and pure oxygen, as well as all other zero-degree-of-freedom equilibria, decrease by 0.059 V per pH unit. Thus, in order to be compared to the potential of the SHE, these calculated values have to be corrected by (14×0.59), or 0.826 V. Then the voltage between lithium and the SHE that is calculated in this way becomes 3.056 V, corresponding to the data in electrochemical tables.

19.5.5   Examples of Phases Containing Lithium That Are Stable in Water

A number of examples can be found in the literature that are consistent with, and illustrate, these considerations. Particularly appropriate are several experimental results that were published by the group of J.R. Dahn some years ago.

They performed experiments on the addition of lithium to $LiMn_2O_4$ in a LiOH-containing aqueous electrolyte using a carbon negative electrode [70] and showed that the two-phase system $LiMn_2O_4$–$Li_2Mn_2O_4$, that is known to have a potential of

2.97 V vs. Li in non-aqueous cells [71] is stable in water containing LiOH. They used a Ag/AgCl reference electrode, referred their measurements to the SHE, and then converted to the lithium scale, assuming that the potential of the lithium electrode is −3.05 V vs. the SHE. They found that lithium began reacting with the $LiMn_2O_4$ at a potential of −0.1 V vs. the SHE, which is consistent with the value of 2.97 V vs. Li mentioned above.

As lithium was added beyond the two-phase composition limit, the potential fell to that of hydrogen evolution. Their data showed hydrogen evolution at a potential 2.2 V vs. pure Li, and found oxygen evolution on a carbon negative electrode at 3.4 V vs. pure Li. It can readily be seen that these experimental results are consistent with the results of the Gibbs triangle calculations shown in Fig. 19.24.

It was also found that the phase $VO_2(B)$ reacts with lithium at potentials within the stability range of water [72]. Electrochemical cell experiments were performed in which $Li_xVO_2(B)$ acted as the negative electrode, and $Li_xMn_2O_4$ as the positive electrode. These aqueous electrolyte cells gave comparable results to those with the same electrodes in organic solvent electrolyte cells.

19.5.6   Materials That Have Potentials Above the Stability Window of Water

At normal pressures, materials with potentials more positive than that of pure oxygen will tend to oxidize water to cause the evolution of electrically neutral molecular oxygen gas. For this to happen, there must be a concurrent reduction process. One possibility is the insertion of positively charged ionic species, along with their charge-balancing electrons, into the material in question. The insertion of protons or lithium ions and electrons into high-potential oxides is one possible example of such a reduction process. When this happens, the potential of the material goes down toward that of pure oxygen.

19.5.7   Absorption of Protons from Water Vapor in the Atmosphere

A number of materials that are used as positive electrode reactants in lithium battery systems have operating potentials well above the stability range of water. Cells containing these materials and carbon negative electrodes are typically assembled in air in the uncharged state. It is generally found that the open circuit cell voltage at the start of the first charge is consistent with lithium-air equilibrium, that is, along the Li_2O/O_2 edge of the ternary phase stability diagram in Fig. 19.24. This can be calculated to be 2.91 V vs. pure lithium. This can be explained by the reaction of these materials with water vapor in the atmosphere. Protons and electrons enter the crystal structures of these high potential materials, reducing their potentials to that value. This is accompanied by the concurrent evolution of molecular oxygen.

19.5.8   Extraction of Lithium from Aqueous Solutions

An analogous situation can occur if a material that can readily insert lithium, rather than protons, has a potential above the stability range of water. If lithium ions and electrons enter the material's structure, the potential will decrease until the value in equilibrium with oxygen is reached. Such a material can thus be used to extract lithium from aqueous solutions. This was demonstrated by experiments on the use of the λ-MnO_2 spinel phase that absorbed lithium when it was immersed in aqueous chloride solutions [73].

References

1. Y.F.Y. Yao and J.T. Kummer, J. Inorg. Nucl. Chem. 29, 2453 (1967)
2. N. Weber and J.T. Kummer, Proc. Annu. Power Sources Conf. 21, 37 (1967)
3. J. Coetzer, J. Power Sources 18, 377 (1986)
4. R.C. Galloway, J. Electrochem. Soc. 134, 256 (1987)
5. R.J. Bones, J. Coetzer, R.C. Galloway, D.A. Teagle, J. Electrochem. Soc.134, 2379 (1987)
6. R.A. Huggins, J. Power Sources 81–82, 13 (1999)
7. D.R. Vissers, Z. Tomczuk, and R.K. Steunenberg, J. Electrochem. Soc 121, 665 (1974)
8. M.S. Whittingham, Science, 192 1126 (1976)
9. M.S. Whittingham, J. Electrochem. Soc. 123, 315 (1976)
10. M.S. Whittingham, in *Fast Ion Transport*, ed. by B. Scrosati, A. Magistris, C.M. Mari and G. Mariotto, Kluwer Academic, Dordrecht (1993), p. 69
11. P. G. Dickens, S.J. French, A.T. Hight and M.F. Pye, Mat. Res. Bull. 14, 1295 (1979)
12. K. Mizushima, P.C. Jones, P.J. Wiseman and J.B. Goodenough, Mat. Res. Bull. 15, 783 (1980)
13. J.B. Goodenough, K. Mizushima and T. Takada, Jap. J. Appl. Phys. 19, Suppl 19–3, 305 (1980)
14. T. Nagaura and K. Tozawa, in *Progress in Batteries and Solar Cells*, ed. by A. Kozawa, JEC Press, Inc. 9, 209 (1990)
15. T. Nagaura, in *Progress in Batteries and Solar Cells*, ed. by A. Kozawa, JEC Press, Inc. 10, 218 (1991)
16. A. Yamada, M. Hosoya, S.C. Chung, Y. Kudo and K.Y. Liu, "Concepts in Design of Olivine-Type Cathodes", Abstract No. 205, Electrochemical Society Meeting, San Francisco (2001)
17. K.S. Nanjundaswamy, A.K. Padhi, J.B. Goodenough, S. Okada, H. Ohtsuka, H. Arai, and J. Yamaki, Solid State Ionics 92, 1 (1996). [Nanjundaswamy, 1996 #42]
18. A.K. Padhi, K.S. Nanjundaswamy, C. Masquelier and J.B. Goodenough, J. Electrochem. Soc. 144, 2581 (1997). [Padhi, 1997 #9]
19. S. Okada, H. Ohtsuka, H. Arai and M. Ichimura, Electrochem. Soc. Ext. Abstracts 93-1, May, 1993, p. 130
20. S. Okada, T. Takada, M. Egashira, J. Yamaki, M. Tabuchi, H. Kageyama, T. Kodama and R. Kanno, "Characteristics of 3D Cathodes with Polyanions for Lithium Batteries", presented at Second Hawaii Battery Conference, Jan. 1999
21. I. Sadadone and C. Delmas, J. Mater. Chem. 6, 193 (1996)
22. P.G. Bruce, A.R. Armstrong and R. Gitzendanner, J. Mater. Chem. 9, 193 (1999)
23. Y. Grincourt, C. Storey and I.J. Davidson, J. Power Sources 97–98, 711 (2001)
24. J.M. Paulson, R.A. Donaberger and J.R. Dahn, Chem. Mater. 12, 2257 (2000)
25. M.E. Spahr, P. Novak, B. Schneider, O. Haas, R.J. Nesper, J. Electrochem. Soc. 145, 1113 (1998)

26. T. Ohzuku and Y. Makimura, Chem. Lett. 8, 744 (2001)
27. Z. Lu, D.D. MacNeil and J.R. Dahn, Electrochem. Solid-State Lett. 4, A191 (2001)
28. K. Kang, Y.S. Meng, J. Breger, C.P. Grey and G. Ceder, Science 311, 977 (2006)
29. Z. Liu, A. Yu and J.Y. Lee, J. Power Sources 81–82, 416 (1999)
30. M. Yoshio, H. Noguchi, J-I. Itoh, M. Okada and T. Mouri, J. Power Sources 90, 176 (2000)
31. M.S. Whittingham, Chem. Rev. 104, 4271 (2004)
32. M.M. Thackeray, W.I.F. David, P.G. Bruce, and J.B. Goodenough, Mat. Res. Bull. 18, 461 (1983)
33. M.M. Thackeray, P.J. Johnson, L.A. de Piciotto, P.G. Bruce, and J.B. Goodenough, Mat. Res. Bull. 19, 179 (1984)
34. M.M. Thackeray, in *Handbook of Battery Materials*, ed. by J.O. Besenhard, Wiley-VCH, New York (1999), p. 293
35. D. Guyomard and J.M. Tarascon, Solid State Ionics 69, 222 (1994)
36. G. Amatucci and J.-M. Tarascon, J. Electrochem. Soc. 149, K31 (2002)
37. R.J. Gummow, A. De Kock and M.M. Thackeray, Solid State Ionics 69, 59 (1994)
38. D. Guyomard and J.-M. Tarascon, US Patent 5,192,629, (March 9, 1993)
39. D. Guyomard and J.-M. Tarascon, Solid State Ionics 69, 293 (1994)
40. C. Sigala, D. Guyomard, A. Verbaere, Y. Piffard, and M. Tournoux, Solid State Ionics 81, 167 (1995)
41. Y. Ein-Eli and W.F. Howard, J. Electrochem. Soc. 144, L205 (1997)
42. Y. Ein-Eli, W.F. Howard, S.H. Lu, S. Mukerjee, J. McBreen, J.T. Vaughey, and M.M. Thackeray, J. Electrochem. Soc. 145, 1238 (1998)
43. Y. Ein-Eli, S.H. Lu, M.A. Rzeznik, S. Mukerjee, X.Q. Yang, and J. McBreen, J. Electrochem. Soc. 145, 3383 (1998)
44. T. Ohzuku, S. Takeda and M. Iwanaga, J. Power Sources 81–82, 90 (1999)
45. K. Ariyoshi, Y. Iwakoshi, N. Nakayama and T. Ohzuku, J. Electrochem. Soc. 151, A296 (2004)
46. K.M. Colbow, J.R. Dahn and R.R. Haering, J. Power Sources 26, 397 (1989)
47. T. Ohzuku, A. Ueda and N. Yamamoto, J. Electrochem. Soc. 142, 1431 (1995)
48. J.B. Goodenough, H.Y-P. Hong and J.A. Kafalas, Mat. Res. Bull. 11, 203 (1976)
49. K.S. Nanjundaswamy, A.K. Padhi, J.B. Goodenough, S. Okada, H. Ohtsuka, H. Arai, J. Yamaki, Solid State Ionics 92, 1 (1996)
50. A.K. Padhi, K.S. Nanjundaswamy, C. Masquelier, S. Okada and J.B. Goodenough, J. Electrochem. Soc. 144, 1609 (1997)
51. J. Barker and M.Y. Saidi, US Patent 5,871,866 (1999)
52. M.Y. Saídi, J. Barker, H. Huang, J.L. Swoyer and G. Adamson, Electrochem. Solid-State Lett., 5, A149 (2002)
53. A.K. Padhi, K.S. Nanjundaswamy and J.B. Goodenough, J. Electrochem. Soc. 144, 1188 (1997)
54. N. Ravet, J.B. Goodenough, S. Besner, M. Simoneau, P. Hovington an M. Armand, Electrochem. Soc. Meeting Abstract 99–2, 127 (1999)
55. S.-Y. Chung, J.T. Bloking and Y.-M. Chiang, Nat. Mater. 1, 123 (2002)
56. R. Amin and J. Maier, Solid State Ionics 178, 1831 (2008)
57. P.S. Herle, B. Ellis, N. Coombs and L.F. Nazar, Nat. Mater. 3, 147 (2004)
58. S.P. Ong, L. Wang, B. Kang and G. Ceder, presented at the Materials Research Society Meeting in San Francisco, March, 2007
59. N. Meethong, H.-Y.S. Huang, S.A. Speakman, W.C. Carter and Y.-M. Chiang, Adv. Funct. Mater. 17, 1115 (2007)
60. J. Barker, M.Y. Saidi and J.L. Swoyer, J. Electrochem. Soc. 151, A1670 (2004)
61. J. Barker, R.K.B. Gover, P. Burns and A.J. Bryan, Electrochem. Solid State Lett 9, A190 (2006)
62. J. Barker, R.K.B. Gover, P. Burns and A.J Bryan, Electrochem Solid-State Lett 10, A130 (2007)

63. J. Barker, R.K.B. Gover, P. Burns and A.J Bryan , J. Electrochem. Soc. 154, A882 (2007)
64. J.R. Dahn, E.W. Fuller, M. Obrovac and U. von Sacken, Solid State Ionics 69, 265 (1994)
65. N.A. Godshall, I.D. Raistrick and R.A. Huggins, J. Electrochem. Soc. 131, 543 (1984)
66. S. Stotz and C. Wagner, Ber. Bunsenges. Physik. Chem. 70, 781 (1966)
67. C. Wagner, Ber. Bunsenges. Physik Chem. 72, 778 (1968)
68. R.A. Huggins, Solid State Ionics 136–137, 1321 (2000)
69. I. Barin, Thermochemical Data of Pure Substances, VCH Verlag, Wienheim (1989)
70. W. Li, W.R. McKinnon and J.R. Dahn, J. Electrochem. Soc. 141, 2310 (1994)
71. J.M. Tarascon and D. Guyomard, J. Electrochem. Soc. 138, 2864 (1993)
72. W. Li and J.R. Dahn, J. Electrochem. Soc. 142, 1742 (1995)
73. H. Kanoh, K. Ooi, Y. Miyai and S. Katoh, Sep. Sci. Technol., 28, 643 (1993)

Chapter 20
Primary, Nonrechargeable Batteries

20.1 Introduction

Except for the discussions of the lithium/iodine cell in Chap. 10, all of the discussion concerning batteries for energy storage has been oriented toward understanding the properties of individual cell components and systems. The emphasis has been upon those that are most interesting for use in rechargeable batteries.

There are, however, a number of types of batteries that are very common and important, even though they cannot be readily recharged. They are often called "primary batteries", and are typically discarded when they become discharged. Several of these will be discussed in this chapter.

Because some primary cells have higher values of specific energy than current rechargeable systems, there is continual interest in finding methods to electrically recharge them, rather than having to refurbish them chemically by reprocessing one or more of their components.

20.2 The Common Zn/MnO$_2$ "Alkaline" Cell

A prominent member of this group is the very common Zn/MnO$_2$ "alkaline" cell that is used in many relatively small electronic devices. These cells are available in great numbers in standard AA and AAA sizes.

Elemental zinc is the negative electrode reactant, and the electrolyte is a solution of KOH. The open circuit voltage is initially 1.5 V, but it decreases as energy is extracted and the residual capacity becomes reduced. This reduction in cell voltage is due to the change of the potential of the positive electrode due to the insertion of protons from the electrolyte. This can be described as changing the value of x in the composition H$_x$MnO$_2$ from zero to about 1. The proton content can be increased until the value of x becomes 2. However, the second proton reaction occurs at a cell voltage of about 1 V, which is too low to be of practical use. This topic was discussed in Chap. 18, and will not be repeated here.

R.A. Huggins, *Energy Storage*,
DOI 10.1007/978-1-4419-1024-0_20, © Springer Science+Business Media, LLC 2010

Although these Zn/MnO_2 cells are generally considered to be nonrechargeable, there have been some developments that make it possible to recharge them a modest number of times, and a small fraction of the alkaline cell market has been oriented in that direction. It involves modifications in the design and proprietary changes in the composition of the materials. Their rechargeability depends upon the depth to which they have been discharged, and there is a gradual reduction in the available capacity.

20.3 Ambient Temperature Li/FeS_2 Cells

Another type of consumer battery that is gradually becoming more popular in a number of consumer markets is the Li/FeS_2 cell. In this case, the negative electrode is lithium metal. The electrolyte is a lithium salt dissolved in an organic solvent similar to that used in rechargeable lithium batteries.

The potential of the elemental lithium negative electrode is constant, but that of the positive electrode varies with the state of charge. At very low current drain, it shows two voltage plateaus, at about 1.7 V and 1.5 V versus lithium, as seen in Fig. 20.1. This indicates the formation of an intermediate phase, and therefore a sequence of two different reactions.

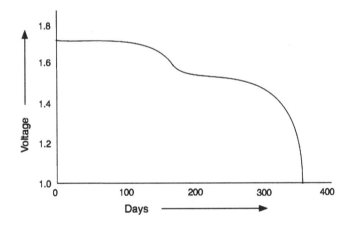

Fig. 20.1 Variation of the cell voltage with state of charge for Li/FeS_2 cells

At moderate, or greater, currents, the plateau structure disappears, and the output voltage drops steadily from about 1.6 to 1.5 V with the state of charge. This means that the intermediate phase does not form under those conditions.

The voltage of this type of battery is about 0.1 V higher than that of the common alkaline cells, and there is less fade as the cell becomes discharged. The primary advantage of the Li/FeS_2 cells over the less expensive Zn/MnO_2 cells is their ability to handle higher currents. This is shown in Fig. 20.2. This property makes them especially useful for pulse applications, such as in cameras.

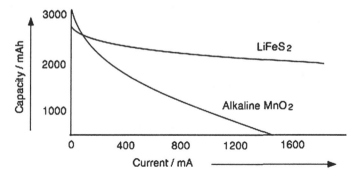

Fig. 20.2 Variation of the capacity of typical Li/FeS$_2$ cells with the current drain

20.4 Li/I$_2$ Batteries for Heart Pacemakers

There was a discussion of the nonrechargeable Li/I$_2$ batteries that are commonly used to provide power for heart pacemakers in Chap. 10. There is no need to repeat that material here, other than to point out the unusual situation that the reaction product, LiI, is actually the electrolyte in this case.

20.5 Lithium/Silver Vanadium Oxide Defibrillator Batteries

Another type of implantable primary cell that is now used to provide power for medical devices, such as defibrillators, is the lithium/silver vanadium oxide cell. The attractive features of this chemistry were first recognized in 1979 [1, 2]. The person responsible for the commercial development of these batteries, Esther Takeuchi, received the National Medal of Technology and Innovation from President Obama in October, 2009.

The negative electrode in these cells is elemental lithium, and the electrolyte is the lithium salt LiBF$_4$ in an organic solvent, propylene carbonate. The positive electrode starts as AgV$_2$O$_{5.5}$, a member of the family of electronically conducting oxides called "vanadium bronzes" [3].

Lithium reacts with this positive electrode material by an insertion reaction that can be written as

$$x\text{Li} + \text{AgV}_2\text{O}_{5.5} = \text{Li}_x\text{AgV}_2\text{O}_{5.5} \qquad (20.1)$$

This reaction occurs over several steps, with corresponding values of x. This can be seen from the plot of the cell voltage as a function of the extent of this reaction shown in Fig. 20.3.

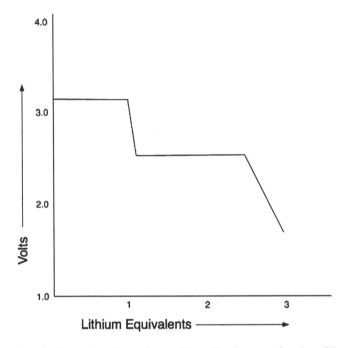

Fig. 20.3 Schematic illustration of the variation of the cell voltage as a function of the amount of lithium reacted

Charge balance is accomplished by a change in the effective charge of the cations originally in the vanadium bronze. In $AgV_2O_{5.5}$, all of the vanadium ions have an effective charge of 5^+. Upon adding Li^+ ions, the system moves into a constant-potential 2-phase regime, where both $AgV_2O_{5.5}$ and $LiAgV_2O_{5.5}$ are present. When the overall composition reaches the end of that voltage plateau, it moves into a variable-potential composition range in which only the phase $LiAgV_2O_{5.5}$ is present, and half of the vanadium ions have a charge of 5^+, and the other half a charge of 4^+. The addition of another two lithium ions causes the composition to move into another two-phase plateau in which the phase $LiAgV_2O_{5.5}$ is in equilibrium with a composition that is nominally $Li_3AgV_2O_{5.5}$. The effective charge of the vanadium ions in this latter phase is still 4^+, but the nominal charge upon the silver ions has become zero. This means that there must be some particles of elemental silver present in the microstructure in addition to a phase of composition $Li_3V_2O_{5.5}$. Experiments have shown that the lower-lithium reactions are reversible, but the last, which involves the precipitation of a new phase, is not.

The decrease of voltage as the cell is discharged allows the state of charge to be readily determined by voltage measurements. This is important when such power sources are used in implantable medical devices. These cells exhibit a very low rate of self discharge, have a long shelf life, and store a large amount of energy per unit volume, 930 Wh/l. The latter feature is attractive for applications in which battery size is important.

20.6   Zn/Air Cells

Primary cells based upon the reaction of zinc with air have been available commercially for a number of years. This chemistry can produce a rather large value of specific energy, is relatively inexpensive, and presents no significant environmental problems. One of the first applications was as a power source for small hearing aids.

A cell with metallic zinc as the negative electrode and oxygen (or air) on the positive side is shown schematically in Fig. 20.4.

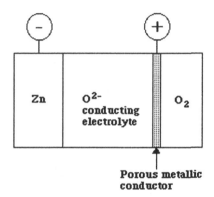

Fig. 20.4 Schematic representation of a Zn/O_2 cell

There must be a mechanism for the flow of electrons into and out of an external electronic circuit from both electrodes. This is accomplished on the negative side by contact with metallic zinc. On the positive side, there is a porous metallic conductor in contact with both the oxygen reactant and the alkaline electrolyte. Although this metal plays no role in the overall cell reaction, the three-phase contact allows the electrochemical reaction that converts neutral atoms into ions and electrons.

The discharge reaction mechanism involves the transport of oxygen across the cell from the positive electrode to the negative electrode, with the formation of ZnO on top of the Zn. A cell that is partially discharged is shown schematically in Fig. 20.5.

ZnO is an electronic conductor, so the electrochemical interface, where the electrical charge transport mechanism is converted from ions to electrons, is at the interface between the ZnO and the electrolyte. It is the electric potential at that interface that determines the externally measurable electrical potential of the negative electrode.

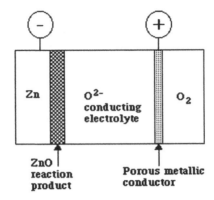

Fig. 20.5 Schematic representation of a Zn/O$_2$ cell that is partially discharged

The reaction that determines the potential is generally assumed to be the formation of ZnO

$$Zn + \frac{1}{2}O_2 = ZnO \tag{20.2}$$

so that the voltage would be determined by the standard Gibbs free energy of formation of ZnO from zinc and oxygen.

$$E = \frac{-\Delta G_f(ZnO)}{zF} \tag{20.3}$$

where $z = 2$, and F is the Faraday constant, 96.5 kJ/volt equivalent. The value of $\Delta G_f(ZnO)$ at 298 K is -320.5 kJ/mol so the equilibrium voltage E is 1.66 V at that temperature.

However, these cells operate in air, rather than pure oxygen. Therefore, the chemical potential of oxygen is lower, and the electrical potential of the positive electrode is reduced. The chemical potential of oxygen in the positive electrode can be expressed as

$$\mu(O_2) = \mu^0(O_2) + RT \ln p(O_2) \tag{20.4}$$

where $\mu^0(O_2)$ is the chemical potential of oxygen in its standard state, a pressure of 1 atmosphere at the temperature in question, and $p(O_2)$ is the actual oxygen pressure at the electrode.

In air, the oxygen partial pressure is approximately 0.21 atmospheres, so that the cell voltage is reduced by

$$\Delta E = \frac{RT}{zF} \ln(0.21) \tag{20.5}$$

The result is that the equilibrium voltage of the Zn/O_2 cell when air is the reactant on the positive side should be reduced by 0.02 V. Thus, a Zn/air cell should have an open circuit voltage of 1.64 V. If the oxygen pressure is maintained at a constant value, the voltage will be independent of the state of charge, that is, it will have the characteristics of a plateau in a battery discharge curve.

The value of the maximum theoretical specific energy can be calculated from this information using the weights of the reactants. As discussed in Chap. 9, the value of the MTSE is given by

$$\text{MTSE} = 26,805\ (zE)/W_t\ \text{Wh/kg} \qquad (20.5)$$

The value of the reactant weight, W_t, is the weight of a mol of Zn (65.38 g) plus the weight of 1/2 mol of oxygen (8 g), or a total of 73.38 g per mol of reaction. The value of z is 2, the number of elementary charges involved in the virtual cell reaction.

Using this value and a cell voltage of 1.64 V for the case of air at the positive electrode, the MTSE is 1,198 Wh/kg. If pure oxygen were used, it would be 1,213 Wh/kg.

But there is a problem. The measured open circuit voltage of commercial Zn/air cells is about 1.5 V, not 1.64 V. The reason for this has to do, again, with what is actually going on at the positive electrode. The normal assumption is that the positive electrode reactant is oxygen, and therefore the potential should be that of pure oxygen at the partial pressure of air.

Experiments have shown the presence of peroxide ions at the positive electrode in alkaline aqueous cells. Instead of a conversion of oxygen from O_2 in the gas phase to O^{2-} ions in the electrolyte, there is an intermediate step, the presence of peroxide ions.

In peroxide ions, O^-, oxygen is at an intermediate charge state between neutral oxygen, O^0, in oxygen molecules in the gas, and oxide ions, O^{2-}, in the KOH electrolyte. In such aqueous systems, this can be written as two steps in series

$$O_2 + H_2O + 2e^- = HO_2^- + OH^- \qquad (20.6)$$

and

$$HO_2^- + H_2O + 2e^- = 3OH^- \qquad (20.7)$$

The result is that the electrical potential in the positive electrode is determined by the presence of hydrogen peroxide, which is formed by the reaction of oxygen with the KOH electrolyte.

This is also the case with aqueous electrolyte hydrogen/oxygen fuel cells, where the open circuit voltage is determined by the presence of peroxide, rather than oxide, ions [4–6]. This is shown in Fig. 20.6. High temperature proton or oxide ion-conducting fuel cells have open circuit voltages that correspond to the assumption that the positive electrode reactant is oxygen.

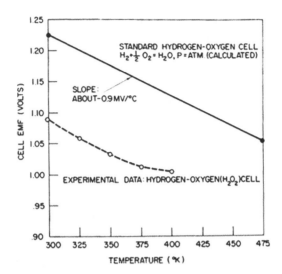

Fig. 20.6 Experimental data on the voltage of aqueous fuel cells, showing the influence of the presence of peroxide ions at the positive electrode. After [6]

Zn/O$_2$ cells are sold with a removable sealing material that prevents access of air to the positive electrode structure so that there is no self-discharge before they are used.

The specific energy is very large, about 30 times the value of the maximum theoretical specific energy of a typical Pb/PbO$_2$ cell, so it is obvious why there is an interest in finding a way to make this system reversible. To do so, three general problems must be solved; the rechargeability of the zinc oxide product, the reversibility of the air electrode, and the sensitivity of the KOH electrolyte to contamination from CO$_2$ in the ambient air. CO$_2$ reacts with hydroxides to form solid carbonates, which can block the ionic transport through the electrolyte.

Development efforts toward the alleviation or avoidance of these problems have been undertaken in a number of laboratories, but this has not yet led to large-scale applications.

A large effort undertaken with the support of the German Post some years ago ran into several problems, the major one being the cost and logistics of the chemical regeneration of the zinc oxide product back into useful zinc electrodes.

20.7   Li/CF$_x$ Cells

Lithium reacts with poly(carbon monofluoride), CF$_x$, at ambient temperatures.

The value of x in CF$_x$ can vary from about 0.9 to 1.2, depending upon its synthesis parameters.

These cells are generally used in situations in which low to moderate rates are required.

Elemental lithium is used as the negative electrode reactant, and the electrolyte is typically $LiBF_4$ in propylene carbonate. The reactant in the positive electrode is powdered CF_x. Although this material has a lamellar structure that can be thought of as analogous to graphite, its structure consists of an infinite array of cyclohexane "boats" [7] instead of thin graphene sheets. Lithium does not readily move between these layers, and therefore the electrode reaction mechanism does not involve insertion, as in the case of lamellar graphite.

Instead, a polyphase reaction occurs during discharge that can be written as

$$xLi + CF_x = xLiF + C \qquad (20.8)$$

Since this is a simple displacement reaction, the voltage remains constant, at 2.75 V, during discharge.

Because the reactants have low weights, the maximum theoretical specific energy, the MTSE, of these cells is very high, 1,940 Wh/kg. This would be attractive for use for implantable medical applications. However, because the voltage remains constant, it is difficult to determine when the capacity is almost consumed. An indication that a power source is soon going to reach its end-of-life is especially important when it is used for such purposes, and this problem is currently receiving a considerable amount of attention.

20.8 Reserve Batteries

20.8.1 Introduction

The discussion of positive electrodes in lithium batteries thus far has assumed that the reactants are either solids or a gas. However, this is not necessary, and there are two types of primary batteries that have been available commercially for a number of years in which the positive reactant is a liquid, the Li/SO_2 and $Li/SOCl_2$ (thionyl chloride) systems. They both have very high specific energies. But because of safety considerations, they are not in general use, and are being produced primarily for military and space purposes.

These are examples of *reserve batteries*, in which some method is used to prevent their operation until their stored energy is needed. There are two general ways in which this can be done.

One is to prevent the electrolyte from contacting one or both of the electrodes. This prevents self discharge, as well as the operation of any unwanted side reactions. A way to do this is to contain the electrolyte in a glass container that can be broken when cell operation is desired. A second method involves the use of an electrolyte that does not conduct current until it melts at an elevated temperature. When battery operation is desired, the electrolyte is heated to above its melting point. Examples of both of these strategies are discussed below.

20.8.2   The Li/SO$_2$ System

High energy Li/SO$_2$ cells are generally constructed with elemental lithium negative electrodes and large surface area carbon electrodes on the positive side. X-ray experiments have shown that Li$_2$S$_2$O$_4$ is formed upon the positive electrode upon discharge. The discharge curve is very flat, at 3.0 V, as shown in Fig. 20.7.

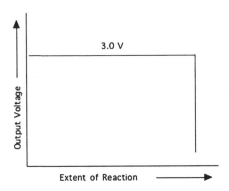

Fig. 20.7 Discharge curve for a Li/SO$_2$ cell

As discussed earlier, this type of behavior indicates that the cell operates by a reconstitution reaction. It should be possible to calculate the voltage by consideration of the thermodynamic properties of the phases involved in this system at ambient temperature. These are shown in Table 20.1.

Table 20.1 Gibbs free energies of formation of phases in the Li–S–O system at 25 °C

Phase	Gibbs free energy of formation (kJ/mol)
Li$_2$O	−562.1
SO$_2$	−300.1
Li$_2$S	−439.1
Li$_2$S$_2$O$_4$	−1,179.2

From this information, the stable tie lines in the ternary phase stability diagram for this system can be determined, as discussed earlier. The reaction equations relevant to each of the subtriangles can also be identified, and their potentials calculated. The resulting diagram is shown in Fig. 20.8.

It can be seen that the Li$_2$S$_2$O$_4$–SO$_2$–O subtriangle triangle has a potential of 3.0 V versus lithium. Since the SO$_2$–Li$_2$S$_2$O$_4$ tie line on the edge of that triangle points at the lithium corner, no oxygen is formed by the reaction of lithium with SO$_2$ to produce Li$_2$S$_2$O$_4$.

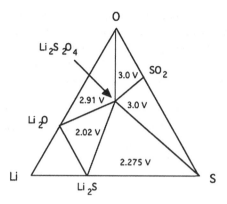

Fig. 20.8 Phase stability diagram for the ternary Li–S–O system at ambient temperature

The formal reaction for this cell is therefore

$$2Li + 2SO_2 = Li_2S_2O_4 \tag{20.9}$$

The theoretical specific energy of this cell can be calculated to be 4,080 kWh/kg, a high value.

20.8.3 The Li/SOCl₂ System

The lithium/thionyl batteries react at a somewhat higher constant voltage plateau, at 3.66 V.

The formal reaction is known to be

$$4Li + 2SOCl_2 = 4LiCl + S + SO_2 \tag{20.10}$$

This involves the Li–S–Cl–O quaternary system. In order to visualize the behavior of this system in a manner similar to that for the Li/SO₂ cell above, a tetrahedral figure would have to be drawn, and the constant voltage plateaus related to each of the subtetrahedra calculated. While straightforward, this is a bit too complicated to be included here, however.

The theoretical specific energy of this cell can be calculated to 7,250 kWh/kg, which is a very high value.

20.8.4 Li/FeS₂ Elevated Temperature Batteries

A good deal of effort went into the development of a high temperature system that uses FeS_2 as the positive electrode reactant, but has either Li–Al or Li–Si alloys, rather than lithium metal, on the negative side. The electrolyte was a molten Li–K halide salt that has a eutectic temperature of 320°C. These cells operated at temperatures over 400°C, and the open circuit voltage was about 1.9 V, with most of the capacity obtained at 1.7 V. The initial development was aimed at their use to power electric vehicles, where their favorable high power operation is attractive. However, the appearance of other alternatives, such as ambient temperature lithium-ion systems, caused this work to be discontinued in the late 1990s.

Because the molten salt electrolyte is only conductive at elevated temperatures, such cells can be stored at ambient temperature and used as *reserve batteries*. Upon heating the electrolyte melts and the cell becomes operable. This type of reserve batteries, often called "thermal batteries", has been used for military applications in which a long shelf life is very important.

References

1. C.C. Liang, M.E. Boltser and R.M. Murphy, US Patent 4,310,609 (1982)
2. C.C. Liang, M.E. Boltser and R.M. Murphy, US Patent 4,391,729 (1982)
3. E.S. Takeuchi and W.C. Thiebolt III, J. Electrochem. Soc. *135*, 2691 (1988)
4. W.G. Berl, Trans. Electrochem. Soc. *83*, 253 (1943)
5. A.W. Winsel, *Advanced Energy Conversion*, Vol. 3 Pergamon Press, Oxford (1963)
6. K.V. Kordesch, in C. Berger, *Handbook of Fuel Cell Technology*, Prentice-Hall, Inc. (1968), p. 361
7. L.B. Ebert, J.I. Brauman and R.A. Huggins, J. Amer. Chem. Soc. *96*, 7841 (1974)

Chapter 21
Energy Storage for Medium-to-Large Scale Applications

21.1 Introduction

Most of the highly visible applications of advanced energy storage technologies are for relatively small applications, such as in portable computers or implanted medical devices, where the paramount issue is the amount of energy stored per unit weight or volume, and cost is not always of prime importance. Such energy storage components and systems have occupied much of the attention in this text, especially the later chapters related to electrochemical cells and systems.

As discussed in Chap. 1, there are several types of large-scale energy storage applications that have unique characteristics, and thus require storage technologies that are significantly different from the smaller systems that are most visible at the present time. These include utility load leveling, solar and wind energy storage, and vehicle propulsion. They play critical roles in the transition away from the dependence upon fossil fuels.

More than for smaller scale applications, the important factors in large systems are the cost per unit energy storage, that is, per kWh, efficiency of the energy storage cycle, that has a large influence upon operating costs, and the lifetime of the critical components. Investors generally expect large systems to be in operation for 25 years or more. In addition, great attention is paid to safety matters.

Several of the storage technologies that are particularly interesting and important for larger-scale applications were described in the early chapters of this book. Some others will be discussed in this chapter.

21.2 Utility Load Leveling, Peak Shaving, Transients

The requirements of the large-scale electrical distribution network, or grid, were discussed in Chap. 1. The major problem is to match the energy available to the needs, which typically undergo daily, weekly, and seasonal variations. In addition, there are short-term transients that can lead to instabilities and other problems in the

R.A. Huggins, *Energy Storage*,
DOI 10.1007/978-1-4419-1024-0_21, © Springer Science+Business Media, LLC 2010

grid. The amelioration of these problems requires not only better technology, but also an intelligent grid control system to couple energy generation, transmission, and storage. Major factors include cost, reliability, lifetime, efficiency, and safety.

The energy storage method that is most widely used to reduce the longer-term variations in some areas involves the use of the pumped-hydro facilities discussed in Chap. 6. However, this is only possible in specific locations, where the required geological features are present. Large-scale underground compressed air storage systems also have related requirements.

Other technologies are useful in reducing the impact of short-term transients, which are now handled by the variation of the AC output frequency. One of these, also discussed in Chap. 6, involves very large flywheels, which are now available with power values up to 100 kW.

The integration of a number of such units to provide total power up to 20 MW is being investigated. The Department of Energy has estimated that 100 MW of flywheel storage could eliminate 90% of the frequency variations in the State of California.

Additional approaches that are being explored at present involve reversible high power electrochemical systems. Here, the amount of energy stored per unit cost is of prime importance. In contrast to other uses of electrochemical systems, the size and weight are often not important. Several of these are discussed later in this chapter.

21.3   Storage of Solar- and Wind-Generated Energy

Solar and wind energy sources are often viewed as technologies that can be employed to both satisfy transient local needs and supply energy into the electricity distribution grid. However, their output generally only roughly matches the time-dependent requirements of the grid. Thus, energy storage mechanisms are required to assist their integration into that large-scale system. Short-term transients in their output, such as when a cloud passes over a solar collection system, or the wind drops in velocity, are generally not of great importance.

As is the case with matching the time dependence of the needs and supplies of energy in the large-scale electricity grid, some electrochemical systems that can have relatively low costs, but are not interesting for portable applications because of their size or weight, can be advantageous.

21.4   Storage Technologies that are Especially Suited to these Applications

21.4.1   Lead-Acid Batteries for Large Scale Storage

Lead-acid batteries, due to their relative ease of manufacture, and favorable electrochemical characteristics, such as rapid kinetics and good cycle life under controlled conditions, are being used in large groups to support solar and wind generation systems. The cost per unit energy stored is a particularly attractive feature of this approach, which was discussed in Chap. 15.

21.4.2 Sodium/Sulfur Batteries

A second type of battery that is beginning to be used for storing energy in large scale systems is the so-called *sodium/sulfur system* that operates at 300 to 350°C. As discussed in Chap. 11, this electrochemical system is best described as a Na/Na_xS cell. These batteries are different from the common systems that most people are familiar with, for the electrodes are liquids, instead of solids, and the electrolyte is a sodium ion-conducting ceramic solid, $NaAl_{11}O_{17}$, called "beta alumina". This is thus a L/S/L, rather than the conventional S/L/S, configuration. The sodium ion conductivity in this ceramic material, discovered by Yao and Kummer, is remarkably high at the operating temperature [1–3], resulting in a resistivity of only about 4 Ω cm at 350°C. The possibility that this material could be used to construct the revolutionary sodium/sulfur battery was soon pointed out by Weber and Kummer [4]. The general construction of such batteries is shown schematically in Fig. 21.1.

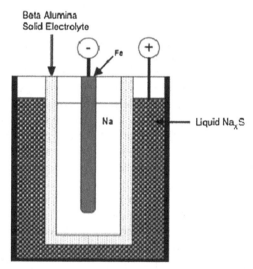

Fig. 21.1 Schematic view of Na/Na_xS cell

Sodium from the negative electrode passes through the surrounding beta alumina cylinder, and reacts with a liquid solution of sodium in sulfur. The capacity is determined by the composition range of this sodium–sulfur liquid phase, as shown in Fig. 11.16. The Na_xS liquid, which is not a good electronic conductor, is contained in a porous carbon "sponge." The cell voltage is somewhat over 2 V, as can be seen in Fig. 11.17. A general reference that contains a lot of information about sodium/sulfur cells is [5].

Early work in both the United States and Europe on this type of cell was aimed toward its potential use for vehicle propulsion. In that case, safety is especially important, and extensive testing relative to the use of these cells in vehicles under crash conditions was performed in Europe in the 1990s. The results were discouraging, and all of these development programs were discontinued.

On the other hand, activities in Japan were aimed at a different application, utility power storage, where they can be kept in a protective environment so that safety considerations can be minimized. A large effort was undertaken by a consortium of NGK Insulators and Tokyo Power (TEPCO) in 1983, and after extensive large-scale testing, this technology became commercially available in 2000.

Large individual cells are enclosed in steel casings for safety reasons, and they can be arranged into parallel and series groups in order to provide the required voltages and capacities. Systems with power values up to 6 MW at 6.6 kV are being used in Japan. This technology is also beginning to be installed in the United States, with facilities currently up to a 1 MW size.

21.4.3 Flow Batteries

21.4.3.1 Introduction

Except for the Na/Na_xS cell, and the Zebra cell, that was discussed briefly in Chap. 12, all of the electrochemical cells that are generally considered have electrodes that are solids. In those two cases, liquid electrodes could be used because the electrolytes are solid, resulting in a L/S/L configuration. There is another group of cells that have liquid electrode reactants, although their electrode structures contain porous solid current collectors. These are generally called "flow batteries", since the liquid reactant is stored in tanks and is pumped (flows) through the cell part of the electrochemical system. Thus, such systems can be considered to be rechargeable fuel cells.

A number of chemical systems have been explored, and in some cases rather fully developed. However, most of them have not been commercially successful to date. As will be seen below, this could well change in the near future.

The general physical arrangement is shown in Fig. 21.2, whereas the configuration of the cell portion of the system is shown schematically in Fig. 21.3.

It can be seen that this is also a type of L/S/L configuration. The electrolyte is a proton-conducting "solid polymer", and the electrode reactants are liquids on its two sides. In the Zebra cell, the reactants are both electronically conducting, whereas in flow cells, the electrode reactants are ionic aqueous solutions that are electronic insulators. In order to get around this problem and provide electronic contact to an external electrical circuit, the liquid reactants permeate an *electronically conducting graphite felt*. This felt provides contact, both to the polymer electrolyte and to a *graphite current collector*.

The electrode reactants are typically acidic, for example 2 M H_2SO_4 aqueous solutions of ions that can undergo *redox reactions*. The function of the polymer

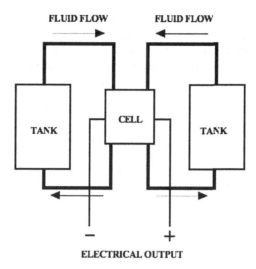

Fig. 21.2 General physical arrangement of a flow battery

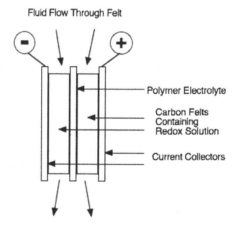

Fig. 21.3 The cell portion of the system. In some cases, there are multiple bipolar cell configurations

electrolyte is to transport protons from one side to the other, thus changing the pH and the charges on the dissolved *redox ions.*

An important difference from the Na/Na$_x$S and Zebra cells is that the reactant materials, the redox ion solutions, can be pumped into and out of the electrode compartments. This means that the capacity is not fixed by the cell dimensions, but is determined by the size of the liquid electrode reactant storage tanks. This can result in very large capacities, and is one of the potential advantages of flow battery systems. Thus, flow batteries deserve consideration for relatively large stationary applications, such as remote solar or wind installations, whose outputs are dependent upon the time of day and/or the weather.

The open circuit voltage across the electrolyte is determined by the difference in the chemical potentials on its two sides. As current passes through the cell protons are transferred, changing the pH, so that the ionic compositions of the two electrode reactant fluids gradually change. Thus, the cell potential varies with the state of charge. The change in the voltage with the amount of charge passed depends, of course, upon the size of the tanks.

Some of the redox systems that have been explored are indicated in Table 21.1.

Table 21.1 Various redox systems used in flow batteries

System	Negative electrode reactant	Positive electrode reactant	Nominal voltage (V)
V/Br	V	Bromine	1.0
Cr/Fe	Cr	Fe	1.03
V/V	V	V	1.3
Sulfide/Br	Polysulfide	Bromine	1.54
Zn/Br$_2$	Zn	Bromine	1.75
Ce/Zn	Zn	Ce	<2

A general discussion of these various alternatives can be found in reference [6].

There is some confusion in the terminology used to describe these systems, for the liquid reactants on the two sides are sometimes called "electrolytes", even though they do not function as electrolytes in the battery sense. Additionally, the liquid reactant on the negative side of the cell is sometimes called the "anolyte", and that on the positive side of the cell the "catholyte". In batteries the negative electrode is generally called the "anode", and the positive electrode the "cathode".

One of the most attractive flow systems involves the vanadium redox system [7–12]. In this case, the negative electrode reactant solution contains a mixture of V^{2+} and V^{3+} ions, whereas the positive electrode reactant solution contains a mixture of V^{4+} and V^{5+} ions. The charge neutrality requirements means that when protons (H$^+$ ions) are added to, or deleted from, such liquids by passage through the polymer electrolyte in the cell, the ratio of the charges on the redox species is varied. This changes the state of charge.

These systems are generally assembled in the uncharged state, in which the chemical compositions of the two liquid reactants are the same. In the vanadium system this is done by adding vanadyl sulfate to 2 M H$_2$SO$_4$, which gives an equal mixture of V^{3+} and V^{4+} ions. The system is then charged by passing current, causing the transport of protons through the polymer electrolyte, so that the ion contents on the two sides become different.

21.4.3.2 Redox Reactions in the Vanadium/Vanadium System

One can write the reactions in the electrode solutions of the vanadium system as

$$VO_2^+ + 2H^+ + e^- = VO^{2+} + H_2O \tag{21.1}$$

or, in terms of the vanadium ions:

$$V^{5+} + e^- = V^{4+} \tag{21.2}$$

in the positive electrode reactant solution, and

$$V^{2+} = V^{3+} + e^- \tag{21.3}$$

in the negative electrode reactant solution.

So that the overall reaction is

$$VO_2^+ + 2H^+ + V^{2+} = VO^{2+} + H_2O + V^{3+} \tag{21.4}$$

or

$$V^{5+} + V^{2+} = V^{4+} + V^{3+} \tag{21.5}$$

The variation of the open circuit cell potential with the state of charge in the case of the V/V system with concentrations of 2 M of each V species is shown in Fig. 21.4. Typical operation would involve cycling between 20 and 80% of capacity and thus at voltages between 1.3 and 1.58 V.

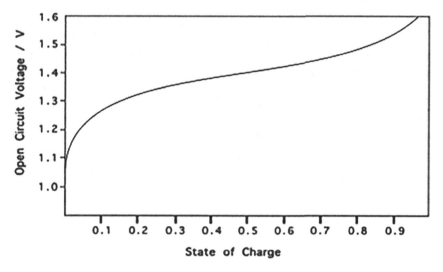

Fig. 21.4 Variation of the open circuit potential versus state of charge for the case of a V/V flow cell at 298 K

Since each cell produces a relatively low voltage, such batteries generally contain a number of cells arranged in series in order to produce a greater overall output voltage. Parallel configurations can be used to provide higher currents.

Depending upon the application, it may be desirable to permit relatively rapid charging, although this may not be necessary during discharge. Thus it may be desirable to include a mechanism to change the number of cells and their series/parallel arrangements during different operating conditions.

If all of the cells are fed from common liquid supplies, this can result in a large voltage applied across the liquid reactants and this can result in the passage of a considerable amount of current. This is a form of self discharge, and is sometimes called "shunt" or "bypass current". This is different, however, from the self discharge that results from neutral species, or neutral combinations of species, traveling through or around the electrolyte in other types of electrochemical systems.

Because there are no solid-state volume changes during charging and discharging, as are typical for electrochemical cells with solid electrodes, the components of the cells, as well as the total system, can have long lives. Thus, long cycle life is not a problem, even with repeated deep charges and discharges.

The vanadium redox system can be used over a temperature range from 10 to 35°C and typically operates at or near ambient temperature. At higher temperatures, the current density increases, but the cell voltage is reduced somewhat. The overall result is that the power is greater at somewhat elevated temperatures.

The electrode kinetics are good, and additional catalysts are not required. The coulometric and voltage efficiencies are high, except for the self-discharge mechanism mentioned above.

The specific energy and energy density are determined primarily by the electrode reactants themselves, which are the major components in these systems. Typical values are 15 Wh/kg and 18 Wh/l, and round trip efficiencies are typically 70–75%.

Since the electrode reactants both consist of vanadium sulfate solutions in aqueous sulfuric acid, only differing by the oxidation states of the vanadium ions, contamination by leakage across the electrochemical cell membranes only results in some capacity loss, and is fully reversible. In flow batteries in which the ions are different on the two sides, this can become a significant and irreversible, problem.

Because the cell voltage is a function of the state of charge, it is possible to determine the state of charge of such systems remotely, which may be an advantage in some cases. The cell design also makes monitoring of the voltage across each cell possible.

Since the cells can be configured in a variety of different series/parallel arrangements, the charging and discharging cycles can operate at different voltages. As a result, such as system can be used as a DC/DC converter.

21.4.4 All-Liquid Batteries

A relatively new concept that could be useful for stationary storage applications is the use of an all-liquid three-component cell [13,14]. In one version, the two electronically-conducting electrode materials have densities different from that of a molten salt electrolyte, one lower, and the other greater. This leads to a three-layer configuration, with the electrolyte in the middle. The container, perhaps with a graphite insert, can act as the current collector for the lower electrode, and the upper electrode material can be contacted by an electronic conductor protruding from above. An inert gas cover, either nitrogen or argon, is needed to prevent reaction with air. This general configuration is illustrated schematically in Fig. 21.5.

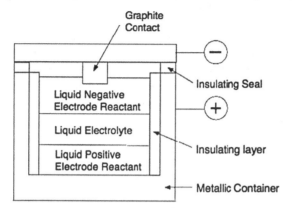

Fig. 21.5 Schematic representation of an all-liquid cell

Such a configuration can operate in the same way as batteries with solid electrodes, with a species leaving one liquid electrode, traveling across the liquid electrolyte, and reacting with the other liquid electrode material.

In a different concept, species from the two electrodes react to form a product that is soluble in the liquid electrolyte. The electrolyte must be more stable (have a greater negative value of the Gibbs free energy of formation) than the soluble product. The stability of the electrolyte limits the possible voltage of the cell, and the capacity is determined by the solubility of the product in the electrolyte phase.

There are several potentially important advantages in this all-liquid configuration.

One is that there are no problems due to the dimensional changes in electrode reactants, which can cause problems in common fine particle battery electrodes. On the other hand, it is necessary to be aware of the potential safety problems. Care must be taken to avoid contact of high temperature molten salts and highly reactive metals, such as magnesium, with air, as the results could be catastrophic.

Molten salt electrolytes can have conductivities much greater than the sulfuric acid and KOH used in aqueous batteries, and the organic solvent electrolytes in lithium batteries.

And with the possibility that all-liquid cells can be designed to have large area electrode contact with high conductivity electrode materials, it is reasonable to expect that such all-liquid cells can operate at very high rates.

The cost will depend to a large extent upon the identity of the materials used in both the electrodes and the electrolyte, which can be much lower than those used in many of the current battery systems. If the product solubility in the electrolyte is appreciable, this concept has an advantage over current flow battery concepts, in which the reactant ions are rather dilute, requiring rather large tanks.

An example of this type of cell is the use of a heavy positive electrode material, such as antimony, with a density of 6.5 g/cm^3, on the bottom, and magnesium, a relatively light negative electrode material with a density of 1.6 g/cm^3, on the top, with a sodium sulfide electrolyte, that has an intermediate density of 4.0 g/cm^3, in the middle. This configuration can be written as a Mg/Na$_2$S/Sb cell. The reaction product is Mg$_3$Sb$_2$, which dissolves in the Na$_2$S electrolyte. The voltage is determined by the Gibbs free energy of formation of Mg$_3$Sb$_2$.

Another design alternative can be used when both electrode materials are heavier than the electrolyte. An example of this type would be the use of Zn (density of 7.1 g/cm^3) as negative electrode, Te (density 6.3 g/cm^3) as positive electrode, and an electrolyte of ZnCl$_2$ (density 2.9 g/cm^3). This is shown schematically in Fig. 21.6.

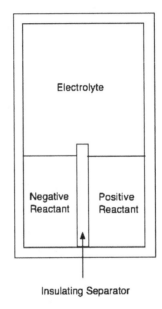

Fig. 21.6 Schematic view of a cell in which both electrode materials have greater densities than the liquid electrolyte

21.5 Storage of Energy for Vehicle Propulsion

21.5.1 Introduction

Most vehicles are propelled by internal combustion motors that consume liquid fuels, either gasoline or diesel fuel. In those cases, the energy storage mechanism is a simple tank to hold the liquid fuel.

Over the years, the commercial introduction of electrically powered automobiles has not generally been successful, due to their high cost and limited performance, compared to what is typical of those with internal combustion motors. This is due primarily to the weight and cost of the batteries required in order to provide what is perceived to be sufficient driving performance and range.

The characteristics needed to meet the requirements for electric vehicle propulsion depend greatly upon the type of duty cycle that is assumed. Extensive measurements of actual vehicle usage were undertaken, and from them, models corresponding to typical usage patterns were established. One of these, known as the ECE-15 cycle, was developed for all-electric vehicles. It was composed of two parts, an urban part that simulated the needs during local travel, and a suburban part that required higher power levels, such as what is needed for travel at higher velocities and greater distances. These were both expressed in terms of power–time profiles, and are shown in Figs. 21.7 and 21.8.

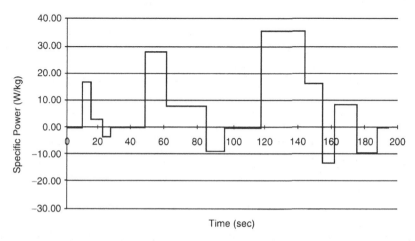

Fig. 21.7 Power demand profile for the ECE-15 reference vehicle in urban travel simulation

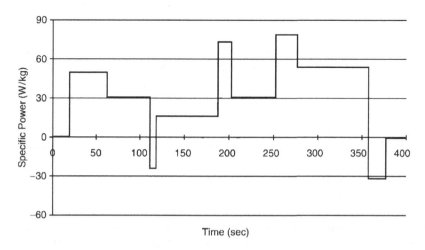

Fig. 21.8 Power demand profile for the ECE-15 reference vehicle in suburban travel simulation

Several auto manufacturers undertook programs to develop electrically powered vehicles in the 1990s. This was in response to a mandate of the California Air Resources Board that required auto manufacturers to develop and produce zero-emission vehicles. The requirement was that at least 2% of the vehicle fleet sold by any manufacturer in the State of California must be a zero-emission type by 1998, and the zero-emission fraction was expected to increase in later years.

One of the most visible responses was the manufacture of almost 1,000 electric vehicles of a model called the EV1 by General Motors. Other auto companies developed prototype vehicles, and some, including Honda, actually put a number on the road.

The EV1 cars were leased, rather than sold. The early ones were powered by lead-acid batteries, and had a range of 80–100 miles under light duty conditions. Hydride/nickel batteries were installed in a later model, which gave a range of 100–140 miles.

After much political pressure from the auto industry and some changes in the composition of the Air Resources Board, the California mandate was eliminated in 2002, and EV1 production was terminated. All the existing cars were repossessed by General Motors and destroyed. The company clearly did not want to have them on the road.

It can be seen in both of the ECE-15 profiles that there are periods in which the required power is negative. This means that the propulsion system can absorb kinetic energy from the motion of the vehicle. Hybrid systems, in which energy can be temporarily stored in a capacitor or battery have been developed for this purpose. This was briefly discussed in Chap. 6.

A number of manufacturers are now selling hybrid vehicles, in which the propulsion is provided by a combination of a modest internal combustion engine and a relatively small battery. The control system allows the engine to operate at relatively high efficiency, and the battery gets some charging during braking, as well as from the motor. Several of these products have proved to be very successful.

Batteries for use in these hybrid vehicles are relatively small, with energy capacities typically about 3 kWh. On the other hand, they are designed to emphasize high power performance so that they can rapidly absorb and deliver energy. Some provide up to 60 kW, giving a power to energy ratio of 15–20. Because of their modest size, they are not particularly expensive.

Another type of hybrid vehicle is beginning to appear, fuel cell-battery hybrids. The most advanced of this type at the present time is the Honda FCX Clarity. These autos, which are mid-size sedans, get their basic propulsion from a fuel cell, which also charges a relatively small high rate lithium battery. When more power is needed for a modest time, the battery kicks in to supplement the fuel cell output. When this is not the case, part of the fuel cell output is used to recharge the battery. A sophisticated control system is used to integrate these two systems to produce a very impressive performance.

It is interesting that the Department of Energy in the United States recently decided to discontinue the support of activities aimed at the use of fuel cells in vehicles, although increasingly successful development efforts have been underway by Daimler Benz in Europe and Honda in Japan for some time.

There is currently increasing interest in another type of hybrid alternative, what is called a "plug-in hybrid" vehicle. In this case, it is expected that the battery part of the system can be electrically recharged, perhaps overnight at home, to provide sufficient energy to propel the vehicle a modest distance – say 20–50 miles – without the use of the internal combustion engine at all. It has been found that a large fraction of the total mileage traveled by vehicle owners in the United States is due to short distance trips, such as shopping, or going to work and back. Thus, a short electrical range will be sufficient to handle much of the total transportation need. This sounds attractive, as the cost of electricity per mile of travel is less than the cost of the equivalent amount of liquid fuel.

The type of battery that is needed for this application must be optimized for its specific energy, rather than for its behavior under transient conditions, as is the case for the other types of hybrid vehicles. This is typically more expensive.

21.5.2 ZEBRA Batteries

There was a brief discussion of the ZEBRA battery, which is based upon the Na–Ni–Cl ternary system, and is sometimes called the $Na/NiCl_2$ battery, in Chap. 12. This system, which evolved from earlier work on the Na/Na_xS battery, was invented in South Africa [15,16], and has had a long and tortuous road toward commercialization [17,18]. This involved work at BETA Research and Development Ltd. in England, and a joint effort of AEG (later Daimler) and Anglo American Corp., AEG Anglo Batteries, GmbH started pilot line production. After the merger of Daimler and Chysler, this activity was terminated, and the technology sold to MES DEA S.A. in Stabio, in southern Switzerland near the Italian border, in the late 1990s. MES DEA was sold to FZ Sonick S.A. in February, 2010. The name ZEBRA initially stood for "Zeolite Battery Research Africa", and is a holdover from the initial idea that the ceramic solid electrolyte would be a zeolite material. General Electric is now beginning to work on this system in the United States.

From the start, it was intended that these cells would be used for vehicle propulsion. Modest numbers have now been produced, and used in the Twingo and the Panda autos in Switzerland and Italy, and the Think City in Norway.

The general configuration is similar to that of the sodium/sulfur cells in that the negative electrode is liquid sodium, and the electrolyte is a solid electrolyte, sodium beta alumina. However, the positive electrode contains both the solid $NiCl_2$ reactant and a second liquid electrolyte, $NaAlCl_4$. The positive electrode is on the inside, and the negative electrode on the outside in this case.

ZEBRA cells are produced in the discharged state, with all of the sodium present as NaCl on the positive side. They are constructed with excess sodium, so the amount of $NiCl_2$ determines the capacity. The operating temperature is kept within the range 270–350°C, and the open circuit voltage is 2.59 V, in accordance with thermodynamic data, as discussed in Chap. 12. The theoretical specific energy of individual cells is 790 W/kg, which is slightly greater than that of Na/Na_xS cells, 760 W/kg.

Groups of batteries are encased in a temperature-controlled container, and the configuration is designed to produce a ratio of power to energy of about two, 50 kW peak power, and 25 kWh energy in one model. On a weight basis, the complete ZEBRA battery system stores about 120 Wh/kg specific energy. An attractive feature is that these cells are fully reversible, with 100% ampere hour efficiency. It is claimed that at this stage of development, the life cycle costs are less than those of lead-acid batteries, despite higher initial costs, due to their much longer lifetime.

Safety tests in Europe have indicated that these batteries are significantly safer than Na/Na_xS cells, and do not represent a significant risk under simulated crash conditions. Both details of the design and several features of the chemistry that provide protection against both overcharge and overdischarge were discussed in [18].

21.5.3 General Comments on Hybrid System Strategies

There is a great variation in the requirements for transient power sources, and in some cases no one type of device, or any one design, will be able to optimally fulfill such diverse needs.

Hybrid systems can include components that meet two different types of needs, a primary energy source and a supplemental source that can meet transient demands for higher power levels than can be handled by the primary source, but has a relatively small energy capacity. This combination can be represented schematically in terms of the commonly used Ragone type of diagram, in which the specific power is plotted versus the specific energy, both on logarithmic scales, as shown in Fig. 21.9.

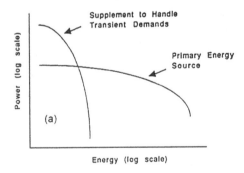

Fig. 21.9 Typical hybrid system characteristics

A possible strategy to consider for accomplishing this is to use a high energy system that operates at a relatively high voltage when the power demand is low. The output voltage of such energy sources typically falls off as the output current is increased. If a second high-power, but lower-energy source that operates at a lower voltage is placed in parallel, it will take over under the conditions that drive the output voltage of the primary high-energy system down into its range of operating voltage, and meet the high-power demand for a short period. When this demand is no longer present, the voltage will again rise, and the high-power component of the system will be recharged by the higher-energy component. This combination is shown schematically in Fig. 21.10.

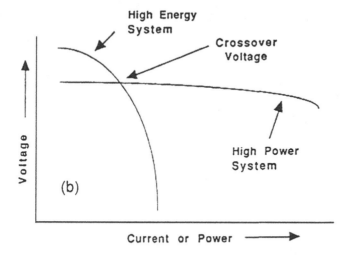

Fig. 21.10 Schematic representation of possible hybrid system strategy

The way that the properties of batteries are typically described, such as by a graphical display of the discharge (voltage vs. state of charge) curves at different constant current densities, or in terms of the change of extractable capacity as a function of the number of discharge cycles, cannot be considered to provide a satisfactory description of behavior in these very different types of applications. Likewise, the value of the capacitance at a single frequency is also certainly not a satisfactory description of the behavior of a capacitor over such a wide range of potential uses.

In order to approach the development of useful devices for this type of application, one should consider the several types of charge storage mechanisms that can be employed, their thermodynamic and kinetic characteristics, and the basic properties of candidate materials, as well as the relationships that determine system performance.

References

1. Y.F.Y. Yao and J.T. Kummer, J. Inorg. Nucl. Chem. 29, 2453 (1967)
2. R.H. Radzilowski, Y.F. Yao and J.T. Kummer, J. Appl. Phys. 40, 4716 (1969)
3. M.S. Whittingham and R.A. Huggins, J. Chem. Phys. 54, 414 (1971)
4. N. Weber and J.T. Kummer, Proc. Ann. Power Sources Conf. 21, 37 (1967)
5. J.L. Sudworth and A.R. Tilley, *The Sodium Sulphur Battery*, Chapman and Hall, London, 1985

6. C. Ponce de Leon, A. Frias-Ferrer, J. Gonzalez-Garcia, D.A. Szanto and F.C. Walsh, J. Power Sources 160, 716 (2006)
7. E. Sum and M. Skyllas-Kazacos, J. Power Sources 15, 179 (1985)
8. E. Sum, M. Rychcik and M. Skyllas-Kazacos, J. Power Sources 16, 85 (1985)
9. M. Skyllas-Kazacos, M. Rychcik, R. Robins, A. Fane and M. Green, J. Electrochem. Soc. 133, 1057 (1985)
10. M. Rychcik and M. Skyllas-Kazacos, J. Power Sources 19, 45 (1987)
11. M. Rychcik and M. Skyllas-Kazacos, J. Power Sources 22, 59 (1988)
12. M. Skyllas-Kazacos and F. Grossmith, J. Electrochem. Soc. 134, 2950 (1987)
13. D. Sadoway, G. Ceder and D. Bradwell, US Patent Application 2008/0044725 A1, Feb. 21, 2008
14. TR10: Liquid Battery, Technology Review, March/April 2009, p. 1
15. J. Coetzer, in Proceedings of the 170th Meeting of Electrochemical Society, San Diego, CA, USA, October 1986, Extended Abstract No. 762
16. R.C. Galloway, J. Electrochem. Soc. 134, 1 (1987)
17. J.L. Sudworth, J. Power Sources 100, 149 (2001)
18. C-H. Dustman, J. Power Sources 127, 85 (2004)

Chapter 22
A Look to the Future

22.1 Introduction

When considering what changes and new developments are possible, or even likely, in the years ahead, it is realistic to consider a quotation attributed to Thomas A. Edison:

> "Making predictions can be rather precarious, especially when they have to do with the future."

Nevertheless, several things are rather obvious. One is that the need for energy storage will certainly grow substantially. This is not just due to the natural growth of all the technologies in which storage is an important component, but also to important changes in the energy production and utilization landscape. There is a greatly increased emphasis upon energy production methods based upon energy sources other than the various fossil fuels. But these are typically periodic, or at least intermittent. Chief among these are the various solar technologies, and those based upon the use of the wind and tidal flows. Their use will surely increase in future years.

But in addition, the pressure for increased efficiency in the use of current energy sources is growing rapidly. An obvious example is the push toward the development of hybrid and plug-in hybrid vehicles.

Official US Department of Energy targets typically assume that new technologies will need to meet the same requirements as those of the current technologies that they are expected to supplant. An example of this is the assumption that electrically and fuel cell-powered vehicles should be expected to meet the long-range ability of current large, internal combustion vehicles. They also often assume that driving habits in the future will be about the same as those at the present time.

On the other hand, more and more of the public's attention is being given to the fact that a very large fraction of the actual vehicle use involves relatively short range daily commuter trips, for which limited range vehicles, whether battery or fuel cell-powered, would be perfectly satisfactory. Occasional longer trips would require the use of a different type of vehicle, of course. It is not unreasonable to

R.A. Huggins, *Energy Storage*,
DOI 10.1007/978-1-4419-1024-0_22, © Springer Science+Business Media, LLC 2010

think in terms of either two-car families, or the occasional rental of a long-distance vehicle when necessary.

The picture is not the same in all parts of the world. As can be seen in the discussion of several technologies in previous chapters, there are a number of directions in which progress has been greater in other countries than in the United States.

A large fraction of the government financial support of research and development activities related to energy technologies comes through, or is greatly influenced by, the Department of Energy in the United States. This leads to a concentration of work in a relatively small number of directions.

Recently, the US Department of Energy decided to terminate efforts to develop hydrogen fuel cell-powered vehicles. In contrast, significant progress in that direction has been made in Japan and Europe, with significant numbers of demonstration vehicles now on the road. This was mentioned in Chap. 21.

Some time ago, it was also decided to terminate work in the United States on both solid electrolyte and molten salt electrolyte elevated temperature batteries. As mentioned in Chap. 21, large sodium/sulfur solid electrolyte batteries are now being produced in Japan for use in large-scale storage facilities connected to the electrical distribution grid, and ZEBRA cells, which also have solid electrolytes, are now being produced in Switzerland for use for vehicle propulsion. Catch-up development efforts are now currently being undertaken in the United States.

A large fraction of the long-range research in both the government and university laboratories in the United States is now aimed at advanced energy storage technologies that seem to be more applicable to small, high-tech portable, rather than larger-scale stationary applications. Despite the great concern about the limitations of the large electrical energy distribution grid, work on the latter is primarily concentrated on a relatively few demonstration projects.

22.2   Recently Discovered Large Natural Gas Source

Although the subject of this book is energy storage, not energy sources, a recent development in the latter area deserves brief mention here. It may have a significant influence upon the energy supply, distribution, and storage in the United States.

Advances in drilling techniques have recently resulted in the expectation that it will be possible to obtain large amounts of natural gas from Devonian black shale rock that is found in the Marcellus basin in the Appalachian area that stretches from New York to West Virginia. It has been estimated that as much as 500 trillion cubic feet of natural gas, the equivalent of 80 billion barrels of oil, can be recoverable. This has resulted in a dramatic increase in the estimated natural gas reserves in the United States, and has been called the most important new natural resource find in the United States since the discovery of oil. There are also large shale gas basins in other areas, Texas, Wyoming, Arkansas, and Michigan, but they are not now getting as much attention as the Appalachian area.

Although the existence of this shale, which lies more than a mile below the surface, and that it contains natural gas, has been known for some time, it was previously considered to be too expensive to extract significant quantities of gas using normal drilling methods.

The change in this picture resulted from the development of a new drilling strategy, involving drilling horizontally into the shale, rather than through it. This is illustrated schematically in Fig. 22.1.

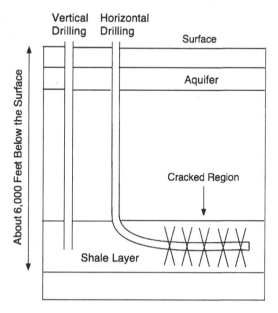

Fig. 22.1 Schematic representation of the difference between normal vertical drilling and horizontal drilling

A mixture of water and sand is forced into the formation at very high pressure. This causes many small fractures, or cracks, in the rock. They would normally close as soon as the water pressure is decreased, but the fine sand goes into the cracks, and holds them open. This lets much larger quantities of the gas become accessible.

Regular production, using this method, first started in 2005. The oil and gas industry recently began to make very large investments to get access to this potentially large source of natural gas, and the rate of production should increase rapidly.

As was discussed earlier, the combustion of natural gas produces only about half as much carbon dioxide as the combustion of oil. Thus, it is often considered to be a "bridge" or "transition" fossil fuel during the transition from the dependence upon oil to increased use of more environmentally-friendly energy sources.

22.3 Emerging Technological Directions

Although most of the attention given to energy storage technology at the present time seems to be focused upon needs related to portable devices, such as computers and telephones that operate at ambient temperature, further development and increased use of larger systems, with different requirements is imperative. Several of these were discussed in Chap. 21. One example is a regeneration of interest in elevated temperature battery systems for use in both large stationary applications and vehicles. Recent technological progress in both of these directions has been made in Japan and Europe.

Flow batteries, discussed in Chap. 21, are also attracting increased attention, but there has not been very much real activity to date. The most attractive vanadium/vanadium system was developed in Australia, and the major company producing such facilities in the United States recently went bankrupt. The leading activity is now in Austria.

New alternatives are also emerging. One of these, that is actually still in the research stage, is the concept of the use of multilayer liquid battery systems that was also briefly mentioned in Chap. 21. It is too early to judge its significance.

Another approach to very large scale energy storage that may become very important has begun to get a lot of attention in the last few years, initially in Europe, but now also in the United States [1] . It involves the use of the sensible heat in relatively inexpensive molten salts as thermal storage media in conjunction with large solar systems. It can be used to periodically supply large amounts of energy to the electrical distribution grid so as to reduce the time-dependent variations in the demand placed upon the major electrical utilities. This is a type of load leveling, and could have a major effect on the cost of electrical energy, especially in areas such as the state of California, in which the demand varies by up to 50%, depending upon the time of day.

Such a system involves the use of long parabolic reflectors to focus the sun's radiation upon tubes that carry a moving fluid. This fluid transfers the heat to a large molten salt bath, whose sensible heat acts as the storage system. This heat is then fed into a Rankine cycle steam turbine to produce electricity when needed.

The material that is initially heated by sunlight is sometimes called "solar oil," and is typically a synthetic organic material, a 50/50 mixture of the organic materials diphenyl oxide and biphenyl oxide. It has a low freezing point, 12°C, so there is little danger that it might solidify, and it can be used up to about 400°C. It transfers heat to a less expensive molten salt, such as the 50/50 eutectic mixture of $NaNO_3$ and KNO_3, which is sometimes called "solar salt." This salt melts at 221°C, is stable up to about 500°C, and has a heat capacity about half of that of water. It can be stored in large tanks, and supplies heat as needed to the steam turbine.

Typical prices are $0.5–1 per kg of nitrate salts, and $3–4 per kg for the low-melting organic heat transfer oils. As might be expected, efforts are being undertaken to find less expensive heat transfer media to replace the organic *solar oil*, or even a single material that can be used to handle the total thermal transfer and storage system in order to avoid the need for oil-to-salt heat exchangers. In addition, it would be desirable to be able to operate at higher temperatures, where the steam turbine is more efficient. It is important that the heat transfer material does not freeze inside the solar collector system or associated piping, of course. These nitrate salts are not corrosive, and can be readily contained in a number of metals and alloys.

Data on the compositions and minimum operating temperature of some of the nitrate molten salt materials that have been investigated are included in Table 22.1.

Table 22.1 Compositions and liquidus temperatures of several nitrate salts

Mol% Li	Mol% Na	Mol% K	Mol% Ca	Liquidus temp. (°C)
	66	34		238
	50	50		221
	21	49	30	133
30	18	52		120
31		58	11	117

The important factors in the consideration of new technological approaches and systems related to large-scale applications are different from those that are important in the smaller and perhaps more high-tech applications. Both initial and lifetime costs are of great importance. In addition, as systems get larger, there will inevitably be more emphasis on safety, for larger problems can evolve into major disasters.

22.4 Examples of Interesting New Research Directions

22.4.1 Organic "Plastic Crystal" Materials

The use of organic phase-change materials for the storage of thermal energy was discussed in Chap. 3. The examples that were mentioned all involved the use of their heat of fusion. There are also some organic materials that undergo solid-state reactions, and exhibit "plastic crystal" behavior. They include some amines and polyalcohols that have large values of solid state phase transition enthalpy and low enthalpies of fusion [2, 3]. This topic is discussed in [4].

22.4.2 Organic Electrode Materials for Lithium Batteries

Present approaches to lithium-ion batteries involve the use of metal alloys and inorganic materials as electrode reactants, as discussed in Chaps. 18 and 19. There have been several recent investigations of the potential of the use of organic materials in this application [5–7]. A recent example is the use of polycarbonyl materials [8]. One of the advantages of these materials is that it is possible to tune the reaction potential. On the other hand, their solubility in electrolytes can be a problem. However, it is believed that this can be alleviated by increasing the molecular weight and increasing the magnitude of negative charge.

22.4.3   *New Materials Preparation and Cell Fabrication Methods*

As in a number of other areas of both science and technology, there is currently a lot of interest in the synthesis of nanosized materials, and their potential use in connection with energy storage technologies.

The advantages of the use of small-dimensioned particles as electrode reactants in batteries are quite obvious in situations in which either the large surface area or the solid-state diffusion distance play an important role in controlling the kinetic behavior of electrodes.

But small nanowires can have an additional advantage in the case of some electrode materials with very large capacities. A particularly interesting example is the lithium–silicon alloy system. Under equilibrium conditions at elevated temperatures up to 4.2 lithium atoms can react per silicon atom [9], resulting in a theoretical electrode capacity of 4,200 mAh g^{-1}. But the volume changes by about 400% upon insertion and extraction of so much lithium, and this results in decrepitation (pulverization) and capacity fading [10].

However, synthesizing silicon in the form of nanowires that are spaced apart makes it possible to accommodate such large volume changes without mechanical damage. This has been done by [11] and [12], who used the vapor–liquid–solid (VLS) method.

Using this method, it is possible to grow silicon nanowires on metallic substrates, such as stainless steel, so that each wire is attached to the current collector, avoiding the problem of the loss of electronic contact often found with particulate reactants.

The VLS method was first used in connection with the growth of whiskers for entirely different purposes [13]. It has subsequently been used for the growth of a number of other materials [14–19].

Another method that can be used for the formation of large numbers of nanowires employing a special chemical etching procedure has also been recently reported [20]. This is done by electrochemically etching silicon to form macropores, followed by uniform chemical etching to increase their diameter to the point that adjacent pores touch. The result is the formation of a large number of parallel fine nanowires, or pillars, of silicon. The galvanic deposition of copper onto the substrate results in a structure in which the wires are encased in copper at the bottom.

Innovative methods are also being pursued in a number of laboratories for the synthesis of positive electrode materials and their incorporation in novel electrode structures in high-energy batteries. These often involve variants of wet chemistry. One particular interesting method involves the formation of very fine particle oxides by using a polymer precursor decomposition method [21–23], and a modification involving the use of citric acid [24].

In addition to the synthesis of fine reactant materials, there is a significant interest in methods to coat them with protective, yet ionically-transparent layers. Another variant is the use of nanofibers to support thin layers of reactant material. One example is the deposition of amorphous silicon coatings onto carbon nanofibers [25].

22.4.4 Alternate Electrolytes

New electrolytes are also being investigated. One group of these that is drawing a lot of attention includes materials called "ionic liquids". These are molten salts that have low melting temperatures. This is accomplished by making one or both of the ions have complicated high entropy structures that are hard to crystallize. They typically contain large organic groups with rather low symmetry. It appears that some of these materials are stable in the presence of lithium battery electrode components.

There is also a growing interest in the use of aqueous electrolytes in lithium systems, primarily for application in moderate-to-large systems in which low cost, high rate, and safety are of particular interest [26–30].

22.5 Final Comments

Energy storage is becoming increasingly important. There are two general reasons for this. One is the recognition of the inevitable depletion of nonrenewable fossil fuels such as oil, and the need to shift, at least partially, away from today's dependence upon them as the primary energy source, and toward the use of alternate energy sources. Some of the most important alternative technologies provide energy only intermittently.

In addition, there is growing concern about the pollution resulting from the use of the major current sources. This may be relieved, at least in part, by the use of some of the alternative sources.

On the smaller scale, there is an increasing number of relatively small portable electrically-powered devices that have to carry their energy sources with them. This results in the need for improvement in electrochemical battery or portable fuel cell technology.

The author hopes that this book will be helpful in providing an understanding of the different methods by which energy can be stored.

He also wishes to applaud, and cheer on, all those who have contributed to the current state of knowledge of energy storage science and technology.

References

1. R.W. Bradshaw and N.P. Siegel, ES2008-54174, in Proc. of Conf. on Energy Sustainability 2008, ASME (2008).
2. E. Murrill and L. Breed, Thermochim. Acta *1*, 239 (1970).
3. D.K. Benson, W. Burrows and J.D. Webb, Sol. Energy Mater. *13*, 133 (1986).
4. D. Chandra, W-M. Chien, V. Gandikotta and D.W. Lindle, Z. Phys. Chem. *216*, 1433 (2002).
5. T. Umemoto, US Patent 6737193 B3 (2004).

6. H. Chen, M. Armand, G. Demailly, F. Dolhem, P. Poizot and J-M. Tarascon, ChemSusChem. *4*, 348 (2008).
7. H. Chen, et al., J. Am. Chem. Soc. *131*, 8984 (2009).
8. M. Armand, S. Grugeon, H. Vezin, S. Laruelle. P. Ribiere, P. Poizot and J-M. Tarascon, Nat. Mater. *8*, 120 (2009).
9. J. Wen and R.A. Huggins, J. Solid State Chem. *37*, 271, (1981).
10. B.A. Boukamp, G.C. Lesh and R.A. Huggins, J. Electrochem. Soc. *128*, 725 (1981).
11. C.K. Chan, H. Peng, G. Liu, K. McIlwrath, X.F. Zhang, R.A. Huggins and Y. Cui, *Nat. Nanotechnol. 3*, 31 (2008).
12. B. Laïk, L. Eude, J-P. Pereira-Ramos, C.S. Cojocaru, D. Pribat, and E. Rouviere, Electrochim. Acta *53*, 5528 (2008).
13. R.S. Wagner, in: A.L. Svitt (Ed.), Whisker Technology, Wiley Interscience, New York, 1970, p. 47.
14. A.M. Morales and C.M. Lieber, Science *279*, 208 (1998).
15. M.H. Huang, et al., Adv. Mater. *13*, 113 (2001).
16. K.A. Dick, et al., Adv. Funct. Mater. *15*, 1603 (2005).
17. Z.W. Pan, Z.R. Dai, and Z.L. Wang, Science *291*, 1947 (2001).
18. Y. Wang, V. Schmidt, S. Senz, and U. Gosele, Nat. Nanotechnol. *1*, 186 (2006).
19. J.B. Hannon, S. Kodambaka, F.M. Ross, and R.M. Tromp, Nature *440*, 69 (2006).
20. H. Föll, H. Hartz, E. Ossei-Wusu, J. Carstensen, and O. Riemenschneider, Phys. Status Solidi RRL, *9999*, 4 (2009).
21. B.H. Hamling, U.S. Patent 3,385,915 (1968).
22. B.H. Hamling, U.S. Patent 3,736,160 (1973).
23. H.D. Deshazer, F. LaMantia, R.A. Huggins, and Y. Cui, presented at 15th Int. Mtg. on Lithium Batteries, Montreal (June 29, 2010).
24. P. Shen, D. Jia, Y. Huang, L. Liu and Z. Guo, J. Power Sources *158*, 608 (2006).
25. L-F. Cui, Y. Yang, C-M. Hsu and Y. Cui, Nano Lett. *9*, 3370 (2009).
26. W. Li, J.R. Dahn and D.S. Wainwright. Science *264*, 1115 (1994).
27. W. Li, W.R. McKinnon and J.R. Dahn, J. Electrochem. Soc. *141*, 2310 (1994).
28. W. Li and J.R. Dahn, J. Electrochem. Soc. *142*, 1742 (1995).
29. M. Zhang and J.R. Dahn, J. Electrochem. Soc. *143*, 2730 (1996).
30. C. Wessells, F. LaMantia, R. Ruffo, R.A. Huggins, and Yi Cui, presented at 15th Int. Mtg. on Lithium Batteries, Montreal (July 1, 2010).

Index

Unit cell dimension changes, 158
US Department of Energy hydrogen storage
      targets, 108
US Dept. of Energy targets, 109, 383
US oil production peak, viii
Uses of energy, v, 1–3, 7
Utility load leveling, 367–368

V
V/Br, 372
V/V, 372, 373
V_2O_5, 218, 219, 320–322
V_6O_{13}, 217, 218, 320
Vacancy motion, 348
Valence band, 221
Valve-regulated lead-acid technology, 237, 246
Van der Waals forces, 209, 210
Van't Hoff equation, 260
Van't Hoff Plot, 262, 263
Vanadium bronzes, 357, 358
Vanadyl sulfate, 372
Vapor-liquid-solid method (VLS), 388
Variation during the week, 4, 367
Variation of cell voltage with state of charge,
      239–240, 356
Variation of voltage during discharge and
      recharge, 133
Variations in energy needs with time, 5
Vario-stoichiometric phases, 159, 271, 272, 282
Vegetable oil, xiv, 49
Virtual chemical reactions, 122, 145
Virtual reaction, 123, 147, 157, 166, 188–192
VLS method, 388
$VO_2(B)$, 350
Voltage calculation, 190–192
Voltage instability, 7
Valve-regulated lead-acid (VRLA) cells, 246

W
Wagner, C., 157, 269, 348
Water
    consumption, 255
    decomposition, 29, 39, 100, 102–103, 247,
        251, 269, 276, 277, 284, 285
    electrolysis, xiv, 95, 100–102, 107,
        269, 283

introduction into solids, 348
    storage in dams, 2, 60
Water-gas shift reaction, 96–98
Waterwheels, 60
Weber, N., 369
Weekly storage, 3
Weppner, W., 158
Whisker growth, 388
Whittingham, M.S., 320
Wind energy storage, 367, 368
Windows, 79, 211, 240, 278, 307, 349, 350
Wood, v–vii, xi, xiv, 2, 3, 22, 49, 52, 53
Wood's metal alloys, 304
Woodall, J.M., 104
World oil production peak, viii
World oil production per person, x
Worldwide energy consumption, 1

Y
Yao, Y.F.Y., 369
$YBa_2Cu_3O_7$, 91, 92
Young's modulus, 55, 56

Z
Zebra battery, 319, 379, 380
ZEBRA cells, 186–188, 335, 370, 371, 379,
      380, 384
Zebra system, 186–188
Zero-strain insertion reaction, 334
Zinc electrode, 113, 251–253, 255, 362
    problems, 253
Zincate ions, 251, 253
Zn, 150, 252, 255, 262, 359, 361, 372
$Zn(NO_3)_2 \cdot 6H_2O$, 40
$Zn(OH)_4^{2-}$, 253
Zn/air batteries, 134, 359–362
Zn/Br_2, 372
Zn/MnO_2
    "alkaline" cell, 355–356
    cell, 250, 251, 269, 270, 355–356
    primary cell, 251
Zn/O_2 cells, 127, 361
$ZnCl_2$, 376
ZnO, 113, 127, 150, 252, 253, 255, 359, 360